NEW TECHNOLOGY-BASED FIRMS IN THE NEW MILLENNIUM VOLUME III

New Technology-Based Firms in the New Millennium Volume II (2002)
Edited by Ray Oakey, Wim During and Saleema Kauser

New Technology-Based Firms in the New Millennium Volume I (2001)
Edited by Wim During, Ray Oakey and Saleema Kauser

New Technology-Based Firms at the Turn of the Century

New Technology-Based Firms at the Turn of the Century (2000)
Edited by Ray Oakey, Wim During and Michelle Kipling

New Technology-Based Firms in the 1990s

New Technology-Based Firms in the 1990s Volume VI (1999)
Edited by Ray Oakey, Wim During and Syeda-Masooda Mukhtar

Previous titles in this series published by Paul Chapman Publishing

New Technology-Based Firms in the 1990s Volume I (1994)
Edited by Ray Oakey

New Technology-Based Firms in the 1990s Volume II (1996)
Edited by Ray Oakey

New Technology-Based Firms in the 1990s Volume III (1997)
Edited by Ray Oakey and Syeda-Masooda Mukhtar

New Technology-Based Firms in the 1990s Volume IV (1998)
Edited by Wim During and Ray Oakey

New Technology-Based Firms in the 1990s Volume V (1998)
Edited by Ray Oakey and Wim During

Other titles of interest from Elsevier

Management of Technology, the key to Prosperity in the Third Millennium
Edited by T. Khalil

Management of Technology, Growth through Business Innovation and Entrepreneurship
Edited by M. Van Zedtwitz

Management of Technology, Sustainable Development and Eco-Efficiency
Edited by Louis A. Lefebvre, Robert M. Mason and Tarek Khalil

Forthcoming
Silicon Valley North: A High-Tech Cluster of Innovation and Entrepreneurship
Edited by Larisa V. Shavinina

Related Journals – Sample copies available on request

Journal of Business Venturing
Journal of Engineering and Technology Management
Journal of High Technology Management Research
Technological Forecasting and Social Change
Technovation

NEW TECHNOLOGY-BASED FIRMS IN THE NEW MILLENNIUM VOLUME III

EDITED BY

WIM DURING

University of Twente, Enschede, The Netherlands

RAY OAKEY

Manchester Business School, Manchester, UK

SALEEMA KAUSER

Manchester Business School, Manchester, UK

2004

ELSEVIER

Amsterdam – Boston – Heidelberg – London – New York – Oxford
Paris – San Diego – San Francisco – Singapore – Sydney – Tokyo

ELSEVIER B.V.
Sara Burgerhartstraat 25
P.O. Box 211
1000 AE Amsterdam
The Netherlands

ELSEVIER Inc.
525 B Street, Suite 1900
San Diego
CA 92101-4495
USA

ELSEVIER Ltd
The Boulevard, Langford
Lane, Kidlington
Oxford OX5 1GB
UK

ELSEVIER Ltd
84 Theobalds Road
London
WC1X 8RR
UK

First edition 2004

Library of Congress Cataloging in Publication Data
A catalog record is available from the Library of Congress.

British Library Cataloguing in Publication Data
A catalogue record is available from the British Library.

ISBN: 0-08-044402-4

⊗ The paper used in this publication meets the requirements of ANSI/NISO Z39.48-1992 (Permanence of Paper). Printed in The Netherlands.

Contents

Contributors

Tamar Almor	Jerusalem School of Business Administration, The Hebrew University, Israel
Paul Benneworth	Centre for Urban and Regional Development Studies (CURDS), Newcastle University, Newcastle-upon-Tyne, UK
Felix Bellido	Comunidad de Madrid, Spain
D. Jane Bower	Division of Economics and Enterprise, Glasgow Caledonian University, City Campus, Cowcaddens Road, Glasgow, Scotland, UK
Nettie Buitelaar	BioPartner Network, the Netherlands
David Charles	Centre for Urban and Regional Development Studies (CURDS), Newcastle University, Newcastle-upon-Tyne, UK
Sarah Cooper	Senior Lecturer in the Hunter Centre for Entrepreneurship, University of Strathclyde, Glasgow, Scotland, UK
Mario Cugini	Accenture, London
Per Davidsson	Jönköping International Business School, Jönköping, Sweden
Wim E. During	Faculty of Technology and Management, Dutch Institute for Knowledge Intensive Entrepreneurship, University of Twente, Enschede, the Netherlands
Tom Elfring	Free University Amsterdam, Amsterdam, the Netherlands — also affiliated at Wageningen University
Abby Gholbadian	Middlesex University Business School, The Burroughs, London, UK
Aard Groen	Faculty for Technology and Management, University of Twente, Enschede, the Netherlands

Paula D. Harveston	Campbell School of Business, Berry College, USA
Niron Hashai	Jerusalem School of Business Administration, The Hebrew University, Israel
Haifen Hu	BioPartner Network, the Netherlands
Willem Hulsink	Rotterdam School of Management, Erasmus University Rotterdam, Rotterdam, the Netherlands
Saleema Kauser	Manchester Business School, Booth Street West, Manchester, UK
Ben L. Kedia	University of Memphis, Memphis, USA
Magnus Klofsten	Center for Innovation and Entrepreneurship (CIE), Linköping University, Linköping, Sweden
Manuel Laranja	Instituto Superior de Economia e Gestão — CIRIUS, Universidade Técnica de Lisboa, Portugal
Andy Lockett	Nottingham University Business School, Jubilee Campus Wollaton Road, Nottingham, UK
Cheo Machin	(Work Psychology) and Instituto Universitario Euroforum Escorial (IUEE), Spain
Ray Oakey	Professor of Business Development, Manchester Business School, Booth Street West, Manchester, UK
Nicholas O'Regan	Centre for Interdisciplinary Strategic Management Research [CISMR], Middlesex University Business School, The Burroughs, London, UK
David Osborne	Campbell School of Business, Berry College, USA
Thelma Quince	ESRC Centre for Business Research, Cambridge University, Cambridge, UK
Davena Rankin	Division of Economics and Enterprise, Glasgow Caledonian University, City Campus, Cowcaddens Road, Glasgow, Scotland, UK
Sami Saarenketo	Telecom Business Research Center, Lappeenranta University of Technology, Lappeenranta, Finland
Olav R. Spilling	Department of Innovation and Economic Organisation, Center for Industrial Development and Entrepreneurship, Norwegian School of Management BI, Sandvika, Norway

Jartrud Steinsli

Department of Innovation and Economic Organisation, Center for Industrial Development and Entrepreneurship, Norwegian School of Management BI, Sandvika, Norway

António Teixeira

Coordinator of Intellectual Property Office, GAPI.UP, University of Porto, Portugal

Beatrice I. J. M. Van Der Heijden

Maastricht School of Management, Department Organizational Behavior, Maastricht, the Netherlands and University of Twente, School of Business, Public Administration and Technology, Department Human Resource Management, Enschede, the Netherlands

Peter Van Der Sijde

Dutch Institute for Knowledge Intensive Entrepreneurship, University of Twente, NIKOS, Enschede, the Netherlands

Marijke Van Der Veen

Faculty of Technology & Management, Dutch Institute for Knowledge Intensive Entrepreneurship (NIKOS), University of Twente, Enschede, the Netherlands

Jaap Van Tilburg

Top Spin International and BTC-BIC Twente, Enschede, the Netherlands

Ajay Vohora

Nottingham University Business School, Jubilee Campus Wollaton Road, Nottingham, UK

Hugh Whittaker

Doshisha University, Kyoto 602-8580, Japan

Mike Wright

Nottingham University Business School, Jubilee Campus Wollaton Rd, Nottingham, UK

Part I

Introduction

Chapter 1

The Emerging Research Agenda

Wim During, Ray Oakey and Saleema Kauser

This tenth volume of papers emanating from the Annual High Technology Small Firms Conference completes a full decade of papers on the innovation and growth problems of High Technology Small Firms (HTSFs). Because the series has become a major vehicle for publication among the best international researchers working in this area, it represents an authoritative voice on HTSF development problems. Moreover, the international credentials of this latest 2002 Conference are confirmed by papers from Canada, Norway, Finland, Portugal and the USA, together with several contributions from The Netherlands and the United Kingdom.

Over the past ten years research into HTSFs reported in this series has fallen into two broad areas; policy towards HTSFs, and the business management processes within HTSFs (where there has been a strong emphasis on networking). Major topics under the policy heading are financing instruments, incubators, Science Parks, and policy comparisons between regions and countries. More recent policy research shows an emphasis on the actual development and use of such instruments. Here the focus is on impact research, benchmarking and best practise studies. The main development in this context appears to be from signalling market imperfections to the proper use of policy in correcting such market failures.

Within business management processes, major topics have included innovation, small versus large firms, strategic alliances, technology transfer and the cooperation process. Concerning networking, the focus has evolved over the past ten years from measuring the importance of networking to the development of HTSFs, into research concerning the actual process and outcome of cooperation. While early studies often had regional or national contexts, later studies have been more focussed on international networks.

Following this brief introduction (Part I), the current volume is divided into *five* main topic areas. The second and largest of these Parts is that of Strategy (Part II), a management subject within NTBF studies that has key relevance, both to performance, and consequently, to the suitability of NTBFs for external investment. In Chapter 2, Nicholas O'Regan and Abby Ghobadian emphasise the importance of senior management in their paper "Leadership — Realising the Potential in High Technology Small Firms." Marijke van Der Veen

New Technology-Based Firms in the New Millennium, Volume III
Copyright © 2004 by Elsevier Ltd.
All rights of reproduction in any form reserved.
ISBN: 0-08-044402-4

takes a more technological approach in Chapter 3 by "Measuring E-Business-Adoption in SMEs." "The Business Platform Model: A Practical Tool for Understanding and Analysing Firms in Early Development" is investigated by Per Davidsson and Magnus Klofsten in Chapter 4, while "Unravelling the Capital Structure Puzzle in Small Firms: Matching Instruments to Investments" is considered by Guy Gellatly, and Stewart Thornhill in Chapter 5.

Four papers in Part III contribute to the growing body of knowledge on HTSF "*Spin offs*," their contribution to HTSF growth, and the problems they encounter during the traumatic "spin off" process. Such problems are reflected in a paper by Willem Hulsink and Tom Elfring in Chapter 6, where the consider "Entrepreneurs, New Technology Firms & Networks: Experiences from Lone Starters, Spin-Offs & Incubatees in the Dutch ICT Industry 1990–2000." The national location changes in Chapter 7 when "Virtual Incubation of Research Spin-Offs and the Comunidad de Madrid Case" is discussed by Jaap van Tilburg, Cheo Machin, Peter van Der Sijde and Felix Bellido. The "spin off" theme is further pursued in Chapter 8 when Jay Vohora, Mike Wright and, Andy Lockett explore "The Formation of High-Tech University Spinouts Through Joint Ventures." Finally this section is concluded in Chapter 9 by a paper from Holland on "The Netherlands Life Sciences Sector Biopartner: Stimulating Entrepreneurship in the Life Sciences" by Haifen Hu, Nettie Buitelaar, Wim During and Aaard Groen.

Part IV, comprising three papers, is concerned with the current "hot topic" of *clusters* and their role in enhancing HTSF formation and growth. In an initial offering (Chapter 10) Paul Benneworth, and David Charles contribute findings on "Overcoming Learning Uncertainties in the Innovation Process: The Contribution of Clustering to Firms' Innovation Performance." This is followed in Chapter 11 by a paper on a related theme of "Close Encounters: Evidence of the Potential Benefits of Proximity to Local Industrial Clusters" by Thelma Quince and Hugh Whittaker. The final paper in this short section is offered in Chapter 12 "On The Role of High Technology Small Firms in Cluster Evolution," by Olav Spilling and Jartrud Steinsli.

The above clustering section is followed in Part V by four papers on related topic of *networking*, in which many of the geographical-based issues raised in clustering literature are pursued at an organisational level. An initial offering on this theme in Chapter 13 concerns "The Competitiveness of New-Technology Based Firms: The Contribution of Trade Associations by Mario Cugnini and Sarah Cooper." This mainly externally oriented paper is then balanced by a paper in Chapter 14 that is more internally directed by Beatrice van Der Heijden on "Organizational Influences Upon the Development of Professional Expertise in SMEs in the Netherlands." "Examining the Mental Models of Entrepreneurs From Born Global and Gradual Globalizing Firms" is a theme explored by Paula Harveston, David Osborne and Ben Kedia in Chapter 15, while Jane Bower and Davena Rankin conclude this section in Chapter 16 with their consideration of "Knowledge Transfer and Policy Imperatives: Science in a Devolved Scotland."

Three papers on global issues in Part VI conclude the book, which is a welcome new research area for this series. Since many HTSFs are truly "born global" in that their early customers (and often suppliers) are international in nature, this topic is very relevant and timely. An initial paper from Finland on "Born Global Approach to Internationalisation of High Technology Small Firms" is provided in Chapter 17 by Sami Saarenketo. This is followed in Chapter 18 by Ttamar Almor, and Niron Hashai who ask "Configurations of International Knowledge-Intensive SMEs: Can the Eclectic Paradigm Provide a Sufficient Theoretical Framework." The final paper comprises an exploration of the "Internationalisation of High Technology Small Firms in Portugal by Antonio Teixeira and Manuel Laranja."

Part II

Strategy

Chapter 2

Leadership — Realising the Potential in High Technology Small Firms

Nicholas O'Regan and Abby Ghobadian

Introduction

Firms of all sizes are experiencing unprecedented change as a result of an increasingly competitive and changing environment — Webber (1988), D'Aveni (1994). Many of the changes such as uncertainty are well documented in the literature — Ghobadian & Gallear (1997). Such changes include the increasing globalisation of markets, advances in telecommunications technology and the growing interdependencies between economies. Other factors impacting on HTSFs include "heightened volatility, hyper competition, demographic changes, knowledge based competition" — Daft & Lewin (1993). Industry structure is changing as knowledge and information achieves greater importance and innovation is now seen as the main driver of competitive advantage. The changing environment and operating structures mean that a change in leadership styles is vital. Consequently, it is fair to say that the "command and control" leadership ethos is becoming rapidly outdated and obsolete.

To-date, most research has focused on factors that contribute to HTSF survival or issues such as the availability and sources of finance, rather than a greater understanding of the drivers of growth and sustainable competitive advantage — Storey (1994). The evidence from the literature suggests that these comments equally apply to the understanding of leadership and its broad impact on strategic planning. Nevertheless, leadership is increasingly seen as an important ingredient in achieving competitive advantage — Bass (1990). Its importance is shown by the various empirical studies indicating a close association with strategic planning — Berkeley Thomas (1988), Hart & Quinn (1993), Keller (1992), Wilderom & van den Berg (1997). There is no shortage of studies on the strategy-performance relationship — Schwenk & Shrader (1993), Baker et al. (1993), McKiernan & Morris (1994), Kargar & Parnell (1996), Joyce et al. (1996), Roper (1997), Naffziger & Mueller (1999).

The main task of HTSF leaders is to prepare their firms for the future. This, by necessity, involves the interpretation of the opportunities and threats that they encounter — Aram

New Technology-Based Firms in the New Millennium, Volume III
Copyright © 2004 by Elsevier Ltd.
All rights of reproduction in any form reserved.
ISBN: 0-08-044402-4

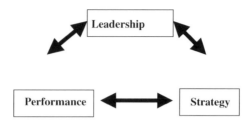

Figure 1: The leadership-strategy-performance relationship.

& Cowan (1990). LeBarre (2002) referring to the air attacks on the World Trade Centre in New York on September 11th 2001, states that "in effect a new economy has emerged which calls for an urgent leadership agenda. Business leaders must be alert to the realities of the here and now and must aggressively pursue the changes and challenges that have resulted from these dramatic events."

The authors contend that the time is now right to assess the impact of leadership on both strategic planning and the performance of HTSFs. Are HTSFs grasping the need to create a "new leadership agenda" and if so, what leadership style and strategic planning approach are appropriate?

Aims of the Research

This study aims to establish the relationship between leadership, strategy and performance as shown in Figure 1.

The majority of studies to date have focussed on large firms. This study focuses on HTSFs — an area that is under researched. The study examines the type of leadership that is likely to lead to enhanced performance and compares the degree of emphasis by HTSFs and firms in a traditional low technology sector — engineering, on the factors crafting strategy and on the extent to which performance criteria have been met. Finally, the study compares the degree of emphasis given to leadership attributes by high and low performing firms.

This study has a number of practical benefits for the Managing directors of HTSFs. Firstly; it provides some guidelines on the leadership styles most likely to be correlated with strategic planning and organisational performance. Second, it will enable HTSFs to compare their approach to both leadership and strategy with low technology firms such as those engaged in engineering. Finally, the study identifies the attributes of leadership emphasised by high performing firms.

Accordingly the following hypotheses were formulated:

Hypothesis 1. Leadership styles are associated with strategic planning and performance.

Hypothesis 2. High technology firms emphasise leadership attributes to a greater extent then low technology firms.

Hypothesis 3. High performing firms emphasise leadership attributes to a greater extent than low performing firms.

Strategy

The development of a strategy consists of three separate but interdependent phases. These are the processes and tools deployed in crafting strategy, the actual content of the strategy and finally, the implementation plan — Hitt *et al.* (2001), Hill & Jones (2001). The authors subscribe to the view that strategy is much more than the content of a written document — it also encompasses the formulation process and the deployment of the strategic plan. The phases of developing the strategic vision and monitoring/control are accepted as pre-requisite activities in strategic management for the purposes of this study.

A small number of research studies indicate that small firms using strategic planning performed better than non-strategic planning firms — Bracker *et al.* (1988). Others found that "strategic" small firms were likely to have significant capability to grow, expand, innovate and introduce new products to the market place — Joyce *et al.* (1996) and achieve greater profitability — Roper (1997).

Overall, the literature suggests that strategic planning processes enhance the ability of the organisation to maintain its competitive advantage. In addition, a formal strategic planning process ensures that there is an awareness of strategic issues throughout the organisation — Porter (1996), Skinner (1996). Strategic planning is also considered to be one of the most effective ways for firms, regardless of size or sector, to cope with the changes in their business environment — Hart & Banbury (1994). However, the literature points out that while most large firms emphasise strategic planning — Ancona & Caldwell (1987), the position in relation to HTSFs is blurred and unclear.

Table 1: Characteristics of strategic planning.

Characteristic	Literature Sources
External orientation	Mintzberg (1994), De Geus (1997: 35), Mills & Friesen (1992), Porter (1980: 88)
Internal orientation	Ramanujam *et al.* (1986), Porter (1990), Fombrun (1992)
Functional integration	Womack & Jones (1994), Hirschorn & Gilmore (1992), Ramanujam & Venkatraman (1987), Ramanujam *et al.* (1986), Hammer (1990)
The use of analytical techniques	Ramanujam *et al.* (1986), Kargar & Parnell (1996), Hayes & Abernathy (1980), De Geus (1997), Ramanujam & Venkatraman (1987)
Resources for the strategic planning process	Hamel & Prahalad (1994), Ramanujam *et al.* (1986), Kargar & Parnell (1996), Bracker *et al.* (1988), Thomas *et al.* (1991), Chan (1993), Ramanujam & Venkatraman (1987), Veliyath & Shortell (1993)
Systems capability/creativity	Porter (1990), De Geus (1997: 22), Sexton & Bowman-Upton (1991), Atkinson *et al.* (1997), Storey (1994), Hamel & Prahalad (1994), Roper (1997)
Focus on control	Goold (1991), Goold & Quinn (1993)

Table 2: Selected definitions of leadership.

Author	Definition
Yuhl (1989)	the process whereby one individual influences other group members towards the attainment of defined group or organisational goals
Weihrich & Koontz (1993)	the art or process of influencing people so that they will strive willingly and enthusiastically toward the achievement of the group's mission
Ackoff (1999)	guiding, encouraging and facilitating others in the pursuit of ends by the use of means, both of which they have either selected or approved

Operationalising Strategic Planning

Researchers have adopted a number of independent characteristics to delineate strategic processes — Ramanujam *et al.* (1986), Ramanujam & Venkatraman (1987), Veliyath & Shortell (1993), Kargar & Parnell (1996). Following an extensive examination of the literature we adopted the characteristics listed in Table 1 to represent the key factors considered in a developing strategy. Table 1 also provides support for the efficacy of each factor. Their relevance was further examined through qualitative interviews with six managing directors of HTSFs.

Leadership

The literature on leadership is rich and fragmented. Managers are continually striving to access the latest thinking in a bid to gain, retain or regain competitive advantage — Moxley (2000). However, despite the growing interest demonstrated by researchers in recent years, and the wide ranging and diverse studies on leadership, there is no agreed definition. More theoretical and empirical work is needed to understand the nature and scope of this phenomenon.

Kouzes & Posner (1997) describe leadership as "a reciprocal relationship between those who choose to lead and those who decide to follow." Other definitions are depicted in Table 2. Bennis & Nanus (1985) suggest that the principle task of leadership is to ensure the effective deployment of the strategic plan by championing and communicating a vision and obtaining commitment to its achievement. Accordingly, leadership is a critical ingredient in the search for sustainable competitive advantage. However, leadership is a highly complex concept — Yuhl (1998), particularly in HTSFs. In relation to HTSFs, Hellstrom *et al.* (2001) state that "the classic leadership par excellence has been brought down from the heights of the corporate ivory tower to the top-management team, or to middle management, where multiple role taking, reflexivity, knowledge brokerage, and the affecting of initiative are now far more important than personality traits and 'unity of command.' "

Table 3: A summary of leadership styles derived by Wilderom & van den Berg (1997).

Leadership Style	Main Emphasis	Literature Sources
Transformational	Shows the degree of inspiration, consideration and assistance given to employees	Tichy & Devanna (1986), Kouzes & Posner (1987), Bass & Avolio (1989), Bass (1990), Yammarino & Bass (1990), Conger (1991), Hirschorn & Gilmore (1992), Wilderom & van den Berg (1997), Lim (1997), Shamir *et al.* (1998), Ackoff (1999)
Transactional	Shows an exchange relationship between the leader and employees based on the benefits of co-operation to each party	Bass (1985, 1990), Wilderom & van den Berg (1997)
Human and results	Provides for the concerns of employees while emphasizing the business goals	Lee (1991), Hampden-Turner (1990), Wilderom & van den Berg (1997)
Laissez faire	Passive management or the absence of active management	Bass (1990), Wilderom & van den Berg (1997)

This study operationalised leadership based on the four leadership styles derived by Wilderom & van den Berg (1997) as shown in Table 3. These dimensions have been tested and validated for use in small firms, are appropriate for use in small firms and are thus appropriate for use in this study.

The Relationship Between Leadership and Strategy

Empirical studies on leadership in large firms has received increasing attention in the literature, reflecting the increasing frequency and importance of leadership and strategy. A review of previous studies indicates that some leadership styles are associated with greater performance than others. For example, Bass (1990) found that transformational leadership has the greatest impact, followed by transactional leadership style and that laissez faire style has the least impact. Lim (1997) in a study of U.K. firms found that transformational style was associated with performance. Both Keller (1992) and Wilderom & van den Berg (1997) suggest that a balanced transformational and human resources orientation is likely to lead to enhanced performance. This finding is consistent with the "new leadership agenda" referred to by LeBarre (2002). Accordingly, it is no surprise to note that Davis *et al.* (1997) argue that a "balanced" style is common sense as it aims to treat employees fairly, in terms of employee satisfaction and organisational productivity.

While the majority of studies focus on leadership in the context of large firms, arguably the role of the Chief Executive in the smaller firm is more significant as s/he is the controlling influence as regards decisions and strategic planning — Begley & Boyd (1986), Day & Lord (1988). The actions of the chief executive can be constrained by their operating environment — Hambrick & Mason (1984).

The literature suggests that radical change can only be achieved through the quality of ideas and leadership commitment to their achievement — Alexander (1985). This implies that transactional style and laissez faire leadership styles will be associated to a lesser extent with both strategy and performance than transformational or human resources leadership styles.

Methodology

The sample consisted of 1,000 small and medium sized manufacturing firms in the U.K. Practical considerations largely guided the choice of the electronics and engineering sectors. First, the contrasting product life cycles. Engineering organisations, by and large, operate in a mature market, whereas electronic firms operate in a market characterised by short product cycles. Second, both sectors are economically important. Third, both sectors have a large number of small and medium sized firms. Small and medium sized firms were defined as having fewer than 250 employees. A survey questionnaire was administered to a random sample based on a directory available from a reputable commercial firm.

Questions concerning the strategic planning process were based on a survey instrument devised and tested on small banks in the U.S. — Kargar & Parnell (1996). In the case of leadership, questions were derived from a survey instrument devised and tested on Dutch SMEs — Wilderom & van den Berg (1997).

The validity of the constructs used and their relevance was tested during the qualitative phase of the research. This involved in-depth interviews with six managing directors of HTSFs and discussions with employer representative bodies. Furthermore, the survey instrument was pre-tested and modified through pilot field-work. Internal consistency was established using Cronbach's Alpha and factor analysis. Factor analysis was used to examine the dimensionality of the constructs and also as a means of data reduction. Canonical correlation analysis was used to test the association between the characteristics of strategic planning and leadership. Canonical correlation is a statistical technique for studying the association between two variables to form linear combinations of the variables that have maximum correlation — Ramanujam & Venkatraman (1987), Kargar (1996).

Responses

From the 1,000 questionnaires mailed 194 valid responses (27%) were received. Non-respondents were contacted to ascertain the reasons for non-response. We compared respondents and non-respondents on demographics and no differences were detected. Thus we believe that the similarities in the response patterns and the range of responses across demographic characteristics suggest the absence of response bias.

Data Analysis

To establish the relationship between the leadership styles and the characteristics of the strategic planning process in manufacturing SMEs, the analysis is preceded in two steps. First, canonical correlation analysis ascertained the degree of correlation between strategic planning and leadership. The results of the canonical correlation analysis are depicted in Table 5. Second, the aggregate score for each leadership style was computed for each of the participating organisations. These scores were used to classify organisations into four quartiles. Correlation analysis was conducted out to ascertain the correlation between firms classified in the upper quartiles of each leadership style with firms classified as being in the lower quartile. To test the importance attached to each characteristic of strategic planning, that is, the upper and lower quartiles, a comparison between the two dimensions was conducted for the purpose of drawing inferences. Table 4 summarises the results of the canonical correlation analysis. The attributes of each leadership style loaded onto two or more separate styles as shown in Table 4. Similarly, the attributes of two characteristics of strategic planning loaded onto more than one factor.

The results indicate that the three main leadership styles account for over 70% of the variance of the association with the characteristics of strategic planning as follows: *Human resources — employee focussed style* 31% of the variance, *transformational — creativity style* 18%, *transformational — competence style* 11.5%, while *transactional — rewards style* accounts for over 10%. The components of a "balanced transformational style" comprising all transformational and human resources styles account for over 60% of the variance. This finding is consistent with the increasing emphasis given to transformational and human resource styles in the literature and reflects the decline in the reward style of leadership.

The results, however, do not show any association between leadership styles and the *resources for strategy, management time, products and performance* or the use of *analytical techniques* characteristics. The lack of association with the *use of analytical techniques* is not surprising as the literature indicates that this characteristic is perceived as being the least important to SMEs — Kargar & Parnell (1996). It may be that the *use of analytical techniques* is not indicative of any of the main leadership styles. It is also possible to ascertain that a lack of an association between leadership styles and *products and performance* is not a cause for concern as product strengths and weaknesses may be implicitly covered in other attributes. This suggests that SME leadership doesn't impact on past performance, but only can be used to draw limited conclusions. The lack of association between leadership styles and the *resources for strategy* and *management time* characteristics is more difficult to explain. One explanation may be that the attributes of these characteristics are included or implied in the remaining characteristics and consequently there is no need to state each attribute separately. Overall, the analysis shows varying degrees of association between the dimensions of leadership and the characteristics of strategic planning process.

Relating Leadership Styles to Strategic Planning

The emphasis placed on each of the strategic planning characteristics by organisations in the upper and lower quartiles of each of the leadership styles was examined. For this

Table 4: Relationship between the canonical functions — leadership and strategic planning.

Variable	L[a]	L2[b]	%[c]	Canonical Cross Loading
Criterion set (leadership)				
Transactional style				
Transactional — rewards	−0.493	0.243	10.65	−0.329
Maintenance of standards	−0.360	0.130	5.70	−0.240
Performance focused	−0.212	0.045	1.97	−0.142
Own interests	−0.113	0.013	0.57	−0.075
Transformational style				
Competence	**−0.514**	**0.264**	**11.57**	**−0.343**
Creativity	**−0.645**	**0.416**	**18.24**	**−0.431**
Vision	−0.134	0.018	0.79	−0.089
Mistakes focused	−0.371	0.138	6.05	−0.248
Human resources style				
Employee focused	**−0.840**	**0.706**	**30.95**	**−0.561**
Organisation based	−0.375	0.141	6.18	−0.250
Laissez faire style				
Reactive	−0.397	0.158	6.93	−0.265
Passive	−0.095	0.009	0.40	−0.064
Total		2.281	100.00	
Predictor set (strategic planning)				
Internal orientation				
Internal capability	**−0.524**	**0.275**	**11.64**	**−0.350**
Products and performance	−0.340	0.116	4.91	−0.227
External orientation	**−0.602**	**0.362**	**15.33**	**−0.402**
Departmental co-operation	**−0.583**	**0.340**	**14.39**	**−0.389**
Resources for strategy				
Resources for strategy	−0.241	0.058	2.46	−0.161
Management time	−0.340	0.116	4.91	−0.227
Use of analytical techniques	−0.175	0.031	1.31	−0.117
Staff creativity	**−0.922**	**0.850**	**35.99**	**−0.616**
Control mechanism	**−0.463**	**0.214**	**9.06**	**−0.309**
Total		2.352	100.00	

Note: Figures in bold indicate associated variables of significance: excess of ±0.300.
[a] Canonical loading — correlation of the canonical variate with each variable in their respective sets.
[b] Measure of variance of observed variables with the canonical (unobserved) variate.
[c] Proportion of explained variance accounted for by each variable in both the criterion and predictor sets.

Table 5: Correlation analysis leadership styles and strategic planning attributes.

Leadership Styles	Strategic Planning Characteristics	
	Upper Quartile	**Lower Quartile**
Transformational — competence	Creativity[**]	External orientation[*]
Transformational — creativity	External orientation[*]	Internal orientation[*]
Human resources — employee focussed	Creativity[**]	External orientation[**] and creativity[*]
Transactional — rewards	Internal orientation[*]	None

[*]Correlation significant at the 0.05 level (2-tailed).
[**]Correlation significant at the 0.01 level (2-tailed).

purpose, the aggregate score for each leadership style was computed for each of the participating organisations. These scores were used to classify organisations into quartiles. Firstly, correlation analysis was conducted to ascertain the relationships between leadership styles and the characteristics of strategic planning for the two quartiles. The results are presented in Table 5.

The analysis in Table 5 is consistent with the results of the canonical correlation analysis in Table 4. For example, *transformational — competence* leadership style is correlated with staff creativity and external orientation, which mirrors the high degree of association shown in Table 4. Similarly, *transformational — creativity* leadership style is correlated with external orientation in the upper quartiles and to a lesser extent, internal orientation in the lower quartiles. This is also consistent with the findings in Table 4 which shows that *transformational — creativity* leadership style (accounting for the 18% of the variance) is associated with external orientation (15% of the variance) and internal orientation (almost 12% of the variance). *Human Resources — employee focussed* leadership style is correlated with staff creativity in the upper quartile and external orientation in the lower quartile. This mirrors the canonical correlation analysis where *human resources — employee focussed* leadership style accounts for the largest proportion of the variance (31%) and is associated with the staff creativity and external orientation characteristics of strategic planning (accounting for 36% and 15% variance respectively — the highest and second highest variance levels).

Finally, firms in the upper quartile of the *transactional — rewards* leadership style correlate with internal orientation in the upper quartile but fail to show any significant correlation in the lower quartile. Again, the findings mirror the canonical correlation analysis where *transactional — rewards style* (fourth highest measure of variance in the criterion set) is associated with internal orientation (fourth highest measure of variance in the predictor set). The analysis in Table 5 is consistent with the findings of Wilderom & van den Berg that suggest that a balanced human resources/transformational leadership style is likely to lead to greater performance. Transformational (both types) and human resources styles are the two strongest leadership styles in the criterion set in the canonical correlation analysis and indicate a strong association with the strongest strategic planning characteristics — creativity and external orientation. The correlation analysis for the upper and lower quartiles of the transformational (competence and creativity), human resources (employee focussed) leadership styles and strategic planning mirrors this finding.

The analysis in Table 4 examined the emphasis placed by firms classified according to predominant leadership style and the factors used in the crafting of strategy.

Table 6 indicates that organisations in the upper quartile placed greater emphasis on each of the strategic planning characteristics than organisations in the lower quartile. These observations suggest that leadership style has a positive influence on the level of emphasis placed on each and every one of the strategic planning attributes.

More specifically, leadership style irrespective of its type resulted in the greatest emphasis being placed on *external orientation and department co-operation. Transactional — rewards* and *human resources — employee focussed* styles resulted in the greatest emphasis placed on *internal orientation*, while *transformational — creativity* style more often lead to a greater emphasis being placed on *staff creativity*. These results are not entirely in line with expectations. On a prioiri, we expected both types of *transformational style* firms to place a greater emphasis on staff creativity. We anticipated less emphasis on *external orientation* where a *transactional — rewards style* was more dominant. Leadership style least influenced the degree of emphasis placed on *strategy — a control mechanism*.

An examination of the degree of emphasis placed on each attribute of strategic planning within each leadership style suggest that *transformational — creativity* style accounted for the highest level of emphasis on each of the planning characteristics. Arguably, this would suggest that strategy development is a craft rather than science. *Transformational — competence* style accounted for the second highest level of emphasis on each of the strategic planning characteristics. These observations lead us to the tentative conclusion that greater attention is being paid to dimensions of strategic planning when both transformational type styles are more prominent.

Leadership-Performance Relationship

The four leadership styles were further examined in order to ascertain their influence on organisational performance. Firstly, correlation analysis was use to ascertain the relationship between firms classified as being in the upper and lower quartiles of each leadership style and the factors used to measure performance. This analysis is depicted in Table 7.

Table 7 indicates that there is a strong correlation between firms emphasising *transformational — competence* leadership to a significant extent, in the upper quartile, and market share whereas firms in the lower quartile tend to focus more on short term performance.

Interestingly, there is also a strong correlation between firms in the upper quartile of *transformational — creativity* leadership and long term performance. This is to be expected as both *transformational (competence and creativity)* styles are normally linked to a longer term outlook that incorporates both performance and market share. *Human resources — employee focussed* leadership style also shows a significant correlation between firms in the upper quartile and market share. This finding is consistent with the work of Wilderom & van den Berg who suggest that both transformational and human resources leadership styles are likely to lead to greater performance in the longer term.

Table 6: Emphasis placed by organisations with scores in the upper and lower quartile of each of the four leadership styles and factors used for crafting strategy.[a]

Strategy Characteristic	Transformational — Competence		Transactional — Rewards		Human Resources — Employee Focussed		Transformational — Creativity	
	Upper	Lower	Upper	Lower	Upper	Lower	Upper	Lower
Staff creativity	63.8	15.4	43.9	24.9	50.9	17.2	100.0	–
External orientation	75.2	43.6	60.6	44.2	64.4	39.9	86.1	18.4
Departmental co-operation	66.5	28.3	45.4	26.5	66.3	22.8	83.4	13.1
Internal orientation	65.9	51.3	65.7	53.0	69.2	57.1	80.6	52.6
Strategy — a control mechanism	22.5	15.7	25.7	4.0	23.6	7.2	27.9	2.6

[a] Excludes resources for strategy and the use of analytical techniques — which are not associated with leadership styles (see Table 5).

Table 7: Correlation analysis — leadership styles and organisational performance.

Leadership Styles	Factors Used to Measure Performance	
	Upper	**Lower**
Transformational — competence	Market share[**]	Short term performance[*]
Transformational — creativity	Long term performance[*]	Innovation[*]
Human resources — employee focussed	Market share[*]	None
Transactional — rewards	None	None

[*]Correlation significant at the 0.05 level (2-tailed).
[**]Correlation significant at the 0.01 level (2-tailed).

Finally, it is no surprise that firms in both the upper and lower quartiles of *transactional — rewards* leadership styles fail to register any significant correlation with any of the factors used to measure performance. This would suggest that in the engineering and electronics sectors, performance is manually driven by factors other than internal orientation.

Table 8 indicates that in the case of each of the performance dimensions, organisations with a more prominent leadership style, (the upper quartile) performed better than the organisations with uncertain leadership styles (scores in the lower quartile). The observed differences varied in magnitude, but the direction, nevertheless, remained the same. These observations suggest that a definitive leadership style, regardless of the style itself, is a significant influence on organisational performance. On the other hand, uncertain leadership irrespective of the actual leadership style is detrimental to high performance.

Examination of figures presented in Table 8 illustrates the impact of each management style on the different facets of performance. Leadership style irrespective of its type has the greatest impact on improving *long-term performance* and *innovation. Transactional — rewards* and human resources styles more often led to improvement in *market share*, while both *transformational type styles (competence and creativity)* more often led to improvements in *short term performance*. These results are not entirely in line with expectations. On a priori, we expected a more dominant relationship between *transactional — rewards* style and *short-term performance*. At the same time, we expected a more dominant relationship between transformational style and *market share*. Leadership style least influenced the *ability to evaluate alternatives* and *avoid problem* areas. *The evaluation of alternatives and avoidance of problem areas* rely on tools and planning and as such weak influence of leadership style on these factors is not surprising.

An examination of each dimension of performance within each leadership style suggests that *transformational — creativity* accounted for highest performance in the attainment of *financial results, evaluation of alternatives, avoidance of problems, and innovation*. On the other hand, *transformational — creativity* style accounted for the highest performance in the following three areas; *market share, improved short-term performance* and *improved long-term performance*. The most profound finding relates to the emphasis on a *transformational creativity* style of leadership. This style includes a number of transformational-type

Table 8: Percentage of firms indicating fulfilled/entirely fulfilled performance objectives.

Leadership Style: Quartile	Transformational — Competence		Transactional — Rewards		Human Resources — Employee Focussed		Transformational — Creativity	
	Upper	Lower	Upper	Lower	Upper	Lower	Upper	Lower
Market share	61.9	55.4	57.4	54.5	61.4	55.4	50.0	39.5
Financial results achieved	56.7	53.9	54.5	48.5	50.7	48.6	61.1	42.1
Evaluate alternatives	45.3	28.2	39.3	32.8	46.2	33.3	61.0	5.4
Avoid problem areas	40.3	25.6	43.9	39.9	40.0	24.6	58.4	32.4
Improve short term performance	68.6	41.0	50.0	47.7	52.3	52.1	63.9	43.2
Improve long term performance	85.1	61.6	69.7	67.2	70.7	66.7	80.5	67.6
Innovation	63.7	61.5	67.7	47.4	71.4	65.6	83.3	57.9

attributes: *encouraging new ideas, introducing new projects and stimulating employees.* The proportion of firms in the upper quartile emphasising the *creativity* style is much higher than the proportion in the lower quartile for all the measures outlined. The most dramatic difference relates to the *evaluation of strategic alternatives.* This finding is not unexpected, as the essence of creativity is the ability to seek alternative ways of achieving a task.

Finally, the largest differences between upper and lower quartiles in the introduction of new products relate to the transactional — rewards style. This is more difficult to explain unless the product type is being rapidly introduced to a changing market place.

Comparing High and Low Technology Small Firms

High technology small firms in the electronics sector were compared with low technology firms in the engineering sector in terms of their emphasis on strategic planning characteristics and performance achieved. The results are depicted in Tables 9 and 10.

The analysis of Table 9 indicates that electronics firms are predominantly characterized in terms of transformational — competence and human resources — employee focus styles whereas engineering firms tend to emphasize a transactional — rewards style. Where engineering firms were found to emphasize the leadership styles other than transactional, the analysis indicated an association with the internal orientation and strategy as controlling mechanism characteristic of strategic planning. This finding is profound and clearly shows that engineering firms are more inward looking and thus may fail to grasp any potential opportunities arising. Moreover, the analysis has also indicted that in cases where HTSFs are outward looking, they have leadership ethos consistent with the achievement of enhanced performance.

The analysis of Table 10 indicates that the transformational — creativity leadership style in electronics firms is associated with many of the measures used to indicate performance. In addition, the correlation between transactional — rewards leadership style and performance measures is in relation to the improvement of short-term performance in engineering firms.

High Performing Firms and Leadership

To ascertain the impact of leadership styles in practice, a comparison was conducted to measure the extent of emphasis given to each of the attributes of leadership by high performing and low performing firms. Firms were classified as high and low performing based on increases or decreases in their market share. For example, Gale & Buzzell (1993: 137–145) state:

> large market share is both a reward for providing better value and a means of realising lower costs. Under most circumstances, enterprises that have achieved a large share of the markets they serve are considerably more profitable than their smaller share rivals. This connection ... is clearly demonstrated in the results of our research over the last fifteen years.

Table 9: Comparison of leadership-strategy correlations in high and low technology firms.

	Transformation — Competence		Transformation — Creativity		Human Resources — Employee Focus		Transactional — Rewards	
	Electronic	Engineering	Electronic	Engineering	Electronic	Engineering	Electronic	Engineering
Internal orientation	✓*	✓**	✓**	–	✓*	✓*	–	✓*
External orientation	✓*	–	✓**	–	✓*	–	–	–
Departmental co-operation	✓**	–	✓**	–	✓*	–	–	–
Analytical techniques	✓*	–	✓**	✓**	–	–	–	–
Staff creativity	✓*	–	✓*	–	✓*	–	–	–
Strategy — a control mechanism	–	✓*	✓*	–	✓*	✓*	–	✓*

*Correlation significant at the 0.05 level (2-tailed).
**Correlation significant at the 0.01 level (2-tailed).

Table 10: Comparison of leadership-performance correlations in high and low technology firms.

	Transformational — Competence		Transformational — Creativity		Human Resources — Employee Focus		Transactional — Rewards	
	Electronic	Engineering	Electronic	Engineering	Electronic	Engineering	Electronic	Engineering
Customer satisfaction	–	–	✓*	–	✓**	–	–	–
Customer retention	–	–	✓**	–	–	–	–	–
Market share	✓*	–	✓*	–	✓*	–	–	–
Predict future trends	–	–	–	–	–	–	–	–
Evaluate alternatives	–	–	–	–	✓**	–	–	–
Avoid problem areas	–	–	✓*	–	–	–	–	–
Short term performance	–	✓*	–	–	–	–	–	✓*
Long term performance	–	–	✓*	–	–	–	–	–
Innovation	–	–	✓*	–	–	–	–	–

*Correlation significant at the 0.05 level (2-tailed).
**Correlation significant at the 0.01 level (2-tailed).

Firms with a perceived increase in market share were classified as high-performing firms, while firms with a perceived decreased market share were classified as low-performing firms. Firms indicating that their market share remained static were omitted from the analysis.

Table 11: The emphasis on attributes of leadership associated with strategic planning in HTSFs high and low performing firms.

Leadership	High Performing		Low Performing	
	Mean	$N = 108$	**Mean**	$N = 35$
Transformational — competence				
Instil perfect trust	**3.88**	**75 (70%)**	**3.32**	**17 (55%)**
Gives employees the feeling that management can overcome any obstacle	**3.45**	**53 (49%)**	**3.06**	**12 (40%)**
Shows an extraordinary ability in everything they undertake	**3.17**	**29 (27%)**	**2.72**	**7 (23%)**
Makes a powerful impression	**3.61**	**56 (52%)**	**3.13**	**9 (29%)**
Transformational — creativity				
Encourage new ideas from employees	3.62	66 (61%)	3.52	17 (55%)
Introduces new projects and challenges	**3.81**	**78 (62%)**	**3.39**	**14 (41%)**
Stimulates employees to support their opinions with good arguments	**3.65**	**50 (46%)**	**3.16**	**17 (55%)**[a]
Transactional — rewards				
Offers the prospects of rewards for good work	3.52	59 (58%)	3.45	16 (52%)
Agrees with the employees on the rewards that they can look forward to if they do what has to be done	**3.06**	**37 (34%)**	**2.61**	**18 (58%)**[a]
Tell employees criteria for performance related benefits	**3.23**	**51 (48%)**	**2.87**	**13 (42%)**
Human resources — employee focussed				
Has an ear for matters that are important to employees	**3.90**	**64 (59%)**	**3.33**	**25 (81%)**[a]
Gives advice to employees when they need it	**3.87**	**73 (68%)**	**3.42**	**23 (74%)**[a]
Creates a feeling of working together on major assignments/missions	**3.80**	**66 (61%)**	**3.32**	**16 (52%)**[a]
Shows employees how to look at problems from new angles	3.32	43 (40%)	3.23	13 (42%)[a]

[a] Attribute has a higher percentage response and a lower mean score than high performing firms.

The choice of perceived change in market share for classifying firms as high or low performing was tested by examining the ability of the firms classified as high and low performing to meet their initial goals, expected financial results, and ability to deploy strategy within the allocated resources. A chi-square test established that the differences are statistically significant ($\chi^2 = 3.83$, df $= 1$, $p < 0.05$). This suggests that change in market share is a good predictor of the level of performance. The analysis showed that high-performing firms had a perceived achievement rate of 75% of their initial goals and objectives, over 58% of the financial results expected and 64% for deployment within the resources allocated, whereas low-performing firms had a perceived achievement of less than 50% of their initial goals, slightly over 34% of the financial results expected and 45.7% perceived achievement within the resources allocated.

To ascertain the differences between the two groups, the mean score for each attribute was compared using the non-parametric Wilcoxon test. The results of the analysis is depicted in Table 11.

Table 11 indicates that the emphasis given by high performing firms to the dimensions of leadership and their attributes is greater than the emphasis given by low performing firms on the same attributes. The analysis shows the mean scores and the number and percentage of firms that emphasised the attributes of leadership "to a great extent" or "to a very greater extent" (mean score 4 or more). A Wilcoxon test is used on the mean scores of both high and low performing firms and indicates that the majority of the attributes in each dimension of leadership are statistically significant at $p < 0.05$ (outlined in bold type in Table 11).

A significant difference between high and low performing firms is in relation to the emphasis placed upon the transactional — rewards leadership style attributes *agrees the rewards that staff can expect . . . and tells employees criteria for performance related benefits*, which are emphasised to a far greater degree by high performing firms. However, the surprise inclusion of the transactional — rewards style attributes means that a combined balance of transformational — competence and human resources — employee focus leadership style must be moderated to include a "rewards" element in manufacturing SMEs. In conclusion, the results confirm that high performing firms place a higher emphasis on the dimensions and attributes of leadership than low performing firms. This finding demonstrates the use of leadership in practice.

Conclusion and Implications of Findings

The findings indicate that leadership is associated with strategic planning to a significant degree. The results of the correlation analysis indicate that the three main leadership styles are associated with five characteristics of strategic planning. The strongest associations are based on the *human resources — employee focussed, transformational — creativity and competence* styles followed by *transactional — rewards* leadership style.

Overall the results indicate that *human resources — employee focussed* style is associated with staff creativity. The analysis indicates that a high proportion of firms meeting performance measures to a significant extent are in the upper quartile of the *transformational — creativity* leadership style. In addition, in excess of 80% of firms in the upper quartile indicate that they emphasise all the strategic planning characteristics with the exception of strategy as a control measure to a strong/very strong extent.

The results also indicate that *transformational (competence and creativity)* leadership styles are associated with an external strategic orientation and departmental co-operation. Managers wishing to pursue an external strategic orientation and seek to encourage greater synergy based on departmental co-operation should emphasise transformational leadership styles.

Finally, *transactional — rewards* leadership style is associated with internal capabilities and strategy as a control mechanism. This means that a strong internal strategic orientation incorporating control mechanisms is best pursued in tandem with a leadership style based on a reward approach. Accordingly, Hypothesis 1 was confirmed.

The results also show that HTSFs place a greater emphasis on the attributes of leadership than low technology firms. This confirms Hypothesis 2. Finally, high performing firms place a greater emphasis on the attributes of leadership than low performing firms. This finding confirms Hypothesis 3.

The analysis suggests that managers would be well advised to give detailed consideration to their leadership style to ensure that it is aligned with their overall strategic plan. The results outlined provide a practical guide towards achieving this. Nevertheless, it is also necessary to note that there is no clear evidence to suggest that any given leadership style is guaranteed to lead to success. Clearly, the uniqueness of each organisation, their product, processes, stakeholders and markets are also important.

References

Ackoff, R. L. (1999). Transformational leadership. *Strategy and Leadership*, 27(1), 20–26.

Alexander, L. D. (1985). Successfully implementing strategic decisions. *Long Range Planning*, 18(3), 91–97.

Ancona, D., & Caldwell, D. (1987). Management issues facing new product teams in high technology companies. In: D. Lewin, D. Lipsky, & D. Sokel (Eds), *Advances in industrial and labour relations* (Vol. 4, pp. 191–221). Greenwich, CT: JAI Press.

Aram, J. D., & Cowan, S. S. (1990). Strategic planning for increased profit in small business. *Long Range Planning*, 23(6), 63–70.

Atkinson, A. A., Waterhouse, J. H., & Wells, R. B. (1997). A stakeholder approach to strategic performance measurement. *Sloan Management Review*, 38(3), 25–37.

Baker, W. H., Addams, H. L., & Davis, B. (1993). Business planning in successful small firms. *Long Range Planning*, 26(6), 82–88.

Bass, B. M. (1985). *Leadership and performance beyond expectation*. New York: Harper.

Bass, B. M. (1990). From transactional to transformational leadership: Learning to share the vision. *Organizational Dynamics*, 18(3), 19–31.

Bass, B. M., & Avolio, B. J. (1989). *Manual for the multifactor leadership questionnaire*. Palo Alto, CA: Consulting Psychologists Press.

Begley, T. M., & Boyd, D. P. (1986). Executive and corporate correlates of financial performance in smaller firms. *Journal of Small Business Management*, 24(2), 8–15.

Bennis, W., & Nanus, B. (1985). *Leaders: The strategies for taking charge*. New York: Harper & Row.

Berkeley Thomas, A. (1988). Does leadership make a difference to organisational performance? *Administrative Science Quarterly*, 33, 388–400.

Bracker, J., Keats, B., & Pearson, J. (1988). Planning and financial performance among small firms in a growth industry. *Strategic Management Journal, 9*(6), 591–603.

Chan, K. C. (1993). World class manufacturing. *Industrial Management and Data Systems, 93*(2), 5–12.

Conger, J. A. (1991). Inspiring others: The language of leadership. *Academy of Management Executive, 5*, 31–45.

Daft, R., & Lewin, A. (1993). Where are the theories for the 'new' organisational forms? An editorial essay. *Organisation Science, 4*(4), i–vi.

D'Aveni, R. (1994). *Hypercompetition: Managing the dynamics of strategic manoeuvring.* New York: Free Press.

Davis, J. H., Schoorman, F. D., & Donaldson, L. (1997). Towards a stewardship theory of management. *Academy of Management Review, 22*(1), 20–47.

Day, D. V., & Lord, R. G. (1988). Executive leadership and organisational performance: Suggestions for a new theory and methodology. *Journal of Management, 14*(3), 453–464.

De Geus, A. (1997). The living company. *Harvard Business Review, 75*(2), 51–59.

Fombrun, C. J. (1992). *Turning points: Creating strategic change in corporations.* New York: McGraw-Hill.

Ghobadian, A., & Gallear, D. (1997). TQM and organisation size. *International Journal of Operations and Production Management, 17*(2), 121–163.

Goold, M. (1991). Strategic control in the decentralized firm. *Sloan Management Review, 32*(2), 69–81.

Goold, M., & Quinn, J. (1993). *Strategic control: Milestones for long-term performance.* London: Pitman Publishing.

Hambrick, D. C., & Mason, P. A. (1984). Upper echelons: The organisation as a reflection of its top managers. *Academy of Management Review, 9*(1), 193–206.

Hamel, G., & Prahalad, C. K. (1994). *Competing for the future.* MA: Harvard Business School Press.

Hammer, M. (1990). Reengineering work: Don't automate, obliterate. *Harvard Business Review, 68*(4), 104–112.

Hampden-Turner, C. (1990). *Creating corporate culture: From discord to harmony.* Reading, MA: Addison-Wesley.

Hart, S., & Banbury, C. (1994). How strategy-making processes can make a difference. *Strategic Management Journal, 15*(4), 251–269.

Hart, S. L., & Quinn, R. E. (1993). Roles executives play: CEO's, behavioural complexity, and firm performance. *Human Relations, 46*(5), 543–574.

Hayes, R. H., & Abernathy, W. J. (1980). Managing our way to economic decline. *Harvard Business Review, 58*(4), 67–77.

Hellstrom, T., Malmquist, U., & Mikaelsson, J. (2001). Decentralising knowledge: Managing knowledge work in a software engineering firm. *Journal of High Technology Management Research, 12*, 25–38.

Hill, C. W. L., & Jones, G. R. (2001). *Strategic management: An integrated approach* (5th ed.). Houghton Mifflin.

Hirschorn, L., & Gilmore, T. (1992). The new boundaries of the 'boundaryless' corporation. *Harvard Business Review, 70*(3), 104–115.

Hitt, M. A., Ireland, R. D., Camp, S. M., & Sexton, D. L. (2001). Strategic entrepreneurship: Entrepreneurial strategies for wealth creation. *Strategic Management Journal, 22*(Special Issue), 479–492.

Joyce, P., Seaman, C., & Woods, A. (1996). In: R. Blackburn, & P. Jennings (Eds), *Small firms: Contributions to economic regeneration.* London: Institute for Small Business Affairs.

Kargar, J. (1996). Strategic planning system characteristics and planning effectiveness in small mature firms. *The Mid-Atlantic Journal of Business, 32*(1), 19–34.

Kargar, J., & Parnell, J. A. (1996). Strategic planning emphasis and planning satisfaction in small firms: An empirical investigation. *Journal of Business Strategies, 13*(1), 42–64.

Keller, R. T. (1992). Transformational leadership and the performance of research and development project groups. *Journal of Management, 18*(3), 489–501.

Kouzes, J. M., & Posner, B. Z. (1987). *The leadership challenge: How to keep getting extraordinary things done in organizations.*

Lee, C. (1991). Followership: The essence of leadership. *Training, 28*, 27–35.

Lim, B. (1997). Transformational leadership in the U.K. management culture. *Leadership and Organizational Development Journal, 18*(6–7), 283–290.

McKiernan, P., & Morris, C. (1994). Strategic planning and financial performance in U.K. SMEs: Does formality matter? *British Journal of Management, 5*(Special Issue), S31–S41.

Mills, D. Q., & Friesen, B. (1992). The learning organization. *European Management Journal, 10*(2), 146–156.

Mintzberg, H. (1994). The rise and fall of strategic planning. *Harvard Business Review, 72*(1), 107–114.

Moxley, R. S. (2000). *Breathing new vitality and energy into individuals and organisations.* San Francisco: Jossey-Bass.

Naffziger, D. W., & Mueller, C. B. (1999). Strategic planning in small businesses: Process and content realities. *Proceedings of the 14th Annual USASBE Conference.* San Diego, CA.

Porter, M. E. (1980). *Competitive strategy: Techniques for analyzing industries and competitors.* New York: Free Press.

Porter, M. (1990). *The competitive advantage of nations.* London: MacMillan.

Porter, M. E. (1996). What is strategy. *Harvard Business Review, 74*(6), 61–78.

Ramanujam, V., & Venkatraman, N. (1987). Planning and performance: A new look at an old question. *Business Horizons, 30*, 19–25.

Ramanujam, V., Venkatraman, N., & Camillus, J. (1986). Multi-objective assessment of effectiveness of strategic planning: A discriminant analysis approach. *Academy of Management Journal, 29*(2), 347–372.

Roper, S. (1997). Strategic initiatives and small business performance: An exploratory analysis of Irish companies. *Entrepreneurship and Regional Development, 9*, 353–364.

Schwenk, C. R., & Shrader, C. B. (1993). The effects of formal strategic planning on financial performance in small firms: A meta analysis. *Entrepreneurship in Theory and Practice, 17*(3), 53–64.

Sexton, D. L., & Bowman-Upton, N. (1991). *Entrepreneurship: Creativity and growth.* New York: MacMillan.

Shamir, B., Zakay, E., Breinin, E., & Popper, M. (1998). Correlates of charismatic leadership behavior in military units: Subordinates attitudes, unit characteristics, and superior' appraisal of leader performance. *Academy of Management Journal, 41*, 387–409.

Skinner, W. (1996). Three yards and a cloud of dust: Industrial management at century end. *Production and Operations Management Society, 5*(1), 15–24.

Storey, D. (1994). *Understanding the small business sector.* London: Routledge.

Thomas, A. S., Litschert, R. J., & Ramaswamy, K. (1991). The performance impact of strategy-manager co-alignment: An empirical examination. *Strategic Management Journal, 12*(7), 509–522.

Tichy, N. M., & Devanna, M. A. (1986). *The transformational leader.* New York: Wiley.

Veliyath, R., & Shortell, S. M. (1993). Strategic orientation, strategic planning systems, characteristics and performance. *Journal of Management Studies, 30*, 359–381.

Webber, A. (1988). *Harvard Business Review, 66*(1), 4–5.

Weihrich, H., & Koontz, H. (1993). *Management: Global perspectives* (10th ed.). New York: McGraw-Hill.

Wilderom, C., & van den Berg, P. (1997). A test of the leadership-culture-performance model within a large Dutch financial organisation. Working Paper 103952, Tilburg University.

Womack, J., & Jones, D. (1994). From lean production to the lean enterprise. *Harvard Business Review*, *72*(4), 93–103.

Yammarino, F. J., & Bass, B. M. (1990). Transformational leadership and multiple levels of analyses. *Human Relations*, *43*(10), 975–996.

Yuhl, G. A. (1989). *Leadership in organisations* (2nd ed.). New York: Academic Press.

Yuhl, G. A. (1998). *Leadership in organizations* (4th ed.). Upper Saddle River, NJ: Prentice-Hall.

Chapter 3

Measuring E-Business-Adoption in SME

Marijke Van Der Veen

Introduction

In this paper we describe the process and result of designing an instrument to measure the level of electronic business (e-business) adoption in SMEs. The instrument (a questionnaire) will be used within the context of a national Dutch program "The Netherlands goes Digital" (EZ 2001) in order to monitor the result of assisting SMEs in adopting e-business. In the program the assistance is specifically aimed at Internet access, Internet presence, e-mail facility, and carrying out business transactions via the Internet. We will show in this paper, that measuring the level of ICT applications does not provide insight into the actual use of the applications. We believe there are several dimensions that need to be examined in order to indicate the level of use of ICT applications for electronic business. We propose that electronic business is not limited to the use of Internet-related applications and e-mail. In our view electronic business means that a firm, with the help of applications like e-mail and Internet, electronically:

– offers or gets data;
– exchanges data; and/or
– enters into or concludes an agreement

with another party in order to improve business processes (adapted from EZ 1998). Further, we consider electronic business to be an innovation as expressed by Rogers (1995): an idea, practice, or object that is perceived as new by an individual or other unit of adoption.

The main question we would like to address in this paper is: can we find an instrument capable of assessing the extent of use of electronic business in small companies? The paper is organised as follows.

First, we will review into innovation adoption research and information systems research to identify theoretical clues for measuring the level of adoption in general. Next, we will discuss theoretical and practical requirements for the measuring instrument and determine the dimensions of e-business we want to investigate. In the subsequent section

New Technology-Based Firms in the New Millennium, Volume III
© 2004 Published by Elsevier Ltd.
ISBN: 0-08-044402-4

we review literature specifically investigating e-business adoption to look for existing answers and useful approaches to the measuring problem. Subsequently, we will describe the operationalisation of the different dimensions of e-business adoption. The paper closes with discussing the limitations to this instrument, conclusions and suggestions for future research.

Theoretical Background

In innovation adoption research several authors have discussed the operationalisation of innovation as a dependent variable. Downs & Mohr (1976) observe three principal, interrelated operationalisations of innovation: (1) the time of first adoption or use; (2) dichotomous adoption or non-adoption; and (3) the extent to which an organisation has implemented an innovation, or the degree to which an organisation is committed to it. The first two operationalisations are relatively easy to measure. However, "the extent of implementation comes closer to capturing the variations in behaviour that we really want to explain" (Downs & Mohr 1976: 709). In discussing "ideal" innovation attribute studies, Tornatzky & Klein (1982) focus on both adoption and implementation as the dependent variable instead of a dichotomous yes/no adoption decision. They state that a degree-of-implementation will vary widely across a group of adopting organisations, whereas adoption will not. Gatignon & Robertson (1985) conclude that the concept of adoption has been used in a rather limited way and refers to a single decision. They conceptualise adoption in two dimensions (width and depth) to be able to assess maximum diffusion potential within a social system. Width refers to the number of people adopting or the product. Depth indicates the amount of usage.

In information systems research, Trice & Treacy (1988) investigate utilisation as a dependent variable. They state that the decision about which aspects of use to measure should be guided in part by the purpose of the investigation. For example, in order to guide an organisation's EDI initiatives, Massetti & Zmud (1996) present a multidimensional approach to measure EDI-usage. They investigated EDI-implementation along four distinct dimensions: volume, breadth, diversity, and depth and depicted that mapping implementation contributes to understanding the nature and impact of EDI on an organisation. Broadbent & Weill (1996, 1999, 2000) show that assessing IT infrastructure capability by means of three measures (two types of services, reach, and range) makes it possible to capture this complex concept and make firms comparable. According to Lassila & Brancheau (1999), information technology utilisation is generally defined as the *volume* of technology used (number of hours of technology use), the *reliance* on the technology to get the job done (how much a user depends on the technology), and the *diversity* of different functions put to use (the number of different software features used).

Both innovation adoption research and information systems research suggest using a measure that indicates a *degree* of adoption instead of a dichotomous yes/no adoption variable. The dimensions of adoption suggested particularly important to investigate are:

– diffusion of the adoption of the innovation in the organization;
– the number of different uses for the innovation; and
– the amount or volume of the innovation used.

In the next section, we will consider these theoretical suggestions along with practical prerequisites in order to operationalise the adoption of e-business.

Dimensions of E-Business Adoption

In this section we describe our approach to measuring the e-business adoption level in SMEs. We determine five dimensions of e-business adoption. Theory and empirical findings support our choice. We also take some practical requirements into consideration.

To investigate the level of e-business adoption, we need to consider the extent of implementation and the different uses for innovation. We decide to take a process-based view of electronic business to approach these dimensions. In trying to assess whether business processes are supported by e-business applications, we will: (a) attempt to describe the extent of implementation within a company; and (b) ascertain where e-business applications are utilized.

We also provide empirical support for taking a process-based view on e-business. During interviews we observed, that most entrepreneurs refer to their business processes when they are asked to describe their company (Wessels 2002). Consequently, respondents are easily able to identify a process-based approach to electronic business. In addition, for the instrument to be valuable to consultants and entrepreneurs, it should not only provide an insight into the current diffusion of e-business practice within the company, but also guide future application. When considering possibilities for future use of e-business, SMEs mainly look for improving efficiency and effectiveness of business processes. Improving efficiency has traditionally been the primary use of information technology (Riggins 1999; Riggings & Mitra 2001; Stroeken 1999; Stroeken & Coumans 1998). In a recent survey among Dutch SMEs the majority of companies named ease of use and speed as the main reasons for planning e-business (NIPO 2001). In the evaluation of the Dutch e-business project Sp.OED, SMEs ($n = 522$) expect that doing business electronically will mainly improve the quality of service and save time and costs (Syntens 2002). These findings confirm that SMEs are primarily interested in improving their business processes. By assessing which business processes are supported by e-business applications and which are not, we will also attempt to determine where applications could support business processes in the future.

Dimension 1. The extent of business process support.

In our view electronic business means a firm, which utilizes applications like e-mail and Internet, electronically:

– offers or gets data;
– exchanges data; and/or
– enters into or concludes an agreement

with another party in order to improve business processes (adapted from EZ 1998). From this definition we can distinguish three levels of data-exchange in business process support: (a) information (offering or getting data); (b) interaction (exchanging data);

and (c) transaction (exchanging data leading to an agreement). The levels indicate an increasing "intensity" or complexity of the exchange (see also Boddy & Boonstra 2000; Dekkers 2000). Consequently, in general, the supporting ICT-applications will increase in complexity as well. For example, offering general information to customers via a website is less complicated to facilitate than offering customers the possibility on a website to inquire about their order status on-line. The use of ICT-applications for business process support can therefore be classified according to these levels of data-exchange.

Dimension 2. Level of data-exchange.

Electronic business application is a generic term for ICT applications, enabling data-exchange to support business processes. In practice, entrepreneurs and consultants do not restrict their perception of electronic business to the use of e-mail or Internet (Peelen 2001; Wessels 2002). Consequently, when investigating adoption we need to determine what type of application is used to support the business process.

Dimension 3. E-business applications used.

The dimensions of adoption we have chosen so far will enable us to assess to what extent the exchange of data in business processes is supported by certain applications. Another important dimension to investigate is the amount or volume of the innovation used. How often are the e-business applications used, and how many people use them? To what extent are conventional methods of communication like mail, fax, or telephone, substituted by for example e-mail and Internet?

Dimension 4. Amount of use.

As we noted earlier, e-business can improve the efficiency and effectiveness of business processes. However, Riggins (1999; Riggings & Mitra 2001) distinguishes a third category of value creation for e-business: strategic benefits. For example, an international website could attract new customers or open up a new market. By assessing where and how business processes are supported, we provide insight into the impact of e-business for the company. By adding this fifth dimension of adoption, we have an additional measure of what the innovation is used for.

Dimension 5. Impact of use.

Apart from choosing the dimensions of e-business adoption as described above, we need to make some choices to satisfy practical requirements for the instrument. As stated in the introduction, the main goal of this research is to design an instrument capable of measuring the extent and use of electronic business in small and medium sized companies. The instrument will be used in a project to monitor the result of assisting SMEs in adopting e-business. During the project's procedure a first assessment of the e-business adoption needs to be taken before the actual advice is given. We decided to choose the form of a questionnaire for the instrument because it can be sent in advance and filled out independently by the entrepreneur.

To guard the usefulness and practicality of the questionnaire, the wording of the questions and terms used, need to be understandable to a wide variety of SME entrepreneurs. The participants in the project work in different lines of business and overall heterogeneity of

SMEs is a well-known fact. A lot of attention should be given to translating theoretical concepts into unambiguous questions, which correspond with daily SME reality. Finally, the time needed to fill out the questionnaire is limited to about 5–10 minutes, as it is part of a larger questionnaire.

In the next section, with the above dimensions of e-business adoption and the practical considerations in mind, we review existing work on measuring e-business adoption. We look for useful approaches to facilitate operationalisation of the dimensions.

Literature Review

Recently several researchers have attempted to measure the extent of use of e-business applications in one way or the other. We collected 28 studies on e-business, e-commerce and Internet-related innovation-adoption regardless of their precise definition, to get an idea of the alternative ways to operationalise these innovations. Basically, two groups of studies emerge. One group of studies is clearly explorative or descriptive and focuses on mapping the extent of use (upper half of Table 1). Usually several dimensions of use are investigated. The other group of studies focuses on explaining the use or extent of use through a set of independent variables (lower half of Table 1). Prime goal is to understand what influences the use of Internet or e-commerce, often to make recommendations to policy makers eager to stimulate diffusion of Internet-related innovation. The extent of use (the dependent variable) seems of less importance in these studies and is therefore often operationalised in a simple way. In both explorative and explanatory studies we find different approaches to operationalise the use or adoption of e-business. The measure of use is often operationalised along several dimensions. We were able to distinguish six categories of dimensions:

Applications

Here, use is operationalised as the use in the company of certain applications like e-mail, Internet, website, intranet and the like. Ticking applications comes close to measuring adoption/non-adoption of a list of applications. Sometimes the variable of use is a dichotomous variable (marked as X (d) in Table 1) referring to the adoption or non-adoption of only one application. A lot of studies (15 out of 28) apply an application-based measure of use.

Activity

An activity-based type of measure offers more insight into the way that the company is supported by e-business applications. Usually several business processes are listed like sending purchase orders to suppliers or offering information to customers. In a lot of activity based measures it is unclear how the list of business processes or activities was put together. This measure of use was also frequently applied in the studies examined (15 out of 28).

Table 1: Studies of e-business adoption.

Author(s)	Measure of Use	Dimensions Explored	Categorization of Dimensions			Impact	First Time of Use	Other
			Application	Activity	Stage of Development			
Nature of study: descriptive								
Abell & Lim (1996)	Business use of internet	Internet resources	X					
		Internet uses		X				
		Current perceived benefits				X		
Poon & Swatman (1996)	Internet-usage	Communication technologies used	X					
		Usage of company web pages		X				
Colarelli O'Connor & O'Keefe (1997)	Role and use of the web	Categorization of on-line marketing			X			
Poon & Strom, Poon & Swatman (1997)	Status of internet-use	Function of the internet		X				
		Benefits gained				X		
		Stage in business relationship development			X			
Teo *et al.* (1997)	Internet usage	Uses of internet		X				
		Parties communicating with						X
		Usage experience					X	
		Frequency of use						X
		Daily usage time						X
Ng, Pan & Wilson (1998)	Business use of WWW	Access to internet (# years)					X	
		Web presence (# years)					X	
		Business use of website		X				
		Cost of setting up/maintaining website						X
		Perceived effectiveness of website				X		
Teo & Tan (1998)	Internet adoption	Use of email		X				
		Rank of internet applications used	X					
		Benefit evaluation				X		

Study	Construct	Measure						
Poon & Swatman (1999)	Internet commerce	Experienced benefits				X	X	
		Function of the internet		X		X		
		Stage of integration internet-internal systems			X			
Riemenschneider & McKinney (1999)	Web-based ecommerce adoption	Employment of web-based ecommerce	X(d)			X		
Walczuch et al. (2000)	Internet-adoption	Current perceived benefits		X				
		Ways of using internet		X				
		Applications/possibilities on website	X	X				
Ellis-Chadwick et al. (2002)	Internet adoption	Registered site (URL)	X(d)	X		X		
		Services offered		X				
Deeter-Schmelz & Kennedy (2002)	Use of internet	Internet utilization		X	X			
		Internet usefullness		X	X			
Descriptive total = 12		Number of studies descriptive	6	9	3	6	2	4
		Total (12 = 100%)	50%	75%	25%	50%	17%	33%
Nature of study: explanatory								
LaRose & Hoag (1996)	Internet adoption	Internet connection	X(d)			X		
Sillince et al. (1998)	Email adoption	Email adoption	X(d)					
		Time of adoption					X	
		Email use	X	X				
		% of employees using email						X
		External vs. internal email use						X
		Email use by trading partners						X
Busselle et al. (1999)	Internet uses	Effect on job				X		
		Frequency of use (# days in last week)						X

Table 1.   (Continued)

Author(s)	Measure of Use	Dimensions Explored	Categorization of Dimensions			Impact	First Time of Use	Other
			Application	Activity	Stage of Development			
Premkumar & Roberts (1999)	Adoption of ICT technologies	Communication technologies used	X					
		Business transactions supported		X				
Teo et al. (1999)	Internet usage	Frequency of usage (# times per day)						X
		Daily usage (# time spent)						X
		Diversity of usage		X				
Nambisan & Wang (2000)	Web technology adoption	Adoption time corporate website					X	
		Adoption time corporate internet					X	
Beatty et al. (2001)	Web adoption	Number of months website operational					X	
Eder & Igbaria (2001)	Internet diffusion and infusion	Departments with min. 1 internet application						X
		Internet deployment level			X			
		Earliness of adoption					X	
Griffin (2001)	Internet-usage	Extent of e-mail use	X					
		Extent of internet browser use	X					
		Internet impact on business value				X		
Katz & Dennis (2001)	Internet use	Level of internet technology	X					
		Centrality of the internet				X		
Kendall et al. (2001)	Ecommerce adoption	Internet activities		X				
		Usage of internet	X					
		Willingness to adopt		X				

Study	Focus	Components of internet adoption	Use of WWW	Use of email	Intensity of internet/WWW use	Diversity of applications used	Frequency of use	Frequency of adoption of e-commerce activities	Firm website
Mehrtens, Cragg & Mills (2001)	Internet adoption	X				X			
Wei et al. (2001)	Uses of internet		X	X				X	
Cheng et al. (2002)	Internet & WWW usage				X			X	
Eastin (2002)	Overall adoption of e-commerce						X	X	X
Sadowski et al. (2002)	Strategic use of internet								X(d)
Number of studies explanatory		9	6	1	6	3	5	7	
Total (16 = 100%)		56%	38%	6%	38%	19%	31%	44%	

Explanatory total = 16

Stage of Development

In only a few studies we see attempts to assess the use of Internet using a stage or level of development model. This is in contrast with literature on Internet- or e-business-strategy where the use of multi-stage business models is very common to characterise companies and their use of the Internet (e.g. Amit & Zott 2000; Earl 2000; Fischer 1998; Holland *et al.* 2000; Timmers 2000; Venkatraman & Henderson 1998). For measuring the level of adoption, stage models can only offer help in roughly categorising the use of e-business in companies or departments.

Impact

Another category of studies characterises utilisation based on the impact the Internet-based applications have on the business. Usually, the respondents are asked about the (perceived) benefits gained by using Internet. Remarkably, negative outcomes of use like higher cost are often left out.

First Time of Use

A classic measure of diffusion is based on the notion, that it is possible to classify organisations into adopter categories based on the point in time when they adopt the innovation relative to other organisations (Rogers 1995). The measure can also be used inside organisations to measure diffusion (e.g. Eder & Igbaria 2001).

Other

Of course other measures of use can be applied. In this category we find measures representing some sort of intensity or frequency: How much, how often or how widespread is the innovation being used? For example, the number of employees using e-mail or the

Table 2.   Comparison of dimensions of e-business adoption.

Dimensions of E-Business Adoption	
Chosen	**From the Literature Review**
Dimension 1. The extent of business process support	Activities
Dimension 2. Level of data-exchange	Stage of development
Dimension 3. E-business applications used	Applications
Dimension 4. Amount of use	Other (some)
Dimension 5. Impact of use	Impact
	First time of use

number of departments with an intranet application can be considered as measures of width of adoption (Gatignon & Robertson 1985).

Overlooking Table 1 we distinguished several dimensions comparable to the dimensions we chose to investigate. In Table 2 we compare the dimensions we chose with the dimensions found in the literature review. The fact that all dimensions chosen by us can be found in literature confirms and grounds our choice. In addition, the studies reviewed can assist in operationalising our dimensions with the exception of the studies reviewed the first time of use of an innovation is frequently used to investigate adoption. We did not select this dimension, because our instrument intends to record the extent of use of e-business at a certain moment. For our purpose, the history or experience with the innovation is not relevant.

Method of Operationalisation

The process of translating the conceptual dimensions into usable questions is iterative in nature. Based on our conceptual ideas and suggestions from the literature review, a questionnaire was designed for the study. During the design process we utilized expert opinion to guide our practical decisions. The questionnaire was reviewed in various stages by fellow academics, Syntens consultants, commercial consultants, a small business research institute (EIM) and SME entrepreneurs. Finally, the questionnaire was formally tested with 15 companies. The test included a structured interview to check validity and comprehensibility (Wessels 2002).

The Proposed Instrument

To measure the extent of use of electronic business in SME we chose to investigate five dimensions of e-business adoption (see Table 2). In terms of operationalisation for each dimension. Firstly, we considered usable operationalisations we found in the literature. Secondly, we evaluated these definitions against our needs and practical requirements, to arrive at measurable indicators for our dimensions of e-business adoption.

Dimension 1. The Extent of Business Process Support

In the literature review we found several studies that investigated the support of business activities as a measure of e-business adoption (Table 1: Activities). However, there are no studies that: (a) consider the company in its entirety i.e. cover more business processes than e.g. sales; and (b) systematically put together a list of activities for these business processes. Therefore we decide to look for a model that can be used to map several business processes in a company.

Porters value chain (Porter & Millar 1985) seems an obvious and well-accepted choice to map business processes in a company. Porter recently showed, on the basis of his value chain model, how application of the Internet can support business processes (Porter 2001).

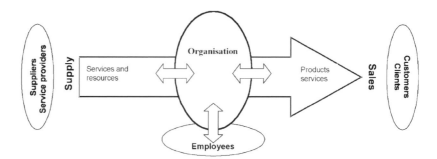

Figure 1: Generic company model.

For two reasons we prefer to use a simplified version of his model. First, Porters' original model is oriented towards traditional production companies. We also found that some companies had difficulty in recognising their business in Porters' model. For example, a wholesaler does not consider procurement as a supporting process and a transport company has trouble understanding separated inbound and outbound logistics. Second, small companies do not identify with all business processes in the model, simply because well-defined "departments" often do not exist. To be able to compare the level of e-business in companies operating in different lines of business, we need a more generic model.

In MIS publications the business processes of a company are sometimes grouped by the different parties with which information is exchanged (e.g. Broadbent & Weill 2000; Kambil & Van Heck 1998). For the business use of Internet a categorisation is possible in three main areas: customer relations, dealing with suppliers, and internal company operations (see also Applegate *et al.* 1996; Chappell & Feindt 1999; Den Hertog, Holland & Bouwman, 1999; IDC 2001; Shaw *et al.* 1997). Another way to roughly divide business processes in categories is to use a general transformation model where input is transformed by an organisation into output (e.g. Daft 1998: 59; Laudon & Laudon 1997: 6; Slack 1998: 10). Combining Porters' model with these views, we have worked out a simple generic company model that proved to be representative of companies in various lines of business. The model is depicted in Figure 1.

To arrive at an instrument that can assess the extent of support for each of these processes by e-business applications, we need to identify what activities take place within each category of business processes, and how they are supported. We will approach the identification of business activities by taking into account the second dimension.

Dimension 2. Level of Data-exchange

The use of e-business applications for business process support can be classified according to three levels of data-exchange: (a) information (offering or getting data); (b) interaction (exchanging data); and (c) transaction (exchanging data leading to an agreement). In the literature, some studies take a stage or level approach to the use of Internet (Table 1: Stage of development). However, we did not find a usable suggestion to operationalise the level of

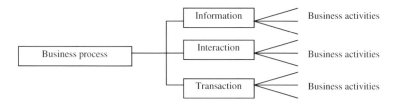

Figure 2: Structure for operationalisation.

data-exchange, as we perceive it. To come to a usable indicator, we decided to group busi-ness activities for each category of business processes according to these levels of increasing complexity. Combining this perspective with the three categories of business processes from our generic company model (supply, organisation and sales), we arrive at a guiding structure to operationalise the extent of business support and the level of data-exchange (Figure 2).

Next, on the basis of the complexity of data-exchange we want to specify business activi-ties for each category of business processes. In the literature review, several authors specify activities that are supported by Internet or e-mail. The activities, usually for the supply and sales process, are often derived from commercial, exchange or trade processes. However, a systematic approach to the specification of activities is often lacking. Peelen (2001) considers three models to represent the activities of the trade process in general: the business market process model by Anderson & Narus (1999), the model of exchange processes by Kambil & Van Heck (1998) and, the general commerce model by Nissen (1997). From these models Peelen deduces five main activities in the trade process: search, specification, fulfilment, payment and after sales service. These main activities are split up into 40 sub-activities. As they are both trade processes, we can use the list to identify the activities of the supply and sales process. Because of the constraint on the available time to fill out the questionnaire, we use the list as a basis to arrive at a smaller list of activities for the supply and sales process. Expert opinion from consultants and the EIM and, relevant literature (a.o. Kambil & Van Heck 1998; Porter 2001; Van Doorne & Verheijen 2000) helped us to make a selection. The activities are described in universal terms, so they are recognisable for producing companies and service providers alike. For the supply process, the list of activities is depicted in Table 3.

There is a lack of literature to help specify our third category of business processes, organisation. In the literature we mostly find that electronic business is limited to the support of supply (e-procurement) and sales processes (e-commerce) by e-business applications. Although the support of intra-organisational processes by e-business applications may not strictly be regarded as "business," we nevertheless want to include the support of the organisation for two reasons. First, benefits are just as well to be gained by supporting the internal organisation with for instance Internet (e.g. Applegate *et al.* 1996; Porter 2001). Second, we observed that positive experience with e.g. e-mail and Internet within the organisation can motivate SMEs to support processes with external parties as well (Wessels 2002).

Within each organisation important processes like human resources management, operations, general management and, quality management take place. Yet, for different

Table 3. Business activities for supply.

Dimension 2: Level of Data-Exchange	Dimension 1: Business Process Support (Supply)
Information	Searching information about suppliers/serviceproviders and their products or services
Interaction	Specifying /selecting products/services
	Requesting quotations for products or services
	Inquiring about logistics (availability, progress, status)
	Electronically exchanging data with suppliers/service providers
Transaction	Electronically ordering a product or service
	Electronically receiving invoices
	Electronic payment of supplies (incl. Electronic banking)
	Electronically receiving products/services

types of organisations different business processes will be relevant. For example, a machine manufacturer relies heavily on its R&D-department, whereas a consultancy firm will not. In order to guard usefulness of the questionnaire, we discussed the relevance of various intra-organisational processes for different lines of business with consultants. We also asked them to particularly consider, for which business activity electronic-business could be beneficial. Again, the recognition of the different level of data-exchange helped us to specify activities. This pragmatic approach yielded a list of business activities for the organisation.

Dimension 3. The E-business Applications Used

Business processes can be supported by various ICT-applications. Overlooking Table 1, with regard to electronic business most authors look at the use of Internet-related applications. The use of various terminologies like WWW, "the" Internet, e-mail, Intranet, can be confusing. According to Applegate *et al.* (1996) we look at different levels of electronic commerce. In our view, worrying too much about correct technical terminology diverts our attention from the purpose of this questionnaire. In assessing *how* business processes are supported we need to know what applications are used. But if we find, for example, a company using e-mail facilities on an internal network (LAN) for its internal communication, we accept the fact that the network may not be based on the Internet-protocol. Consequently, we chose spoken language to identify applications instead of technical terms. In addition to e-mail and Internet, we chose to look at mobile data-communication. We described this category as "electronic data exchange (no speech) using applications on wireless devices like SMS, WAP, GPRS on a mobile phone, applications on handheld terminals, etc." Finally we decided for a fourth miscellaneous category, to give entrepreneurs an opportunity to show any other applications for the electronic exchange of data like EDI. We decided to look at four categories of applications for two reasons. First, in practice, electronic business is not restricted to e-mail and Internet. Second, mobile data-communication is expected to grow in the coming years and consultants observe that within SMEs the use of especially SMS is increasing. During the tests we found respondents understood all four categories of applications well enough for our purpose.

Recapitulating, to investigate dimension 1, 2 and 3 we grouped the questions resulting into a "question table" for each business process (see Appendix 1).

Dimension 4. Amount of Use

Reviewing the measures used so far, we are able to assess to what extent business processes are supported by certain applications. What is lacking is a measure of intensity of use in terms of frequency or volume. In the literature we find some measures that relate to frequency or volume of use (e.g. Busselle *et al.* 1999; Teo *et al.* 1997, 1999; Teo & Choo 2001). However, the fact that, for example, a salesperson uses Internet every day does not render into a measure of relative importance of Internet for her job. Neither does it give us information about the amount use of Internet compared to more conventional methods of communication like mail, fax, or telephone. Research showed that carefully measuring the amount of use of e-business applications is possible, but very complicated and laborious (Peelen 2001). To limit the length of the questionnaire, we decided not to answer these questions in detail. We actually follow the suggestion of Lassila & Brancheau (1998) to measure the *reliance* on the technology to get the job done. To assess to what extent a business process relies on e-business applications, we simply ask what happens to the business process when the applications are out of order (Appendix 2).

Dimension 5. Impact of Use

An additional measure to characterise the use of e-business is assessing the impact of the e-business applications on the company. In the literature review we find several studies investigating the perceived benefits gained by using e-business (Table 1, Impact). In most publications it is unclear how the authors put together the lists of benefits.

To approach the investigation into possible impacts of use in a structured way, we propose to use the work of Riggins. Riggins (1999; Riggings & Mitra 2001) distinguishes three categories of value creation for e-commerce and e-business:

– improving efficiency (time and cost-related);
– improving effectiveness (related to communication); and
– strategic benefits (related to products, markets and services).

We used these categories of value creation as a starting point for operationalisation. Using several other sources of literature and empirical research (e.g. Chappell & Feindt 1999; NIPO 2001; Riggins 1999) we composed a list of possible impacts of e-business (see Appendix 2). This list was evaluated and supplemented by consultants. Also, during the test interviews we made some changes to the list (Wessels 2002). The final list is in Appendix 2 (Dimension 5).

Besides investigating the *type* of impact a company experiences from the application of e-business, we want to investigate the *extent* of the impact. For example, a company indicates that, as a result of e-business, a new market can be served. This, however, is no indication of the organisational changes the company underwent to achieve this. Therefore we ask the respondent to give one's opinion about the extent of the changes as a result of e-business, to obtain a sense of *overall* impact (see Appendix 2, Dimension 5). Again, the limit to the length of the questionnaire kept us from a more detailed question.

Limitations

Overlooking the various categories of dimensions of use in Table 1, and considering the theoretical suggestions, we are able to identify the limitations to the proposed instrument. First, a categorisation based on the first time of use is missing. As a consequence we cannot identify the adopting firm or part of a firm as a pioneer or laggard regarding e-business. Second, as stated earlier, we do not assess intensity of use in terms of volume (number of hours), or frequency (times per period). This implies that, in this questionnaire, a company using e-mail once a month to discuss conditions with one of its suppliers resembles one that uses e-mail every day with all of its suppliers. Adding a question about the companies' reliance on the technology only partly solves this problem. Third, the instrument focuses on the company itself. However, communicating through e-business applications requires some other party to participate as well. We do not assess to what extent and in what way business processes of the focal firm and the communicating parties are integrated or connected.

The application of e-business can support: (a) operations; and (b) strategy (Applegate *et al.* 1999: 16–24). Our measuring instrument examines in what way the operations are supported. For our purpose this is an acceptable limitation. Strategy support by e-business applications however, can offer great opportunities (Riggings & Mitra 2001). By investigating the impact of e-business we nevertheless gain some insight in to what extent entrepreneurs aim for strategic benefits.

Conclusions and Further Research

The main question in this paper was: can we find an instrument capable of assessing the extent of use of electronic business in small companies? We explored the concept of e-business adoption by considering theoretical suggestions and, empirical findings. We found five dimensions of e-business adoption important to investigate: the extent of business process support, the level of data-exchange, the e-business applications used, the amount of use and, the impact of use. We found that in the existing literature a systematic and integral approach to measuring e-business adoption, as we see it, is missing. Therefore, we developed our own instrument covering the five dimensions we identified. It appeared to be possible to operationalise each dimension into usable questions using expert opinion and theoretical suggestions. The result is a questionnaire capable of assessing the extent of e-business adoption.

Consultants are currently using the questionnaire to support and monitor the impact of their work. In the next few months we expect that a sufficient number of questionnaires will be returned to enable further validation of the questionnaire. We expect that the results will indicate significant sector-differences in e-business adoption.

Adoption is a process (Rogers 1995) where actual implementation of, in this case, e-business applications is preceded by awareness of its benefits, and preparation for its application. As Downs & Mohr (1976) suggest, adoption should be measured as the extent of use or commitment. By "ticking applications" and looking at business process support by e-business we do not assess the extent to which a firm is aware of the possibilities of e-business, and "ready" for application. By using the instrument described in this paper, we only observe implemented applications and do not capture raised awareness or preparation for implementation like an improved IT-infrastructure or knowledge level. Den Hertog *et al.* (1999) distinguish three stages in the adoption of e-commerce: readiness, intensity and impact. The instrument described in this paper is an attempt to investigate intensity and impact of electronic business. To fully capture the level of adoption as awareness and use, we need to measure a level of awareness as well. This paper is published as part of ongoing research on the topic of measuring e-business application in SME. A second article on how to assess the extent of use including the level of awareness is being prepared.

References

Abell, W., & Lim, L. (1996). Business use of the Internet in New Zealand: An exploratory study. *Proceedings of the Second Australian World Wide Web Conference, AusWeb96*, Gold Coast, Australia.

Amit, R., & Zott, C. (2000). Value drivers of e-commerce business models. Insead Working Paper 2000/54/ENT/SM.

Anderson, C., & Narus, J. A. (1999). *Business market management: Understanding, creating, and delivering value*. Upper Saddle River, NJ: Prentice-Hall.

Applegate, L. M., Holsapple, C. W., Kalakota, R., Radermacher, F. J., & Whinston, A. B. (1996). Electronic commerce: Building blocks of new business opportunity. *Journal of Organizational Computing and Electronic Commerce*, *6*(1), 1–10.

Applegate, L. M., McFarlan, W., & McKenney, J. L. (1999). *Corporate information systems management: The challenges of managing in an information age*. Chicago: Irwin.

Beatty, R. C., Shim, J. P., & Jones, M. (2001). Factors influencing corporate web site adoption: A time-based assessment. *Information and Management*, *38*, 337–354.

Boddy, D., & Boonstra, A. (2000). Doing business on the Internet: Managing the organizational issues. *Journal of General Management*, *26*(1), 18–35.

Broadbent, M., & Weill, P. (2000). Managing IT infrastructure: A strategic choice. In: R. W. Zmud (Ed.), *Framing the domains of IT management* (pp. 329–459). Cincinatti, OH: Pinnaflex Educational Resources.

Busselle, R., Reagan, J., Pinkleton, B., & Jackson, K. (1999). Factors affecting Internet use in a saturated-access population. *Telematics and Informatics*, *16*, 45–58.

Chappell, C., & Feindt, S. (1999). *Analysis of E-commerce practise in SME's*. Research Report, Knowledge and information transfer on electronic commerce (KITE).

Cheng, C. H., Cheung, W., & Chang, M. K. (2002). The use of the Internet in Hong Kong: Manufacturing vs. service. *International Journal of Production Economics*, *75*, 33–45.

Colarelli O'Connor, G., & O'Keefe, B. (1997). Viewing the Web as a marketplace: The case of small companies. *Decision Support Systems, 21*, 171–183.

Daft, R. L. (1998). *Organization theory and design*. Cincinatti, OH: South-Western College.

Deeter-Schmelz, D. R., & Kennedy, K. N. (2002). An exploratory study of the Internet as an industrial communication tool: Examining buyers' perceptions. *Industrial Marketing Management, 31*, 145–154.

Dekkers, M. (2000). Een gouden plak voor integratie — De vier stadia van ebusiness. *Computable, 44*, 45.

Den Hertog, P., Holland, C., & Bouwman, H. (1999). *Digitaal Zaken Doen: Bouwtekening voor een E-commerce Monitor*. Research Report, Ministry of Economic Affairs, The Hague.

Downs, G. W., & Mohr, L. B. (1976). Conceptual issues in the study of innovation. *Administrative Science Quarterly, 21*(4), 700–714.

Earl, M. J. (2000). Evolving the e-business. *Business Strategy Review, 11*(2), 33–38.

Eastin, M. S. (2002). Diffusion of e-commerce: An analysis of the adoption of four e-commerce activities. *Telematics and Informatics, 19*, 251–267.

Eder, L. B., & Igbaria, M. (2001). Determinants of intranet diffusion and infusion. *Omega, The International Journal of Management Science, 29*, 233–242.

Ellis-Chadwick, F., Doherty, N., & Hart, C. (2002). Signs of change? A longitudinal study of Internet adoption in the U.K. retail sector. *Journal of Retailing and Consumer Services, 9*, 71–80.

EZ (1998). *Actieplan electronic commerce*. Policy Report, Ministry of Economic Affairs, The Hague.

EZ (2001). *SMEs in the Dutch digital delta*. Policy Report, Ministry of Economic Affairs, The Hague.

Fischer, G. (1998). Electronic commerce: Business models for SMEs. In: P. Timmers, B. Stanford-Smith, & P. T. Kidd (Eds), *Electronic commerce: Opening opportunities for business* (pp. 229–236).

Gatignon, H. T., & Robertson, T. S. (1985). A propositional inventory for new diffusion research. *Journal of Consumer Research, 11*, 849–867.

Griffin, J. (2001). Internet usage in small Irish firms. In: W. E. During, R. P. Oakey, & S. Kauser (Eds), *New technology-based firms in the new millennium* (pp. 57–70). Oxford: Elsevier.

Holland, C., Bouwman, H., & Smidts, M. (2000). *Back to the bottom line; onderzoek naar succesvolle e-businessmodellen*. Research Report, Zoetermeer, ECP.NL.

IDC (2001). *Fastrackers: The Adoption of the Internet by SMEs in Western Europe*. Uxbridge: Cisco Systems.

Kambil, A., & Van Heck, E. (1998). Reengineering the Dutch flower auctions: A framework for analysing exchange organisations. *Information Systems Research, 1*(9), 19.

Katz, J. A., & Dennis, D. (2001). Innovation strategy and business strategy in e-businesses. *Proceedings of the Babson-Kauffman Entrepreneurship Research Conference 2001*. Jönkoping, Sweden: Babson College-Kaufmann Foundation.

Kendall, J. D., Tung, L. L., Chua, K. H., Hong, D., Ng, D., & Tan, S. M. (2001). Receptivity of Singapore's SMEs to electronic commerce adoption. *Journal of Strategic Information Systems, 10*, 223–242.

LaRose, R., & Hoag, A. (1996). Organisational adoptions of the Internet and the clustering of innovations. *Telematics and Informatics, 13*(1), 49–61.

Lassila, K. S., & Brancheau, J. C. (1999). Adoption and utilization of commercial software packages: Exploring utilization equilibria, transitions, triggers and tracks. *Journal of Management Information Systems, 16*(2), 63–90.

Laudon, K. C., & Laudon, J. P. (1997). *Essentials of management information systems: Organization and technology in the networked enterprise*. Upper Saddle River, NJ: Prentice-Hall.

Massetti, B., & Zmud, R. W. (1996). Measuring the extent of EDI usage in complex organizations: Strategies and illustrative examples. *Management Information Systems Quarterly* (September), 331–346.

Mehrtens, J., Cragg, P. B., & Mills, A. M. (2001). A model of Internet adoption by SMEs. *Information & Management, 39*(3), 165–176.

Nambisan, S., & Wang, Y. (2000). Web technology adoption and knowledge barriers. *Journal of Organizational Computing and Electronic Commerce, 10*(2), 129–147.

Ng, H. I., Pan, Y. J., & Wilson, T. D. (1998). Business use of the World Wide Web: A report on further investigations. *International Journal of Information Management, 18*(5), 291–314.

NIPO (2001). *Nederland gaat Digitaal.* Research Report (confidential), NIPO, Amsterdam.

Nissen, M. E. (1997). Reengineering support through measurement driven inference. *Intelligence Systems in Accounting, Finance and Management, 6*(2), 109–120.

Peelen, P. P. T. M. (2001). *E-meten is weten?* Thesis, University of Twente, Twente.

Poon, S., & Strom, J. (1997). Small business' use of the Internet: Some realities. *Proceedings of the Internet Society's Seventh Annual Conference, INET'97.* Internet Society, Kuala Lumpur.

Poon, S., & Swatman, P. M. C. (1996). Electronic networking among small business in Australia — an exploratory study. *Proceedings of the Ninth International Conference on EDI-IOS.* Bled, Slovenia.

Poon, S., & Swatman, P. M. C. (1997). Internet-based small business communication: Seven Australian cases. *Electronic Markets, 7*(2), 15–21.

Poon, S., & Swatman, P. M. C. (1999). An exploratory study of small business Internet commerce issues. *Information & Management, 35*(1), 9–18.

Porter, M. E. (2001). Strategy and the Internet. *Harvard Business Review, 79*(3), 62–79.

Porter, M. E., & Millar, V. E. (1985). How information gives you competitive advantage. *Harvard Business Review* (July/August), 149–160.

Premkumar, G., & Roberts, M. (1999). Adoption of new information technologies in rural small businesses. *Omega, The International Journal of Management Science, 27*, 467–484.

Riemenschneider, C. K., & McKinney, V. (1999). Assessing the adoption of web-based e-commerce for businesses: A research proposal and preliminary findings. *Electronic Markets, 9*(1).

Riggins, F. J. (1999). A framework for identifying web-based electronic commerce opportunities. *Journal of Organisational Computing and Electronic Commerce, 9*(4), 297.

Riggings, F. J., & Mitra, S. (2001). A framework for developing e-business metrics through functionality interaction. Working Paper, currently being revised.

Rogers, E. M. (1995). *Diffusion of innovations.* New York: Free Press.

Sadowski, B. M., Maitland, C., & Van Dongen, J. (2002). Strategic use of the Internet by small- and medium-sized companies: An exploratory study. *Information Economics and Policy, 14*(1), 75–93.

Shaw, M. J., Gardner, D. M., & Thomas, H. (1997). Research opportunities in electronic commerce. *Decision Support Systems, 21*(3), 149–156.

Sillince, J. A. A., Macdonald, S., Lefang, B., & Frost, B. (1998). E-mail adoption, diffusion, use and impact within small firms: A survey of U.K. companies. *International Journal of Information Management, 18*(4), 231–242.

Slack, N. (1998). *Operations management.* London: Pitman Publishing.

Stroeken, J. H. M. (1999). Informatietechnologie, innovatie en ketenstructuur. *Bedrijfskunde, 71*(2), 67–78.

Stroeken, J. H. M., & Coumans, J. (1998). The actual and potential use of information technology in small and medium sized enterprises. *Prometheus, 16*(4), 469–483.

Syntens (2002). *Evaluatie Sp.OED Advies.* Internal Report.

Teo, T. S. H., & Choo, W. Y. (2001). Assessing the impact of using the Internet for competitive intelligence. *Information and Management, 39*, 67–83.

Teo, T. S. H., Lim, V. K. G., & Lai, R. Y. C. (1997). Users and uses of the Internet: The case of Singapore. *International Journal of Information Management, 17*(5), 325–336.

Teo, T. S. H., Lim, V. K. G., & Lai, R. Y. C. (1999). Intrinsic and extrinsic motivation in Internet usage. *Omega, The International Journal of Management Science, 27*, 25–37.

Teo, T. S. H., & Tan, M. (1998). An empirical study of adopters and non-adopters of the Internet in Singapore. *Information and Management, 34*, 339–345.

Timmers, P. (2000). *Electronic commerce: Strategies and models for business-to business trading.* Chichester: Wiley.

Tornatzky, L. G., & Klein, K. L. (1982). Innovation characteristics and innovation adoption-implementation: A meta-analysis of findings. *IEEE Transactions on Engineering Management, 29*(1), 28–44.

Trice, A. W., & Treacy, M. E. (1988). Utilization as a dependent variable in MIS research. *Data Base* (Fall/Winter), 33–40

Van Doorne, P., & Verheijen, A. (2000). *eProcurement; Inkooprevolutie in drie stappen.* Research Report, ECP.NL, Zoetermeer.

Venkatraman, N., & Henderson, J. C. (1998). Real strategies for virtual organizing. *Sloan Management Review* (Fall), 33–48.

Walczuch, R., Van Braven, G., & Lundgren, H. (2000). Internet adoption barriers for small firms in The Netherlands. *European Management Journal, 18*(5), 561–572.

Wei, S., Ruys, H. F., Van Hoof, H. B., & Combrink, T. E. (2001). Uses of the Internet in the global hotel industry. *Journal of Business Research, 54*, 235–241.

Wessels, P. J. (2002). *@-doptie in het MKB.* Thesis, University of Twente.

Appendix 1

Which ICT-applications do you use in your business processes?

Dimension 2: Level of Data-Exchange	Dimension 1: Business Process Support	Dimension 3: E-Business Applications			
		Email	Internet Intranet Extranet!	Mobile Datacom *No Speech!*	Other
	Supply				
Information	(a) Searching information about suppliers/serviceproviders and their products or services	☐	☐	☐	☐
Interaction	(b) Specifying /selecting products/services	☐	☐	☐	☐
Interaction	(c) Requesting quotations for products or services	☐	☐	☐	☐
Transaction	(d) Electronically ordering a product or service	☐	☐	☐	☐
Interaction	(e) Inquiring about logistics (availability, progress, status)	☐	☐	☐	☐
Interaction	(f) Electronically exchanging data with suppliers/service providers	☐	☐	☐	☐
Transaction	(g) Electronically receiving invoices	☐	☐	☐	☐
Transaction	(h) Electronic payment of supplies (incl. electronic banking)	☐	☐	☐	☐
Transaction	(i) Electronically receiving products/services	☐	☐	☐	☐
	Organisation				
Information	Offering information electronically to employees about:				
	(a) HRM (contracts, regulations, "who's who", etc.)	☐	☐	☐	☐
	(b) Quality management (specifications, manuals, procedures, etc.)	☐	☐	☐	☐
	(c) Management (reports, plans, budgets, etc.)	☐	☐	☐	☐
	(d) Process (progress, troughput-times, hours, etc.)	☐	☐	☐	☐

Appendix 1 (*Continued*)

Dimension 2: Level of Data-Exchange	Dimension 1: Business Process Support	Dimension 3: E-Business Applications			
		Email	Internet Intranet Extranet	Mobile Datacom *No Speech!*	Other
	Other activities				
Interaction	(e) Communication between employees (incl. Document sharing)	☐	☐	☐	☐
	(f) Recruitment of new personnel	☐	☐	☐	☐
	(g) Teleworking	☐	☐	☐	☐
	(h) Knowledge management (internally collecting and exchanging knowledge)	☐	☐	☐	☐
	(i) Attending courses, education	☐	☐	☐	☐
Transaction	(j) Electronically signing/authorising contracts, invoices, certificates, etc.	☐	☐	☐	☐
	Sales				
Information	(a) Searching information about the market (clients/competitors)	☐	☐	☐	☐
Information	(b) Offering information about your organisation/product/service	☐	☐	☐	☐
Interaction	(c) Offering help in specifying or selecting products or services	☐	☐	☐	☐
Interaction	(d) Quoting electronically	☐	☐	☐	☐
Transaction	(e) Receiving orders electronically	☐	☐	☐	☐
Interaction	(f) Offering logistics information (about availability, status of orders, progress, etc.)	☐	☐	☐	☐
Transaction	(g) Electronic invoicing	☐	☐	☐	☐
Transaction	(h) Offering the possibility of electronic payment	☐	☐	☐	☐
Transaction	(i) Electronic delivery of products or services	☐	☐	☐	☐
Interaction	(j) After-sales service (complaints, manuals, helpdesk, FAQ, etc.)	☐	☐	☐	☐

Appendix 2

Dimension 4: Amount of Use

(for the supply process, the questions for organisation and sales are similar)
  Suppose that the e-business applications can not be used (out of order). What would be the consequence for the supply process in your company?
  The supply process . . .

☐ is idle
☐ experiences a lot of inconvenience
☐ experiences a little inconvenience
☐ experiences no inconvenience
☐ don't know/ not relevant

Dimension 5: Impact of Use

What did your company achieve as a result of doing electronic business? (multiple answers)

Value Creation	Impact
	☐ Nothing
Efficiency	☐ Saving time/money (efficiency improvements)
	☐ Higher costs
	☐ Higher turnover
Effectiveness	☐ Improved customer service
	☐ Improved company image
	☐ Improved communication with suppliers
	☐ Improved communication with customers
	☐ Improved communication between employees
Strategic benefit	☐ Additional market channel
	☐ Offering a new product/products
	☐ Offering a new services/services
	☐ Opening up a new market/markets
	☐ Offering a new combination of product, service and/or market
	☐ Don't know/Not relevant

What was the impact of these changes on your company?

☐ Major change(s)
☐ Some change(s)
☐ Minor change(s)
☐ Don't know/Not relevant

Chapter 4

The Business Platform Model: A Practical Tool for Understanding and Analysing Firms in Early Development

Per Davidsson and Magnus Klofsten

Introduction

Assessing the state of newly started firms in order to predict their fate, or to initiate appropriate action for increasing their probability of survival and growth is no easy task (Hall 1995). Such firms are in a turbulent phase of development where business activities are carried out in a short-term perspective and where it can be difficult to perceive more fundamental shortcomings that are overshadowed by everyday problems (Adizes 1988). Much research has been devoted to establishing associations between various kinds of presumed causal factors on the one hand, and the ability of a company to attain stability and growth on the other (e.g. Cooper 1981; Cooper *et al.* 1994; Dahlqvist *et al.* 2000; Davidsson 1991; Gimeno *et al.* 1997; Jones-Evans & Klofsten 1997; Kazanjian 1988; Littunen *et al.* 1998; Morris 1998; Van de Ven *et al.* 1984). A general conclusion from this research is that a range of factors on the individual, firm and environmental levels of analysis shape the firm's development. At the same time, the results suggest that there is no individual factor that universally and by itself has a strong determining influence. Further, explanatory models based on additive effects of comprehensive lists of presumed causes provide far from full explanations of the outcome variance. This supports the notion that holistic and to some extent idiosyncratic configurations of sets of factors jointly determine the success probabilities of young firms.

Models for assessing and assisting the development of firms are a key interest for research, teaching, and business practice. Numerous manuals have been written that more concretely treat different aspects of new business development. These provide considerable amounts of practical advice on how a firm can increase its chances for survival and success. In the literature, comprehensive problems in the firm are discussed, but no serious attempts are made to anchor the advice in systematic empirical research, or to develop testable theories. The suggestions are based more on experience and on what — for the moment — is considered to be good management (Hall 1995). Although research in the last ten-year period has grown in scope and our knowledge of early growth and development processes in firms are considerably greater than before, it is still fairly unusual for researchers to try to generate practical tools in this field.

Our attempt to develop a quantifiable and research-based instrument for assessing and assisting the development of young firms is based on Klofsten's Business Platform Model. This model builds on previous research and has been used extensively in qualitative research and business consulting. In the next section we present this model and its underpinnings.

The Business Platform Model

It has been shown that firms that succeeded in surviving 2–3 years and that have gone through a number of crucial phases attain a stable foundation from which they can continue to develop (Freeman *et al.* 1983; Hall 1995; Mayer & Goldstein 1961). Gibb & Scott (1986) introduced the concept "base for potential development" as an expression of this stabilising condition. According to the authors, a development base has been attained when the newly started firm is sufficiently developed concerning resources (capacity), experiences, control, leadership, and idea. With this basis, the firm then has the possibility to develop and manage future environmental changes and can thereby be considered to have achieved stability.

Pursuing a similar idea, Klofsten (1992) conducted a comprehensive literature review and defined eight firm-level cornerstones that determine a firm's early development process. These eight cornerstones are the business idea, the product, the market, the organisation, core group expertise, core group drive/motivation, customer relations and other relations. These cornerstones make up the Business Platform Model (see Table 1). The purpose of the cornerstones is to describe the early development process in a holistic manner at the micro-level. It comprises the development process itself (idea, product, market, and organisation); key actors such as the founders, CEO, and board members (expertise and drive/motivation), and the flow of external resources (customer relations and other firm relations).

The fundamental premise of this model is that firms are vulnerable in their early life and continually run the risk of disappearing from the market. Success in the firm is determined by how well this vulnerability is overcome, and it is the early development process that is one of the most important periods in the life of the firm. The thoughts and the driving forces in the firm and the actions taken can be decisive for the continued growth and development of the firm (cf. Kimberly & Miles 1980).

The theory proposes that success will be determined by how well the firm builds and maintains its business platform. A firm attains a platform by satisfying two criteria: securing an input of resources and developing an ability to manage and utilise such resources (cf. Barney 1997, Chapter 5). After the platform has been achieved, a firm has a good deal of leeway in generating and managing its resources.

Early development in a firm can be defined by the business platform cornerstones' progress and can take as little or as much time as is needed, with some firms never actually getting beyond this stage. Taking steps to realise a business concept by initiating activities intended to lead to the creation of a firm is the beginning of early development. It ends when the firm has established a business platform. At the risk of being categorical,

Table 1: The cornerstones of the business platform model.

Cornerstone	Minimum Levels to Attain
Formulation and clarification of the business idea	The idea must be clarified so that the special know-how that makes up the commercial springboard is understandable and can be communicated internally and externally
Development to finished Product	Once the product is available, it must gain acceptance by one or more reference customers — the firm has then proven that it is capable of satisfying the markets' needs and wants
Definition of market	The firm must define a market that is large enough and profitable enough to ensure survival
Development of an operational organization	The running of business operations requires the existence of an organizational structure that facilitates functional co-ordination - this structure should take advantage of the firm's inherent flexibility and innovative ability, and be fairly effective at internal co-ordination and at maintaining and developing external relations
Core group expertise	A business firm must have technological and commercial competence to develop its products and market — it is crucial to have access to expertise for solving the firm's real problems
Commitment of the core group and the prime motivation of each actor	A basic requirement for development is that at least one person is highly motivated and that the other key actors are committed to the business idea
Customer relations	A customer base must be qualitatively and quantitatively strong enough to generate operating revenue
Other firm relations	The firm may sometimes need additional capital, management know-how, or other 'oil' in its machinery — these relations complement the customer relationships

it could be argued that firms aspiring to grow and become significant actors in the market must sooner or later attain a business platform.

The rationale is that the cornerstones that make up a business platform must reach certain minimum levels if a business platform is to be attained (see Table 1). For example, an idea must be communicable both within and outside of the firm (Kazanjian 1988; Timmons 1994). A hobby firm will not exhibit the same levels of activity and progress as a firm with strong driving forces to grow and develop (Naffziger *et al.* 1994; Shaver & Scott

1991). The resource-based view of firms (Barney 1991, 1997; Penrose 1959) argues that it is often necessary to stand out, to have a real edge, on at least some dimension. Thus, the model is not a simple additive and compensatory model. Reaching high levels on several dimensions is not assumed to necessarily make up for severe shortcomings on another dimension, and sometimes achieving average levels across all cornerstones does not suffice, either.

Based on a number of case studies, Klofsten (1992) showed that the existence or non-existence of a business platform was possible to assess in case study research. The model was originally applied to technology-based firms but has also proven to be applicable to other types of firms (Klofsten 1998). He also found reason to argue that if a business platform it is not attained, the firm will sooner or later go under and disappear from the market, at least as independent actor. Moreover, it was possible to analyse the state of the eight cornerstones and to determine whether a business platform has been attained.

The Business Platform Model has since been disseminated (predominantly in Europe) in research, education, and trade and industry (Klofsten 1992, 1994, 1997, 1998). The experience to date is that the platform model is a useful tool for gauging and assisting the development of young firms. An obvious limitation is that its validity has not yet been proven in broadly based research. A necessary first step towards such validation is to develop a standardised operationalization of firms' standing on the "cornerstones" of the platform model. That is what we set out to do in the present study. It is our hope that the development of such an instrument will facilitate also the adoption and sound application of the Business Platform Model by business consultants.

The aim of the present paper is to develop an instrument that can quantify a firms' state on each cornerstone dimension. It should be clear from the above discussion that simple summation of scores from such a quantification will not suffice for determining whether a firm has attained a business platform or not. What is the appropriate minimum level for each dimension, the possible need for excellent scores on some dimensions, and how this varies by industry or type of firm are questions that will need further research.

Method

The Sample

The sample consisted of technology- and knowledge-based firms located at Swedish technopoles (technopoles are regions which exhibit strong technology- and knowledge-based new business development. They are a combination of a university with a technical profile, a research institute, and science parks as well as a number of large and small firms. In Sweden, Göteborg, Linköping, Lund, Stockholm, and Uppsala are the main regions classified as technopoles, cf. Heydebreck *et al.* 2000). The aim of the paper was to study growth and development in young firms, therefore, firms more than 10 years old were excluded. At the same time, it was considered an advantage if the firms had some history and a lower age limit of two years was set. A questionnaire was sent to 313 firms addressed either to the CEO or senior executives. A total of 114 firms returned the questionnaire, with a response rate of 36%.

While these response rates are low in relation to what is expected from Swedish studies (cf. Davidsson 1989a, b, c; Wiklund 1998) in light of previous research, it is not low in comparison to other studies. We have no indication that the level of non-response would have distorted the reported results.

The average and median ages of the firms were 6.7 and 7.0 years, respectively. During 1998, the average turnover was SEK 9.6 million (USD 1.1 million) and the median SEK 9.0 million (USD 1 million) with a range of SEK 0–71 million (USD 0–7.9 million). An average of 4.6 persons (median 3.0) were employed in the firms (range 0–100). The majority of firms (58%) were oriented towards providing services while some (35%) primarily produced products. Six firms (5%) reported other forms of business but did not specify what. More than a quarter of the firms (26%) reported that they had changed the focus of their business since start-up. A number of respondents (representing 55 firms; 48%) stated that the founders had a university education prior to start-up. Those who reported otherwise (30%) came from firms or research institutes.

The Instrument

This section provides the operational definitions of the variables and explains how they were measured.

In developing the instrument we adopted the standard procedure of generating multiple item measures of questions for each intended dimension. In order to arrive at a manageable instrument we aimed at about five items for each dimension. We first generated a pool of five to ten "raw" items for each dimension. Each suggested item was discussed among the team of researchers from a substantive as well as a technical standpoint. Through this process of selecting, revising and deleting items we developed a set of measures specifically for the study. The final version included a total of 36 items.

Following the recommendations of Converse & Presser (1986) we chose a forced choice format. The respondents were asked to indicate their relative degree of agreement on a five-point scale between two statements that were contrasted. Scales and measures are listed in Table 2. For each item the respondent was asked to give separate responses for the firm's current situation and for its status two years earlier. In that way we were able to formulate two — albeit not two independent — evaluations of the technical properties of the instrument.

The package of questions aimed at capturing the various dimensions of the business platform constitutes the lion's share of the mail/e-mail questionnaire. The remainder of the questionnaire concerned background facts about the firm and its founders as well as a question concerning the relative ease or difficulty of answering the business platform items.

Analysis

To assess the extent to which the measures used for the study are reliable, we used Cronbach's Alpha. In the interpretation of this analysis we apply Nunnally's rule of thumb that a Cronbach's Alpha above 0.70 indicates satisfactory measurement quality from a technical

Table 2: Scale construction results for each cornerstone in the business platform model.

	No. of Items	No. of Cases	Mean	S.D.	Cronbach's Alpha
Business idea					
Sample item: "The idea about what the firm's operations should be is not particularly specified." (1) vs. "There exists a very clearly specified idea for what the firm's operations should be." (5)					
Scale "Now"	3	113	12.68	2.12	0.79
Scale "2 years ago"	3	107	10.77	3.41	0.90
Product					
Sample item: "There is no developed product." (1) vs. "There is at least one well developed product that is entirely ready for sale." (5)					
Scale "Now"	3	101	12.29	3.85	0.94
Scale "2 years ago"	3	95	10.07	4.83	0.96
Market					
Sample item: "The firm has no limitations as to what customers it turns to." (1) vs. "The firm turns itself to a very specific customer category." (5)					
Scale "Now"	4	91	15.96	2.80	0.70
Scale "2 years ago"	4	88	12.81	3.97	0.84
Organization					
Sample item: "All staff do most types of work." (1) vs. "All staff have clearly delimited tasks." (5)					
Scale "Now"	5	108	16.63	5.28	0.83
Scale "2 years ago"	5	104	13.35	5.22	0.84
Competence					
Sample item: "To some extent the firm lacks expert knowledge within its domain." (1) vs. "The firm is very well equipped with expert knowledge within its domain." (5)					
Scale "Now"	6	111	18.56	4.31	0.77
Scale "2 years ago"	6	105	17.69	5.04	0.81
Drive/Motivation					
Sample item: "The founder regards the firm as one of several possible ways of earning his/her living." (1) vs. "The founder is completely geared towards a future as business owner-manager." (5)					
Scale "Now"	4	109	15.29	3.38	0.71
Scale "2 years ago"	4	105	13.52	3.68	0.72
Customer relations					
Sample item: "The firm has as yet not sold any product to a customer." (1) vs. "The firm has a large number of customers who have bought its products." (5)					
Scale "Now"	4	107	15.15	2.77	0.60
Scale "2 years ago"	4	104	13.30	3.65	0.73
Other relations					
Sample item: "There exist no relationships with banks or investors." (1) vs. "There exist very good and stable relationships with banks and investors." (5)					
Scale "Now"	4	112	15.23	3.10	0.63
Scale "2 years ago"	4	106	12.91	4.01	0.80

standpoint (Nunnally 1967). Items were dropped from the respective index if scores fell below the accepted level.

Factor analysis was conducted to assess the convergent and discriminent validity of the study's constructs. We conducted a principal components factor analysis with varimax rotation to confirm the construct validity of the data set and provide reassurance as to the reliability of the research design employed. However, we did not assume a priori that the different dimensions (or "cornerstones") of the Business Platform Model would be uncorrelated (orthogonal). For one thing, they should all to some extent be a function of time. Therefore, a crystal clear factor pattern was not expected. Furthermore, summated indices are not identical to factors in a factor analysis, and the summated indices are what we eventually proposed to use.

Essentially, operationalization should also reflect the empirical reality that the conceptualization aims to capture. To ensure high content validity all measures were developed and evaluated by a team including the creator of the Business Platform Model and thus we feel that face validity is high. Table 2 displays Cronbach Alpha values and other statistics for the resulting indices.

Results

As regards the Business Idea three out of the four original items were retained. One item was discarded because it had a low Alpha coefficient. The resulting three-item scale had coefficients greater than 0.70. The distribution is positively skewed with an item average score of 4.25 out of five, and a full 26% of the sample had the maximum score (15) on the "Now" version of the scale. This suggests that some minor rephrasing of the current items and the additions of one or two new items would help capture the full range of variation better than the current version does. However, the current version performed very well.

Only three items were developed for the second scale, Product, and all three were retained. The Cronbach's Alpha values were extremely high for both versions of the scale, indicating very high reliability. The measurement was not entirely unproblematic. The distributions of responses were bi-modal, with over-representation at both extremes. This suggests that the respondents viewed this dimension as being dichotomous: either you have a developed product ready for sale, or you do not. Again, some further improvement seems possible although this first attempt must be judged as being relatively successful.

For the Market scale satisfactory to high Cronbach Alpha values were obtained after deletion of one of the original items. High reliability is thus attributed to the resulting four-item scale. The distributions also look good despite some positive skewness. Approximately 20% of the sample chose to skip one or several items. We have checked that neither for Market nor for Product are there any clear differences between the manufacturing and service firms' relative propensity to give valid responses.

For the Organisation scale the measurement was most successful. The Cronbach's Alpha values are high, the distribution is very close to normal, and internal non-response is modest. After eliminating one item the five-item Competence scale also was very satisfactory, with low internal non-response, high Cronbach's Alpha values and only a mild positive skewness.

Similarly, the results for Drive/Motivation were satisfactory, although the Alpha values are only slightly above 0.70 and the positive skewness somewhat more pronounced.

The results for the final two dimensions, Customer Relations and Other Relations, are very similar. Both are based on four items in the final analysis and both are somewhat positively skewed with across-item averages of 3.79 and 3.81 (out of five), respectively. Neither had any severe problems with internal non-response. However, the "Now" version of the scale had less than satisfactory Alpha scores. Adding one more appropriate item to each of these two cornerstones is recommended.

Taken together, this initial attempt to create a formal operationalization of the Business Platform Model can be judged as very successful from a reliability point of view. The current version of the instrument seems to work reasonably well, although improvement would be desirable for some of the cornerstones. In terms of the instrument's validity the means were uniformly higher, and the standard deviations uniformly lower, for the "Now" version of the scales relative to the "Two years ago" versions. This development was expected over time from a surviving sample of younger firms. A more serious problem detected was the rather substantial internal non-response for the Market and Product dimensions. A closer look at this problem suggests a number of limitations concerning the types of firm for which these dimensions of the model are applicable. Alternatively, it may suggest a more positive solution to the problem. It is conceivable that with face-to-face or telephone interviewing the uncertainty behind the internal non-responses can be resolved. With regards to mail surveys, improved instructions to the respondent might help.

The results for the control question that were asked showed that few respondents found the questions difficult to answer. As regards the situation "Now," two thirds of the sample agreed completely or partly that it was easy to determine what the answers should be. For natural reasons relatively fewer respondents — 49% — found it easy to give responses for the firm's situation two years earlier. The proportions that refuted that the questions were easy stayed at eight and 12 per cent, respectively.

Having established content validity and reliability we now turn to the issue of discriminant validity. For this purpose we ran an exploratory factor analysis with the default criterion that eigenvalues for extracted factors should be higher than unity. We used, the "Now" version of each item and included only those 32 items that were retained after the reliability analysis. This initial analysis yielded ten factors. These ten factors accounted for 70% of the total variation. However, a couple of factors had only one high loading. In order to better assess the fit between the eight concepts and the factor analysis results we performed a second run extracting eight factors. These eight factors accounted for 65% of the variance. (See Table 3 for factor loadings.) Note that due to listwise deletion of cases with missing data this analysis was based on 72 cases only. Note also that for ease of interpretation the factors have been re-numbered and loadings below |0.30| were suppressed.

All items intended to measure the same dimension load on the same factor, and only in two cases out of 32 does an item have a higher loading on another factor than its "own." The number of "side-loadings" above 0.30 is very low over-all. This clearly suggests that the different item packages measure different dimensions. That is, discriminant validity appears to be high.

Table 3: Factor (principal components) analysis loadings for cornerstone items ($n = 72$).

Items	F1: Idea	F2: Product	F3: Market	F4: Organization	F5: Competence	F6: Drive/ Motivation	F7: Customer Relations	F8: Other Relations
Idea 1	0.75			0.36				
Idea 2	0.84							
Idea 3	0.75							
Product 1		0.80		0.38				
Product 2		0.83						
Product 3		0.90						
Market 1			0.70					
Market 2			0.80					
Market 4			0.43	0.64				
Market 5			0.76					
Org. 1				0.77				
Org. 2				0.81				
Org. 3				0.69				
Org. 4				0.65				
Org. 5				0.77				
Comp. 1					0.75			
Comp. 2					0.77			
Comp. 3					0.56			
Comp. 4					0.78			
Comp. 5					0.63		0.31	
Drive 1						0.75		
Drive 2						0.68		
Drive 3						0.68		
Drive 4					0.67	0.39		
Cust. 1							0.75	
Cust. 2							0.69	
Cust. 3							0.56	0.37
Cust. 4							0.65	
Other 1								0.54
Other 2								0.51
Other 3								0.73
Other 4						0.35		0.85

When the analysis was re-run for the "Two years ago" versions of the items the results were similar. In some ways they were "better," in other ways "worse" than those displayed in Table 3. The analysis yielded eight factors, by the default criterion, and these eight factors accounted for 73% of the variance. The factor pattern is similar to that displayed in Table 3 and thus very clear for most dimensions. However, the Competence and Customer Relations dimensions were blurred with a tendency for the items for both of those to split between two mixed competence/customer relations factors.

The factors in the factor analysis can be regarded as weighed indices of all items included in the analysis, but with greater weight given to items with higher loadings. Summated indices, on the other hand, are based solely on those items that were intended to assess that

Table 4: Pearson product-moment correlation between the cornerstones' summated indices.

	Idea	Product	Market	Organization	Competence	Drive/ Motivation	Customer Relations	Other Relations
Idea	1.00							
Product	−0.07	01.00						
Market	0.05	0.36[b]	1.00					
Org.	0.13	0.44[b]	0.44[b]	1.00				
Comp.	0.13	0.22[a]	0.13	0.31[b]	1.00			
Drive	0.19[a]	0.07	0.15	0.23[a]	0.42[b]	1.00		
Cust.	0.04	0.22[a]	0.00	0.17	0.29[b]	0.19	1.00	
Other	0.11	0.13	0.16	0.07	0.19[a]	0.10	0.25[b]	1.00

[a]$p < 0.05$.
[b]$p < 0.01$.

particular dimension, and these items are given equal weight in the summation. Therefore, a summated index is not identical to its corresponding factor. While the factors are constructed to be uncorrelated, the indices may overlap. We therefore supplemented the factor analysis with a correlation analysis to provide more direct evidence on the relative distinctiveness of the factors. This analysis is displayed in Table 4.

The results show that the correlations are positive for the most part. This was expected, since all dimensions had in common that they are in part a function of time. Only one correlation was not statistically significant. Interestingly, two of the dimensions were related — Market and Customer Relations — are not correlated at all in this sample. Most correlations were modest. The fact that the Organization dimension had a few correlations in the 0.40s does not mean that it is not to distinct enough. A correlation of 0.44 reflects that the two factors have less than 20% of the variance in common. This is very far from being identical, and our conclusion that discriminant validity is satisfactory remains valid.

Discussion

Overall, we believe that our attempt to create a formal operationalization of the Business Platform Model has been successful. We have, arguably, established high face validity, moderate to very high reliability, and high discriminant validity for the different cornerstone indices. This is an important step towards making the Business Platform Model a quantifiable, holistic and action-oriented instrument for assessing and assisting the development of young firms, and hence towards an increased and more well-founded use of the model.

Much remains to be done before a fully satisfactory tool has been developed. As noted above, some revision of the instrument may be needed in order to improve its technical properties. These technical properties need to be tested also for the translated version, so that the applicability of the instrument can be generalized to other countries. Further research on large samples should use the improved version of the instrument to assess the state of firms at various points in time during their early development and relate these assessments to outcomes. This is needed in order to establish predictive validity of the Business Platform

Model as operationalized with our instrument. This also involves the issues of establishing minimum levels for each cornerstone and investigating the possible need of outstanding levels on some dimensions, as well as checking for differences in these regards by industry or type of firm.

The fact that the model is based on an extensive literature review and qualitative research should ascertain some external validity. However, further in-depth work is needed in order to determine whether assessments based on the standardized instrument accord with clinical judgment. Experiments would be the ideal for evaluating the model's and the instrument's suitability not only for prediction but also as a basis for corrective action. Such experiments would be difficult to set up in practice and if possible conducting them would be ethically questionable, as it would involve refraining from giving the advice the model predicts is essential for the firm's survival. However, quasi-experiments should be possible. That is, all cases in the study would get proper advice concerning the model. The experimental "manipulation" should be provided by the firm' themselves, i.e. the extent to which they chose to implement the actions suggested by the advice. Evidence that those who followed the advice fared better than those that did not would provide very strong support for the validity of the model and its operationalizations. Ultimately, that is the type of evidence we need in order to apply this tool with great confidence.

References

Adizes, I. (1988). *Corporate lifecycles: How and why corporations grow and die and what to do about it.* Englewood Cliffs: Prentice-Hall.

Barney, J. B. (1991). Firm resources and sustained competitive advantage. *Journal of Management, 17*(1), 99–120.

Barney, J. B. (1997). *Gaining and sustaining competitive advantage.* Menlo Park, CA: Addison Wesley.

Converse, J. M., & Presser, S. (1986). *Survey questions: Handcrafting the standardized questionnaire.* Sage University Paper series on Quantitative Applications in the Social Sciences, series no. 07–063, Beverly Hills: Sage.

Cooper, A. C. (1981). Strategic management: New ventures and small business. *Long Range Planning, 14*(5), 39–45.

Cooper, A. C., Gimeno-Gascon, F. J., & Woo, C. Y. (1994). Initial human and financial capital as predictors of new venture performance. *Journal of Business Venturing, 9*(5), 371–395.

Dahlqvist, J., Davidsson, P., & Wiklund, J. (2000). Initial conditions as predictors of new venture performance: A replication and extension of the Cooper *et al.* study. *Enterprise and Innovation Management Studies, 1*(1), 1–17.

Davidsson, P. (1989a). *Continued entrepreneurship and small firm growth.* Stockholm School of Economics, Stockholm.

Davidsson, P. (1989b). Entrepreneurship – and after? A study of growth willingness in small firms. *Journal of Business Venturing, 4*(3), 211–226.

Davidsson, P. (1989c). Need for achievement and entrepreneurial activity in small firms. In: K. G. Grunert, & F. Ölander (Eds), *Understanding economic behavior* (pp. 47–64). Dordrecht: Kluwer Academic.

Davidsson, P. (1991). Continued entrepreneurship: Ability, need, and opportunity as determinants of small firm growth. *Journal of Business Venturing, 6*(6), 405–429.

Freeman, J., Carrol, G. L., & Hannan, M. T. (1983). The liability of newness: Age dependence in organizational death rates. *American Sociological Review*, *48*, 692–710.

Gibb, A., & Scott, M. (1986). Understanding small firms growth. In: M. Scott *et al.* (Eds), *Small business growth and development* (pp. 81–104). London: Gower.

Gimeno, J., Folta, T. B., Cooper, A. C., & Woo, C. Y. (1997). Survival of the fittest? Entrepreneurial human capital and the persistence of underperforming firms. *Administrative Science Quarterly*, *42*(4), 750–783.

Hall, G. (1995). *Surviving and prospering in the small firm sector*. London: Routledge.

Heydebreck, P., Klofsten, M., & Maier, J. (2000). Innovation support for new technology-based firms: The Swedish teknopol approach. *R&D Management*, *30*(1), 89–100.

Jones-Evans, D., & Klofsten, M. (1997). *Technology, innovation and enterprise — the European experience*. London: Macmillan.

Kazanjian, R. K. (1988). Relation of dominant problems to stages of growth, in technology-based new ventures. *Academy of Management Journal*, *31*(2), 257–279.

Kimberly, J. R., & Miles, R. H. (Eds) (1980). *The organizational life cycle*. San Francisco: Jossey-Bass.

Klofsten, M. (1992). *Tidiga utvecklingsprocesser i teknikbaserade företag*. Linköping University, Department of Management and Economics, Linköping Studies in Management and Economics, No. 24, Ph.D. Dissertation.

Klofsten, M. (1994). Technology-based firms: Critical aspects of their early development. *Journal of Enterprising Culture*, *2*(1), 535–557.

Klofsten, M. (1997). Management of the early development process in technology-based firms. In: Jones-Evans, & M. Klofsten (Eds), *Technology, innovation and enterprise — the European experience* (pp. 148–178). London: Macmillan.

Klofsten, M. (1998). *The business platform: Entrepreneurship and management in the early stages of a firm's development*. TII, Luxembourg: European Commission.

Littunen, H., Storhammar, E., & Nenonen, T. (1998). The survival of firms over the first 3 years and the local environment. *Entrepreneurship and Regional Development*, *10*(3), 189–202.

Mayer, K. B., & Goldstein, S. (1961). *The first two years: Problems of small firm growth and survival*. Washington, DC: Small Business Administration.

Morris, M. H. (1998). *Entrepreneurial intensity: Sustainable advantages for individuals, organizations and societies*. London: Quorum Books.

Naffziger, D. W., Hornsby, J. S., & Kuratko, D. F. (1994). A proposed research model of entrepreneurial motivation. *Entrepreneurship Theory and Practice*, *18*(3), 23–46.

Nunnally, J. (1967). *Psychometric theory*. New York: McGraw-Hill.

Penrose, E. (1959). *The theory of the growth of the firm*. Oxford: Oxford University Press.

Shaver, K., & Scott, L. (1991). Person, process and choice: The psychology of new venture creation. *Entrepreneurship Theory and Practice*, *16*(2), 23–46.

Timmons, J. A. (1994). Opportunity recognition: The search for higher potential ventures. In: W. D. Bygrave (Ed.), *The portable MBA in entrepreneurship* (pp. 26–54). Toronto: Wiley.

Van de Ven, A. H., Hudson, R., & Schroeder, D. M. (1984). Designing new business startups: Entrepreneurial, organizational and ecological considerations. *Journal of Management*, *10*(1), 87–107.

Wiklund, J. (1998). *Small firm growth and performance: Entrepreneurship and beyond*: Jönköping International Business School (diss.), Jönköping.

Part III

"Spin off" Firms

Chapter 5

Entrepreneurs, New Technology Firms and Networks: Experiences From Lone Starters, Spin-Offs and Incubatees in the Dutch ICT Industry 1990–2000

Willem Hulsink and Tom Elfring

Introduction

The formation and early growth of new (technology-based) firms is less spectacular, visionary and heroic than often thought. The field of entrepreneurship is beset with several myths (Bhidé 2000; NCOE 2001), such as the delusions that risk-taking entrepreneurs take wild risks in starting their companies, that high-technology based innovation-based start-ups begin with breakthrough innovations, and that the entrepreneur is an expert, with a strong track record based upon years of experience in his or her industry. Furthermore, one could also add the myth that entrepreneurs have a well-considered business plan and have researched and developed their visionary ideas before taking action, together with venture capital myth that most companies are backed by venture capitalists who provide them with millions to develop their ideas and build a business. However myths offering inspiring accounts of events and stories generally are not true, and can be easily debunked (Bhidé 2001). Most successful entrepreneurs start as relative amateurs with little background experience and only later, with some of the entrepreneurial lessons learned, do they team up with experienced executives. Also, initially, they are often far from successful, often beginning with a modest product or service offering that puts them on a path to something else which eventually brings them success. Instead of having grand vision and pursuing radical innovations, successful entrepreneurs all have a "Master's Degree" in the obvious. Often they slightly modify someone else's (often ordinary) ideas, execute them very well, and take only calculated steps towards improving such a product or the service through the stages of early growth. Last but not least, only a very small minority of new firms is backed by venture capital (i.e. approximately 1%).

New Technology-Based Firms in the New Millennium, Volume III
Copyright © 2004 by Elsevier Ltd.
All rights of reproduction in any form reserved.
ISBN: 0-08-044402-4

In short, new entrepreneurs short of a complete set of skills and resources, typically pursue small and highly uncertain opportunities, that are highly new and unproven and require little capital. Instead of high-powered venture capitalists and customer-friendly banks, they have to rely upon scarce assets and act creativily in serving customers, often by persuading friends and family to support them. Many of these "bootstrapped entrepreneurs" with modest funds, help incubate new disruptive technologies that at first compete with difficulty in niche markets, while producing revenue streams too small to interest bigger companies. Some of these entrepreneurial growth companies even lack a business plan at the start-up and only later, when some innovative ideas have occurred, and investors have expressed interest, are business plans written and grand ambitions expressed. In summary it might be concluded that the aforementioned myths and the empirical evidence on founders and their start-ups fits well in the Schumpeterian and Kirznerian approaches to entrepreneurship, by supporting notions such as the heroic or alert entrepreneur acting vis-à-vis rivals in a basically unstable and anonymous market.

The inventor-entrepreneur, however, acting on his/her own behalf in the search for new technical combinations, spotting new market opportunities, and striving for profit maximisation, tells only half of the story of (innovative) entrepreneurship. The notion that is also crucial is the "social and institutional embeddedness" of entrepreneurship, as introduced and pioneered in economic sociology and organisation theory (e.g. Nohria & Eccles 1992; Swedberg 2000). Granovetter (1995), for instance, has argued that economic activities are socially situated and cannot be explained by reference to individual motives alone; instead they depend critically on the robustness of the underlying social structure. Economic action usually takes place in complex social situations, where actors are related to each other through ongoing networks of (inter)personal and interorganisational relationships. Their face-to-face interactions and economic transactions are influenced by the larger social, political and cultural context; their pursuit of economic goals, for instance, is typically accompanied by that of such non-economic motives as sociability, recognition and approval, status, and power. Also entrepreneurs are embedded in social networks which provide access to critical resources (e.g. information, capital, customers). Aldrich & Zimmer (1986) have defined entrepreneurship as the situational exchange of resources and opportunities, which are embedded in ongoing social relations. Such emergent economic linkages are channeled and facilitated (or constrained and inhibited) by people's positions in larger social networks. The external networks of research institutes, leading edge customers, key suppliers, and technology transfer agencies can act as both a resource and a constraint to the evolution of the technological competencies and strategies in small technology companies. For instance, in a comparative study on British and French biotechnology firms, Estades & Ramani (1998) identify major differences in the composition of the firms' external network and dominant constituencies: while in the U.K. the interorganisational system was characterised by market incentives and key customers and alliance partners acting as principals, in France public research laboratories and state subventions shaped the biotechnology network and its constituent firms.

Thus, in explaining the success of a (new) company it is not only the qualities of the entrepreneur that play a large role, but also the social networks in which entrepreneurs and their companies operate. A network is one of the most powerful assets that any individual can possess: it provides access to information, opportunities, power and to other networks

(Uzzi 1996, 1997). An alternative term for this whole set of active connections among people and organisations, which seems to be in vogue today, is "social capital." More specifically, for Cohen & Prusak (2001: 4) social capital includes the trust, the mutual understanding, shared values and behaviour patterns, that not only bind the members of interpersonal networks and communities together, but also facilitate cooperative action. The link between the entrepreneur, his contact network, and success is not straightforward. On the one hand, a high level of social capital (i.e. dense social networks) will generate positive results, contributing to more intensive knowledge sharing, lowered transaction costs and turnover rates, and the promotion of greater coherence of action. On the other hand, social capital is not the key to organizational success, or more precisely, it can be even be neutral or detrimental to a firm's success. Some organisations succeed, despite the negative effects of low social capital (e.g. poor university or consultancy services); others collapse because of poor market decisions and strategic errors, despite being known for their collegiality and employee commitment. Or even more to the extreme, the ties that bind can also be the ties that blind (Cohen & Prusak 2001: 14): "a cohesive and tightly integrated community can become a problem if that makes it clannish, insular, or even corrupt."

In short, the network of relations within which the start-up entrepreneur finds himself, and the contacts he relies upon for the growth of his venture, has until now received little attention (Elfring & Hulsink 2003). In order to be successful start-up firms must not only bring about new technological combinations and market their product, but they should develop new networks as well — in addition to their existing information and contact networks, thus mobilising knowledge, experience, monies and other funds. For our research, there are two aspects that are important: (i) Are start-ups that operate in tight clusters more innovative and ultimately more successful than companies that are not or hardly a part of them? (ii) Does the need for specific forms of networks and/or strategic partners change as the company goes through different phases? The Dutch government, through various policy instruments (e.g. Techno-starters fund, Twinning initiative, DreamStart), has attempted to motivate nascent technology-based entrepreneurs to set up their own business, develop innovative products & services, and to build networks (and rely on them). Universities and large companies also see possibilities in, on the one hand to offer graduates and former employees the opportunity to start their own company, while on the other hand trying to maintain close ties with the starter and his company by outsourcing contract research in the vicinity (nearby perhaps, or on a university campus), licensing, participations, etc, and by looking for synergies between an increasing number of new and old technology companies within a keiretsu-like organisational form.

The core question in the following research is "In what way does the entrepreneur's network contribute to the success of a start-up company." This research aims at increasing our insights into the influence of the network on the level of achievement of starters in the initial (growth) phases, following the foundation of a company. The network seems to be important when it comes to acquiring knowledge, complementary means and legitimacy. In the literature, however, it is unclear in what way a certain network configuration influences the success of a start-up company in terms of structure (dense/thin) and the type of relations (strong/weak). The network contribution to a starter's success can, however, be negative as well (e.g. network overload), as mentioned above. The more traditional approaches to entrepreneurship (the psychological "traits" approach, and the Kirznerian and

Schumpeterian perspectives in economics) may insufficiently explain success and would benefit if complemented with the network approach. This research is intended to clarify the contingencies of the network contribution to the success of the starter. An application of the theory and practice of the "networking entrepreneur" was made in the Dutch ICT industry. We constructed a network of 30 ICT start-ups which were set up between 1990 and 2000, on the basis of company documents, and conducted interviews with the founders. A distinction was made between three types of initial network conditions. First, start-ups in incubators, the so-called *incubatees*; secondly, *spin-offs* from established companies and research establishments, and, lastly, more or less independent start-ups (i.e. *lone starters*). On the basis of the variations in the structure of the network and the type of relations they rely upon and/or they develop over time, we intend to develop propositions concerning the contribution of a particular network configuration to the ability of the start-up to discover opportunities, to get resources and to gain legitimacy. Each of these three categories contained a mixture of (relative) successes, failures, and firms that were in a dynamic process of dramatic restructuring or, alternatively, in a small market niche limiting any entrepreneurial growth.

Entrepreneurs and Their Networks: Theoretical Perspectives

Schumpeter (1934, 1947, 1976) was one of the first academics to pay attention to the notion of individual leadership, vision, intuition and the special role played in economic development by entrepreneurs and inventors. Schumpeter (1947: 152) made the distinction between the two perfectly clear: "the inventor produces ideas, the entrepreneurs gets things done." While acknowledging the importance of discovery and the activities of inventors in the process of technical change, his main interest was on the entrepreneur as an innovator and an agent of industrial transformation. For Schumpeter, entrepreneurs were relentlessly driven by a larger vision of achieving the commercial goal they had in mind. Commerical breakthroughs and techno-economic shifts were caused and deliberately triggered by "heroic" entrepreneurs and their incessant rivalry in a dynamic and inherently instable capitalist system. Entrepreneurship, for Schumpeter, involved the capacity to think the new, to grasp the essential quickly, and exploit the new product firmly: it is the "doing of new things or the doing of things that are already being done in a new way" (Schumpeter 1947: 151). While the purpose of entrepreneurs is beyond any dispute (e.g. their visions), the means by which they realise their goal are not clear. Entrepreneurs are, according to Schumpter, also risk seeking innovators, who disrupt and disturb existing markets and sectors in order to reform and revolutionise conventional technologies and products and established organisational practices and patterns. For these entrepreneurial processes of industrial and corporate mutation, and the urge to put existing technologies and components together into new commodities, Schumpeter (1976) coined the by now famous terms "creative destruction" and "the search for new combinations." They are a destabilising force in economic development by introducing new commercial ideas, products and ventures to the market. By coming up with innovations and forming new combinations, entrepreneurs disturb the equilibrium and overhaul established market conditions.

Kirzner (1990, 1991, 1997) has addressed similar fields of study to those of Schumpeter (e.g. entrepreneurship, search behaviour, competitive dynamics). He, however, posed a different research question: how do markets work, and more specifically, how do market participants seize opportunities in competitive conditions? In his "Austrian" approach to economics, Kirzner has conceptualised the market as a dynamic and spontaneous process where conditions of disequilibrium and general instability apply. Kirzner has argued that competitive market processes are essentially entrepreneurial and search-oriented. Markets are, according to Kirzner, inherently unstable and are characterised by imperfect information where human action is partly guided by maximising criteria and rivalry, which fosters alertness, judgment and search processes. In such a dynamic setting, entrepreneurs actively select information, and try to seek out and act upon new opportunities for beneficial exchange and private gain in a trial and error fashion. After many half-hearted attempts and erroneous decisions, entrepreneurs will eventually become alert to hitherto overlooked possibilities for exchange and to opportunities for profit, power, prestige and the like. In Kirzner's world, market participants, acting as competitive price and quality makers, acquire more and more accurate knowledge of present supply and demand conditions and market potential. In the successful search by the entrepreneur, his or her boundaries of sheer ignorance and imperfect information are pushed back, and he or she is able exploit some of the opportunities for arbitrage (of course dependent on the quality of sound judgement). By grasping knowledge that otherwise would remain unexploited in this discovery process in a basically imperfect market, the entrepreneur acts as an equilibrating force. By identifying and mitigating market imperfections and performing creative arbitrage activities, a contribution is made towards re-establishing the equilibrium.

The economic sociologist Burt (1992) has analysed entrepreneurial behaviour and market competition in terms of relations between players (cf. player attributes in traditional economics) and emergent interaction patterns (instead of observed relations) in an imperfect market. His view allows for a process-orientation and substantial degrees of freedom for market players and their networks to leverage information and control (instead of focussing on results and power constraints). The central notions in Burt's interface or "stitched" model of markets are that the "*tertius gaudens*" (the third who benefits') and the availability of "structural holes" or opportunity structures in the economic system. The concept of the "tertius" refers to a gate keeper or intermediary playing off "conflicting demands and preferences against one another, while building value from their disunion (Burt 1992: 34)." He/she is someone who derives benefit from brokering between players. This seems to reflect the historical definition of the "entrepreneur" as a strategic broker who obtains profits by coordinating the activities of others (Burt 1992: 274). Originally, entrepreneurs were the men who contracted and organised labour and materials for large government and military projects such as harbours, fortifications, bridges, roads and buildings. The concept of "structural hole" refers to entrepreneurial opportunities for information access, referrals, timing, and control; it is, "an invisible seam of non-redundancy waiting to be discovered by the able entrepreneur" (Burt 1992: 51). The location and distribution of structural holes in a social network provides incentives for strategic action by the "tertius." Entrepreneurial behaviour is shaped by opportunities and constraints in social relationships. Both Burt (1992: 10) and Aldrich & Zimmer (1986: 20) have come up with a relational interpretation of entrepreneurial achievement in which success is determined less by what you know than by whom you know.

An example of a "tertius" knowing about, taking part in, and exercising control over more rewarding opportunities ("structural holes") is the *impannatore*, a key figure in the Italian textiles industry (Brusco 1982). Responding to the regular requests and contracts from his big clients, he acts as a project developer by building an ad-hoc platform of organisations to carry out a particular job, and managing its execution. Normally, there is a division of labour in the network in which the *impannatore* designs the fabric by himself and commissions the spinning, weaving, and finishing to other local enterprises. His relational power base is based on an exclusive access to market information (i.e. the ever changing demands from fashion houses, new trends in fabrics and design etc.), detailed knowledge about contracting and enforcement, and a capability to manage several overlapping support networks and the intangible asset of institutional legitimacy accumulated over the years of doing the brokerage business. Another example of an institution relevant to facilitating information flows and bringing together anonymous participants within a shared purpose to realise complementary interests, is the 128 Venture Group (Nohria in Nohria & Eccles 1992). The monthly meetings of the 128 Venture Fund, scheduled around a plenary forum and more informal sessions, are meant to identify new techno-economic opportunities in the ICT sector and potential business leads, and bring together the local partners with whom one can collectively build a new venture. The following parties involved in setting up a new venture can be distinguished: an entrepreneur with an innovative idea and/or technology, a provider of venture capital, candidates for the venture's management team, and providers of professional and other support services. The process of matchmaking is organised through allocating different badges to the various participants at the meeting: green for capital, red for technology, yellow for professional services and blue for management team candidates. Normally, the candidates for the partnership, who have not met before, have "weak tie" relationships, which are activated through the Venture Group 128 meetings: they all belong to the bounded community of fate of MIT-alumni and live in the Route 128 area in and around Boston (MA, USA).

Social network analysts claim that social capital pays off, either in terms of particular rewards (bigger value for some of us) or in achieving the common good (bigger understanding among all of us). In the first case, scholars such as Burt (1992) and others have argued that personal and organisational accomplishments (e.g. pay, promotion, profits) are to a large extent determined by the structure and composition of one's personal and business networks. Here, the underlying notion is that networks are both opportunity structures that permit, or even encourages certain "entrepreneurial-like" interactions, and constraint structure requiring, discouraging and prohibiting other interactions. Especially if there are a couple of structural holes or gaps in one's network, by linking two other actors who are not themselves directly connected, the focal actor (ego) may be able to discover interesting opportunities and create value out of his or her social capital. For instance, they may get more information, they may get it first, they may learning about new possibilities and finding alternative resources. These so-called entrepreneurial or brokerage networks are ideal for the analysis of the emergence of new products and new firms; especially in a dynamic market place where certain information advantages and privileged network positions may allow strategic linking and manipulation in order to further one's particular interests and eventually yield higher performance than those of rivals. Others, including Coleman (1988) and Baker (2000a, b) have paid less attention to these ultimate zero-sum games, and have instead focused on non-zero sum phenomena, including altruism and loyalty. Here reciprocity, defined as "we are helped because we help others" (Baker 2000b: 131). Often these networks are "clumpy": close ties of similar people and organisations, which share a common identity and a sense of common purpose. Although they are a useful vehicle for building group loyalty and creating consistent performance expectations among its members, these closed networks are inadequate for obtaining information, other resources or influencing people outside such networks.

The structure of networks may vary from a loose collection of ties to close-knit business groups, in which the focal organization is embedded. In this explorative study, a choice has been made for the effect of a particular mix of strong and weak ties in entrepreneurial networks, because this mix allows for an analysis of support networks in terms of both the depth and width of relationships. Granovetter (1995) has specified the intensity and diversity of relationships, i.e. the difference between strong and weak ties, on the basis of four criteria: namely, the frequency of contacts, the emotional intensity of the relationship, the degree of intimacy and reciprocal commitments between the actors involved. While weak ties provide access to (new) industry information and to new business contacts, strong ties are relationships one can rely upon, both in good times and in bad times. Strong ties tend to bind similar people in longer-term and intense relationships. Affective ties with close friends and family members may provide a shortcut to, or even preclude, the search for useful knowledge and access to critical resources. In other words, strong ties contribute to "economies of time" (Uzzi 1997: 49): involving the ability to capitalize quickly on market opportunities. The manifestation of strong bonds will also reduce the time spent on monitoring and bargaining over agreements, while free-riding will be discouraged and transaction costs lowered. Strong ties are more likely to be useful to individuals in situations characterized by high levels of uncertainty and insecurity (e.g. amidst radical innovations). In such complex settings, individuals rely on close friends and family members for protection, uncertainty reduction and mutual learning. Krackhardt (1992: 238) has elaborated on the affective component of strong ties by arguing that commitment, loyalty and friendship within an organization will be critical to an organization's ability to deal with major crises. In short, a relational governance structure based on strong ties will promote the development of trust, the transfer of fine-grained information and tacit knowledge, and joint problem-solving (Rowley *et al.* 2000; Uzzi 1996, 1997).

Strong ties have shortcomings too. There is the risk of *overembeddedness*, i.e. of stifling economic performance (Uzzi 1996). Close ties within and among business communities are vulnerable to exogenous shocks and may insulate such commitments from information that exists beyond their network. There is the danger of being blind to new developments or being "locked-in" (Johannisson 2000). Weak ties refer to a diverse set of persons working in different contexts with whom one has some business connection and infrequent or irregular contact. These loose and non-affective contacts increase diversity and may provide access to various sources of new information and offer opportunities to meet new people. Weak ties represent local bridges to disparate segments of the social network that are otherwise unconnected and may open the door to new options (Burt 1992; Granovetter 1995). In short, both strong and weak ties are useful and contribute to the emergence and growth of firms, although they are beneficial in different ways and at different stages of a company's development. Therefore, the ideal entrepreneurial network includes a particular mix of strong and weak relationships (Uzzi 1996, 1997). We have distinguished three entrepreneurial processes, the ability to discover opportunities, the ability to secure resources, and the ability to gain legitimacy, in which network ties play a role.

The Networking Entrepreneur at Work

In the existing literature on entrepreneurship the importance of having a solid network, in addition to the personal qualities of the entrepreneur, is emphasized as being one of the factors influencing the achievement of starters. In this research we emphasize the influence of the entrepreneur's network on the achievements of the starting company. The network is important to obtain knowledge, complementary means and legitimacy. Until the mid-1990s, most network studies established a simple causal relation between the size of the network and the success of the starter (Aldrich & Zimmer 1986; Larson & Starr 1993). Recently, however, more and more qualifications are being brought forward to indicate that the relationship is not that simple, nor does it necessarily have to be positive. Steier & Greenwood (2000), for instance, introduced the term "network overload." At a certain size the network no longer has a positive impact on the success of the starter, and may even be negative. The positive effect of a number of relations is cancelled by the amount of extra time needed to maintain new relations. To limit the danger of "network overload" an entrepreneur may benefit from an incubator, since the incubator provides him with access to a new network. Another study (Stuart *et al.* 1999) suggests that it is not so much the size of the network as its quality and reputation that have a positive influence on the success of start-ups. In addition, various studies introduce contingencies, for example, with regard to the branch in which the starter is operating. Research conducted by Rowley *et al.* (2000) shows that a network with strong and close relationships has a positive effect on the achievements of starters in a stable industry, but a negative impact on the success of starters in a dynamic market. Research on "social capital" has yielded similar results. The analysis presented by Gargiulo & Benassi (1999), ominously called *The dark side of social capital*, shows that an existing network with close ties can inhibit the search for new opportunities and therefore have a negative impact on the success of a start-up in a dynamic market.

Opportunities

An important source of new ideas and lucrative opportunities may be the networks, in which the entrepreneur is more or less actively participating. Hills, Lumpkin & Singh (1997) found that about 50% of entrepreneurs identified ideas for new ventures through their social network. In addition, in the process from idea to the actual start of a venture, prior knowledge (Shane 2000) and information (Fiet 1996) are important. According to Fiet (1996: 429): "use of network may be viewed as a way of tapping into an information channel to obtain risk-reducing signals about a venture opportunity." Both variables are closely linked to networks, as network relations can be seen as ways to gain access to knowledge and information. In one of the first studies on this aspect, Birley (1985) carefully documented how often entrepreneurs seek advice and feedback on the core ideas of their business plan, when they turn to friends and family for the solution of local problems, and when they use formal ties to look for financial support. Starting-up was seen as an iterative process in which the number of informal and formal ties affected the success of the entrepreneur in finding a lucrative opportunity. The environment and the opportunities it contains are diverse and uncertain. The network of an entrepreneur is a source of information helping the entrepreneur to locate and evaluate opportunities. Networks and in particular weak ties provide access regarding a diverse set of topics, ranging from potential markets for goods and services to innovations and promising new business practices. Weak ties are supposed to lead to a more varied set of information and resources than strong ties can (Bloodgood *et al.* 1995), and consequently weak ties enhance the ability of entrepreneurs to spot new opportunities.

Resources

Providing access to resources is an important contribution of networks to the venturing process. Entrepreneurs rarely possess all the resources required to seize an opportunity. One of the crucial tasks in a new venture is to access, mobilize and deploy resources (Garnsey 1998). This is a difficult task in the initial stages of a start-up with limited financial resources and hardly any ability to generate internal resources and revenues. Close social support networks (e.g. spouse, family ties) may provide the founder/owner with the resources (e.g. financial and human capital) he or she is lacking, and hence provide stability to the new firm in its early stages (Brüderl & Preisendörfer 1998). Additionally, sparse networks facilitate the search for critical asset providers (e.g. investment and technology partners and key customers), who may offer the start-up further access to financial resources, production know-how and complementary technology, distribution channels, etc. Furthermore, there is initial uncertainty about the growth of the venture and the resources it requires (Chrisman, Bauerschmidt & Hofer 1998). In the case of staged investing by venture capitalists in technology start-ups, the amount of uncertainty about a venture declines as it survives and grows. One of the key survival strategies is "asset parsimony" (Hambrick & MacMillan 1984). The required resources must be secured at minimum cost. Paying the market price for resources, such as labor, materials, advice and commitment is often too expensive. Social transactions through network ties play a critical role in the acquisition of venture resources. These resources can be acquired far below the market price when entrepreneurs (as well as intrapreneurs) employ social assets such as friendship, trust, and obligation (Starr & MacMillan 1990). In particular, network members representing strong ties are more motivated to help the entrepreneur than those with whom the entrepreneur has weak ties. Potential entrepreneurs assess their ability to get hold of the required resources at relatively low cost on the basis of their strong ties.

Legitimacy

The third contribution of a network to the success of a start-up is the way it opens possibilities to gain legitimacy. Gaining legitimacy is imperative in starting something that is considered innovative (DiMaggio 1992). Stinchcombe (1965: 148–150) has introduced the concept of the *liability of newness*, or simply stated, young organizations face higher risks of failure than old ones. Established organizations have a set of institutionalized roles and tasks, stable customer ties, experienced constituents, a surplus of capital and creativity (slack), and a shared normative framework at their disposal, all of which contribute to an effective provision of goods and services and their ultimate survival. New firms and novel organizational forms, on the other hand, are more likely to fail just because they still have to develop and acquire such prerequisites (Baum *et al.* 2000). Faced with the aforementioned "liability of newness," a new venture has to organize institutional support and legitimacy. This appears to apply especially to (relatively) radical innovations, where young technology companies need the endorsement of (some of) the prominent players in their industry (Stuart *et al.* 1999). In order to enhance their visibility and gain recognition, new ventures seek to obtain a prestigious business affiliate to build up a strong link with and eventually hope that, through this key contact, they will have access to new customers and partners. Furthermore, biotechnology companies in particular establish large supervisory boards involving well-known industry experts and academics (Elfring & Hulsink 2003).

Suchman (1995: 574) has defined legitimacy in a broad sense as "a generalised perception or assumption that the actions of an entity are desirable, proper, or appropriate within some socially constructed system of norms, values, beliefs, and definitions." Aldrich & Fiol (1994) draw a distinction between cognitive and socio-political legitimacy. Understanding the nature of the new venture is referred to as cognitive legitimacy. It has to do with the spread of knowledge regarding the new business concept. To overcome this legitimacy barrier, network actors, such as competitors, distributors and universities, must be mobilized to create partnerships in order to achieve a wider understanding of the new concepts. The second, and related, type of legitimacy is labeled socio-political legitimacy and refers to the extent to which key stakeholders accept the new venture as appropriate and conforming to accepted rules and standards. Achieving socio-political legitimacy is particularly difficult when the new venture is very innovative and challenges existing industry boundaries. In those cases changes in the institutional framework are often required. Organizing socio-political legitimacy requires collective action, negotiations with other industrial constituents, and joint marketing and lobbying efforts.

Networks and Corporate Evolution

It is becoming increasingly clear that a network does not always have a positive effect on the achievements of start-up companies. There are contingencies and the question arises what type of network under what circumstances will contribute positively to the success of a starter. In addition, it is important to gain an insight into what the causal chain from network to start-up success looks like. In short, "how exactly is that possible positive network effect brought about," and "what are the sources of that network effect" are two core questions we will address in this research. In theory, for example, there is a distinction between "Burt rents" and "Coleman rents" (Kogut 2000). In the first case the starter can use his position as "broker" between various networks to generate additional income. In terms of Burt (1992) the entrepreneur, thanks to his position, gains access to additional sources of profit by locating "structural holes" in the network. In such a network there exist few duplicate relations, it is spread out and links together various subnetworks. The Coleman rents, on the other hand, derive from the benefits of a close network with strong relations wherein trust is important (Coleman 1988). This kind of relationship enables a "tacit" exchange of knowledge, the benefits of which can be used to generate more intense and trusted ties.

In order to understand the causal mechanisms between the network structure and performance, we will focus on the mix of weak and strong ties, each of them contributing in a particular way to the entrepreneurial process. Strong ties are associated with the exchange of "fine-grained" information and tacit knowledge, trust-based governance, and resource cooptation (Krackhardt 1992; Rowley *et al.* 2000). Their advantages are different from the benefits generated by weak ties. Weak ties are beneficial as they provide access to novel information as they offer linkages to divergent regimes of the network (Burt 1992; Granovetter 1995). As strong and weak ties each have qualities, that are advantageous for different purposes we focus on the mix. Thereby we build on the work of Uzzi (1996, 1997), Hite & Hesterly (2001) and Rowley *et al.* (2000) who conclude that a key issue in the determination of network benefits is the search for the optimal mix of strong and weak ties. A number of researchers have utilized a contingency approach to reconcile the different network benefits. For example, the industry context has been introduced as a contingency factor by Rowley *et al.* (2000), while Hite & Hesterly (2001) show that, as ventures progress from emergence to growth, the evolving resource needs require a shift in network structure. Start-ups based on radical innovations require a different mix of strong and weak ties from those pursuing incremental innovations. We argue that this degree of innovation affects the way in which firms approach their network relationships and seek to benefit from them.

Research Design

The research project contributes to the phenomenon of the "networking entrepreneurs" in our Internet- and ICT-based society and to the institutionalisation of "networking" in our network economy. There are a number of processes that indicate that "networking," defined as the exchange of information and contacts and the "wheeling and dealing" between entrepreneurs, business partners and service providers, is not only on the rise in terms of its popularity, but also in terms of quality. Statements such as "without a good support network innovative entrepreneurs are nowhere" are an indication of the social and economic value attached to existing and new contacts and partners within a strategic network. In addition to the well-established technology transfer offices, large companies, universities and research establishments, and intermediary investors and service providers have set up so-called "incubators" to nurture new ideas, entrepreneurs and/or dynamic firms or to speed up product and service innovation and entrepreneurial growth in a controlled environment, where resources, services and contacts are easily accessible (Richards 2002; Smilor & Gill 1986). However, within the community of high-technology and ICT-starters there are also new rituals and innovative institutions, aimed at bringing together new ideas and entrepreneurial professionals that are not yet known to one another. In addition to the aforementioned "incubators," there are also "virtual incubators," such as Garage.com and Factory Zoo, that are trying to exchange contacts and business plans and establish global companies over the Internet. Also, the regular partner evenings and *ICT-parties* (First Tuesday etc.) and the rise of special media focused on information exchange and networking (e.g. bulletin boards/websites exclusively for starters, new magazines created by and for ICT-starters such as Red Herring and Tornado Insider) can be seen as illustrations of "entrepreneurial networking." A very significant question in this context is, however, whether these networking activities, facilitated in part by ICT and communicated through Internet, do contribute to the success of that new company.

This study is about nascent and actual entrepreneurship in the Dutch Information and Communication Technology (ICT) sector and especially the role that is played by the networks in which starters do or do not participate with regards to the innovativeness and success of these ICT-companies. The population of experts and professionals that are working at large companies or research institutes, recent university or polytechnic graduates, and other specialists in a particular domain (e.g. ICT) produces nascent entrepreneurs (van Gelderen *et al.* 2001). Nascent entrepreneurs are people that are seriously considering starting an ICT-company, whether on their own or with others. In the process of founding and building a company the social and strategic network in which the starter operates or wants to operate (with the partners and means that the company presently lacks) can play a determining role in the growth and eventual success of the company. This research is concerned with the individuals who actually did so, in other words young, small and innovative ICT-entrepreneurs. These we define as companies with a minimum of 2 employees that offer ICT-products or services and were founded between 1990 and 2000. Our research focuses on the entrepreneur and his/her network of various contacts and links. Furthermore, our sample of young and dynamic ICT-companies includes a "mere" 30 start-ups and their linkages with relevant investors, business-partners, customers, other entrepreneurs etc. It is, of course, also possible that the company does not value growth and expansion that much, being relatively content with the share of the market niche within which it is operating.

Through interviews with the founders of 30 ICT/Internet-companies and through desk research we have sought to determine to what extent the presence or absence of such support networks have contributed in a positive way to the success of the start-up (e.g. survival, growth and/or profit). In that way we reconstruct the networks of all start-ups in "mini-cases" and analysed them on the basis of development phases (conceptualisation, foundation, growth, etc.). We have divided the 30 companies into 3 groups of 10 companies each, based on the extent to which these starting companies utilise a strategic network to start and build their ICT-company: (i) *lonesome cowboys*; (ii) *spin-offs/spin-outs*; and (iii) *incubator-driven companies* (see Figure 1). Each of the 3 groups will include at least 2 companies that were unsuccessful and that have faced bankruptcy. Although it is relatively hard to obtain the cooperation of entrepreneurs who did not succeed, their findings are of great importance to our research.

(i) The first category of ICT-start-ups, called the *lonesome cowboys*, includes companies that appear as if from nowhere and develop further without substantial support from a strategic network. These are ICT-starters that are being founded within a constellation similar to traditional companies: the entrepreneur or entrepreneurial team initially sets out without network partners and at a later development stage may look for additional

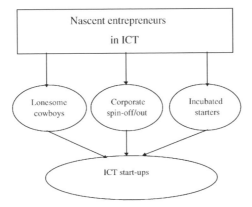

Figure 1: Research design.

knowledge, employees, funds and customer input (next to the conventional commercial and labour relations the company maintains with its customers and employees). Examples of *lonesome cowboys* are Annie Connect (call centre), Euronet (internet service provider), and Ring! (computer-telephony integrator).

(ii) The second category of *spin-offs* and *spin-outs* consists of ICT-start-ups that in some way have been given support when they were founded from their former employer(s) (e.g. in training and coaching, housing, contract research, financing, etc.). Whereas in the case of a spin-off company there is no longer a direct financial relation with the company the entrepreneur has worked for, in the case of a spin-out there does exist a relation with the mother company, for instance in the form of a strategic participation in and/or collaboration with the nascent company. An example of a company that keeps sending new companies into the world is the national research laboratory CWI (Centre for Mathematics and Information Science), which since the early 1990s has created around 10 spin-offs, examples of which are NLnet/Uunet, Data Distilleries and General Design/Satama. Universities and large companies, consciously or unconsciously, can also serve as incubator for innovative ideas and potential entrepreneurs and generate spin-offs and/or spin-outs; examples are HuQ Speech Technologies (University of Groningen) and Carp Technologies (University of Twente). Also established companies can churn out teams of employees that start for themselves seeking to commercialise the technologies they were working on previously. Examples of corporate spin-offs in our sample are Wellance (spin-off from KPN/Planet Internet) and Profuse (spin-off from the Baan Company).

(iii) The third category, that of *incubator-driven ICT-companies*, is created, founded and built *within* a strategic network of (potential) partners and professional service providers, created as such by a specialised *incubator* (e.g. Twinning). Thanks to this closely integrated and varied network or with the help of a strategic partner creating a virtual network, the start-up can develop further. Examples of incubators in the Netherlands are Twinning, Silicon Polder Fund, Gorilla Park and Newconomy. These incubators provide the ICT-start-up (in exchange for a share in the new company)

Table 1: Overview of ICT companies participating in the study.

Lonesome Cowboys	Spin-Offs	Incubatees
Annie Connect	Carp Technologies	Bibit
Co-makers	HuQ Speech Technologies	CareerFever
Euronet	InterXion	FactoryZoo
Keekaboo	Profuse	Gopher
Metrixlab	Proloq	Hot Orange
Nedstat	Tornado Insider	Information Innovation
Planet Internet	Tridion	Oratrix
Rits Telecom	Wellance	Siennax
Ring	Xpertbuyer	Tryllian
The Vision Web	Bitmagic	Punt Edu
Vocognition	Vizigence	
XOIP		

with easier access to a number of important services, such as financing, housing and equipment/infrastructure, counselling and coaching, and information exchange and networking (contacts and referrals to clients, partners, suppliers, research institutions, etc.). Examples of such incubator-driven ICT-starters are Hot Orange, Trylian, and Gopher publishers.

The research was explorative in nature and is aimed at generating hypotheses with regard to the influence of networks on the success of ICT-starters. Thirty ICT-starters were non-randomly selected from the databases of Erasmus University of Rotterdam, EIM, TNO and a number of trade magazines (Automatiserings Grids, Computable, Emerce). In total, 31 entrepreneurs were interviewed (29 individually and one joint interview with the two founders). Twelve lonesome cowboys were selected for the study, 11 spin-offs and 10 incubatees. The interviews were semi-structured and lasted on average between 1 and $1\frac{1}{2}$ hour. The in-depth interviews were taped and transcripts were made of them. On the basis of these transcripts and publicly available company profiles (obtained through desk research), 30 "mini-cases" of the entrepreneur and their firms were made. In the final phase the mini-cases will be analysed and the findings will be discussed (Table 1).

Results

Lonesome cowboys were started without a particular network within the ICT industry. They were relative outsiders and they benefited from some strong ties derived from their backgrounds. These strong ties were in some cases family and friendship ties, but they also profited from relationships in their previous work environments. However, these strong ties appeared to be relatively unimportant. The dominant networking activity was the exploration of weak ties. Most of the founders in this category of start-ups discovered opportunities through their weak ties. In most cases the business model of the start-up changed during the period of emergence. These changes were often inspired by discussion with acquaintances, such as people they recently met or persons they were referred to by relatively "distant" friends. The networking could be characterized as a frantic search for people who could give information on new opportunities and the feasibility of the already spotted opportunities on the one hand. On the other hand, the role of strong ties, although limited in number, was to give "trusted" feedback on the various stages of the business plan (often close friends and family relationships). These strong ties were often outsiders to the ICT community, while the weak ties consisted mostly of insiders. Such weak ties appear to be used to get access to the strategic network of ICT and related firms in the Netherlands.

Some of the weak ties during the opportunity discovery process developed into trusted ties, of which some appeared to play an important role in the process of securing resources. In addition to these "new" strong ties, the older strong ties (i.e. people they know well from their previous activities), were important in getting hold of required resources. On the basis of this limited number of cases we found a balanced mix of strong and weak ties contributing to the resource acquisition process. Some weak ties developed into strong ties and simultaneously some weak ties were dropped because they did not provide sufficient value in the struggle for resources. This selection among weak ties is a process, which appears to be less intense in the resource acquisition process than in the opportunity discovery process. Thus, the wild exploration of network ties in the search for opportunities evolves into a combination of exploration and exploitation of the network in the process of getting resources. Concerning the third entrepreneurial process of gaining legitimacy, the network benefits could be characterized by a mix of strong and weak ties. Although weak ties tended to be dominant, some of the original weak ties developed into strong ties by the time legitimacy was crucial. It was interesting to see that almost all of these entrepreneurs were aware of the importance of legitimacy on the one hand and the impact of association with a well-known player in the field on the other hand. Therefore, searching through their growing network of organisations to be associated with, was high on their priority list. Most of their emphasis was on finding a respectable "launching" customer, but connections to leading venture capitalists or major ICT companies, such as KPN, IBM or Baan were valued as well for their impact on legitimacy.

While the lonesome cowboys in their early growth stages move from exploration to exploitation, the other two categories predominantly focus on exploitation, in the case of spin-offs, and on exploration, in the case of incubatees. Also the development path differs between spin-offs and incubates: while the first is still very close and dependent upon its source (or "mother") organisation (i.e. a strong tie), the latter's use of contacts is less outspoken and sometimes referrals and references are provided by the incubators, and/or alternatively developed by the in-house entrepreneurs themselves. The spin-off entrepreneurs in our sample were "kick-started" for fast early growth due to in-depth industry knowledge of the founding entrepreneurs with many years of experience and the resources provided by their former employer, varying between capital, tangible and intangible assets (easy access to patents and facilities), rolling contracts, and reputational benefits as a consequence of the association with the mother organisation. However, the status of being industry insiders and the almost direct participation in an already established strategic network (e.g. the Baan network), piggybacking on the contacts and resources of the mother organisation, proved to be, in a number of cases, a blessing in disguise. While the spin-off firm had a number of ongoing and strong ties (with a clear industry affiliation), it was relatively unable to develop new weak ties, and as a consequence, unable to break out from the complacent networks, that it already had established. Just by this trained incapacity to pursue weak ties aggressively and cultivate a diverse network, spin-offs lack the drive of the lone starters to take major risks (e.g. experiment with new technologies) and to spot unseen opportunities (work with new customers and partners). They may also lose some of their initial advantages at a latter stage. This could be seen as a lock-in effect, or a path-dependent development.

The category of incubatees and their networking behaviour is more difficult to put into perspective. First of all, the incubator organisations with whom our incubatees were affiliated with, were all young and inexperienced (e.g. Twinning was established in 1998; Gorilla Park, Small Business Link and Newconomy in 1999), and busy with establishing themselves. Instead of offering their incubatees a Rolodex of business contacts instantly, the incubators had to roll out their network of services first, finding business partners and searching for capital and political legitimation, before they could actually help their start-ups. Already during the built-up of their infrastructure, they ambitiously and randomly started to select a large number of start-ups as incubatees, and promised them services, resources and contacts they could not yet fully materialise and deliver. Like their incubatees, the incubators themselves also lacked a track record and standardised procedures. The supply of services, resources, facilities and contacts not only varied between incubators, but also within the portfolio of investments of one incubator: for instance, one Twinning company only marginally benefited from an early investment, and another firm agreed on office space, a whole set of specialised services, and two major co-investments. Some of the incubatees with proven entrepreneurial skills and an extensive industry network were not desperately in need of support from the incubator to seize business opportunities. Others, that were clearly less experienced, could find a shelter and some seed money from the incubator to promote their ideas and consider some market opportunities. In this case the incubator could not really help, since there were not any clear ties (neither weak neither strong) with established companies that could act as a partner or customer for the start-up. In the case of securing resources, most of the incubatees benefited from the services and facilities offered by and through the incubator, and eventually from the new weak ties they now had access to (although they disagreed whether the new contracts with law firms, consultancies, accountancies and investors were worth the money). The relatively unknown incubatees also could benefit from the reputation and the brand name of their well-known incubator, giving them quicker access to banks, investors and other service providers. When the incubators ended up in the stormy economic weather of 2001–2002, the legitimacy benefits offered by incubators evaporated and some incubatees went bust or had to distance themselves from the struggling incubator.

Discussion and Conclusions

The results of the use and development of network ties of the lonesome cowboys category is significantly different from the network benefits to entrepreneurs as suggested by Hite & Hesterly (2001). On basis of a review of the literature Hite & Hesterly (2001) propose that start-ups rely in the emerging phase primarily on their strong ties. And only later in the early growth stage they expand their network to include weak ties as well. The argument for the dependence on strong ties has to do with the high level of uncertainty of the new venture. Strong ties are willing to provide the resources despite the uncertainty, while weak ties tend not to take the risk associated with the uncertain future of the start-up. Furthermore, in the early growth phase, it is necessary to develop a more diverse network in which weak ties may appear to be crucial to discover structural holes (Burt 1992). Structural holes are important in accessing new resource providers in order to fuel further growth. Thus they propose that network benefits develop from exploitation of strong ties to the exploration of weak ties, while we find evidence that it is exactly the reverse; from exploration of weak ties in the search for opportunities to exploitation of emerging strong ties.

In our sample of start-ups weak ties appeared to be very important in the emergence phase and some of them appeared not to have any difficulties concerning the uncertain future of the start-up. There may be two reasons for this different finding. First, in our cases the emergence phase is dominated by the search for the most lucrative opportunity and was not primarily focussed on securing resources, while the focus of Hite and Hesterly seems to be on the resource acquisition process. Secondly, our cases of high-tech start-ups differ in the sense that they indeed take much more time to search for the best business concept (see also Roberts 1991) and thus there is more focus on opportunity discovery, and this process benefits more from weak ties than strong ties. In that process ties are also less committed to the start-up and therefore the uncertainty and the associated risk is not that important as in the situation of being a resource provider. The argument of Hite & Hesterly (2001) for the growing importance of weak ties as the venture evolves from emergence to early growth concerns the need to find structural holes. This use of the structural hole argument is a bit odd. Structural holes, and the role of weak ties, are related to the discovery of new information. This information and thus weak ties may be of importance to spot new opportunities or more specifically for these start-ups to change the business concept on basis of this new information. This process plays in particular a role in the emergence phase. However, once the start-up has reached the early growth stage, they know what they need in terms of resources. Thus there is no reason to discover new information through structural holes, since the search for resources and also legitimacy is straightforward.

Acknowledgments

This study was made possible through a small grant from the Netherlands Organisation for Scientific Research (NWO) in the MES framework (Society & Electronic Highway, project number 014–43–609).

References

Aldrich, H. E., & Zimmer, C. (1986). Entrepreurship through social networks. In: D. Sexton, & J. Kasarda (Eds), *The art and science of entrepreneurship*. Cambridge, MA.

Baker, W. E. (2000a). *Networking smart. How to build relationships for personal and organizational success (1994)*. Backimprint.com.

Baker, W. E. (2000b). *Achieving success through social capital. Tapping the hidden resources in your personal and business networks.* Jossey-Bass.

Baum, J. A. C., Calabrese, T., & Silverman, B. S. (2000). Don't go it alone: Alliance network composition and startups, performance in Canadian biotechnology. *Strategic Management Journal, 21*, 267–294.

Bhidé, A. (2000). *The origin and evolution of new business.* Oxford University Press.

Bloodgood, J. M., Sapienza, H. J., & Carsrud, A. (1995). The dynamics of new business start-ups: Person, context, and process. In: J. A. Katz, & R. H. Brockhaus (Eds), *Advances in entrepreneurship, firm emergence, and growth.* Greenwich: JAI Press.

Brusco, S. (1982). The Emilian model: Productive decentralisation and social integration. *Cambridge Journal of Economics, 3*, 167–184.

Burt, R. S. (1992). *Structural holes. The social structure of competition.* Cambridge: Harvard University Press.

Chrisman, J. J., Bauerschmidt, A., & Hofer, C. W. (1998). The determinants of new venture performance: An extended model. *Entrepreneurship: Theory and Practice, 23*, 5–29.

Cohen, D., & Prusak, L. (2001). *In good company. How social capital makes organizations work.* Harvard Business School Press.

Coleman, J. S. (1988). Social capital in the creation of human capital. *American Journal of Sociology* (Suppl.), S95–S120.

DiMaggio, P. (1992). Nadel's paradox revisited: Relational and cultural aspects of organizational structures. In: N. Nohria, & R. G. Eccles (Eds), *Networks and organizations: Structure, form, and action.* Cambridge: HBS Press.

Elfring, T., & Hulsink, W. (2003). Networks in entrepreneurship: The case of high-technology firms. *Small Business Economics, 21*(4), 409–422.

Estades, J., & Ramani, S. V. (1998). Technical competence and the influence of networks: A comparative analysis of new biotechnology firms in France and Britain. *Technology Analysis and Strategic Management, 10*(4), 483–495.

Fiet, J. O. (1996). The information basis of entrepreneurial discovery. *Small Business Economics, 8*, 419–430.

Gargiulo, M., & Benassi, M. (1999). The dark side of social capital. In: R. Leenders, & S. Gabby (Eds), *Corporate social capital and liability.* Boston.

van Gelderen, M., Frese, M., & Thurik, A. R. (2001). Strategies, uncertainty and performance of small startups. *Small Business Economics* (forthcoming).

Granovetter, M. (1995). *Getting a job. A study of contacts and careers* (2nd ed. 1974). University of Chicago Press.

Hambrick, D., & MacMillan, I. C. (1984). Asset parsimony — managing assets to manage profits. *Sloan Management Review, 25*(Winter), 67–74.

Hills, G. E., Lumpkin, G. T., & Singh, R. (1997). Opportunity recognition: Perceptions and behaviors of entrepreneurs. In: P. Reynolds *et al.* (Eds), *Frontiers of entrepreneurship research 1997* (pp. 168–182). Babson Park: Babson College.

Hite, J. M., & Hesterly, W. S. (2001). The evolution of firm networks: From emergence to early growth of the firm. *Strategic Management Journal, 22*, 275–286.

Johannisson, B. (2000). Networking and entrepreneurial growth. In: D. L. Sexton, & H. Landström (Eds), *The Blackwell handbook of entrepreneurship.* Oxford: Blackwell.

Kirzner, I. M. (1990). The market process: An Austrian view. In: K. Groenveld *et al.* (Eds), *Economic policy and the market process. Austrian and mainstream economics* (pp. 23–39). North-Holland.

Kirzner, I. M. (1991). Market process vs. market equilibrium (1973). In: G. Thompson *et al.* (Eds), *Markets, hierarchies & networks. The coordination of social life* (pp. 53–65). Sage.

Kirzner, I. M. (1997). Entrepreneurial discovery and the competitive market process: An Austrian approach. *Journal of Economic Literature*, *35*, 60–85.

Kogut, B. (2000). The network as knowledge: Generative rules and the emergence of structure. *Strategic Management Journal*, *21*, 405–425.

Krackhardt, D. (1992). The strength of strong ties: The importance of *Philos* in organizations. In: N. Nohria, & R. G. Eccles (Eds), *Networks and organizations: Structure, form and action* (pp. 216–239). HBS Press.

Larson, A., & Starr, J. A. (1993). A network model of organization formation. *Entrepreneurship: Theory and Practice* (Winter), 5–15.

NCOE (2001). *Five myths about entrepreneurs: Understanding how business start and grow*. Washington, DC: National Commission on Entrepreneurship.

Nohria, N., & Eccles, R. G. (Eds) (1992). *Networks and organizations. Structure, form, and action*. Harvard Business School Press.

Richards, S. (2002). *Inside business incubators and corporate venture*. Wiley.

Roberts, E. B. (1991). *Entrepreneurs in high technology. Lessons from MIT and beyond*. New York: Oxford University Press.

Rowley, T., Behrens, D., & Krackhardt, D. (2000). Redundant governance structures: An analysis of structural and relational embeddedness in the steel and semiconductor industry. *Strategic Management Journal*, *21*, 369–386.

Schumpeter, J. A. (1934). *Theory of economic development*. Cambridge: Harvard University Press.

Schumpeter, J. A. (1947). Creative response in economic history. *Journal of Economic History*, *7*(2), 149–159.

Schumpeter, J. A. (1976). *Capitalism, socialism and democracy* (1943). London: Allen & Unwin.

Shane, S. (2000). Prior knowledge and the discovery of entrepreneurial opportunities. *Organization Science*, *11*, 448–469.

Smilor, R. W., & Gill, M. D. (1986). *The new business incubator. Linking talent, technology, capital and know-how*. Lexington.

Starr, A. S., & MacMillan, I. C. (1990). Resource cooptation via social contracting: Resource acquisition strategies for new resources. *Strategic Management Journal*, *11*, 79–92.

Steier, L., & Greenwood, R. (2000). Entrepreneurship and the evolution of angel financial networks. *Organization Studies*, *21*, 163–192.

Stinchcombe, A. L. (1965). Social structure and organizations. In: J. G. March (Ed.), *Handbook of organizations*. Chicago: Rand McNally & Company.

Stuart, T. E., Hoang, H., & Hybels, R. C. (1999). Interorganizational endorsements and the performance of entrepreneurial ventures. *Administrative Science Quarterly*, *44*, 315–349.

Swedberg, R. (Ed.) (2000). *Entrepreneurship. The social science view*. Oxford University Press.

Uzzi, B. (1996). The sources and consequences of embeddedness for the economic performance or organizations: The network effect. *American Sociological Review*, *61*, 674–698.

Uzzi, B. (1997). Social structure and competition in interfirm networks: The paradox of embeddedness. *Administrative Science Quarterly*, *42*, 35–67.

Chapter 6

Virtual Incubation: The Case of the VVE of the Comunidad de Madrid

Jaap Van Tilburg, Cheo Machin, Peter Van Der Sijde
and Felix Bellido

Introduction

The incubation of knowledge intensive spin-off companies from research institutes and universities have been the focus of many reports by the European Commission (e.g. EC 2002; Centre for Strategy & Evaluation Sevices 2002) and theme issues of academic journals (e.g. Van der Sijde 2002a). An aspect that receives much attention in all these reports and articles is how the incubation process can best be better supported. Very few authors pay attention to the perspective of the entrepreneur; most consider the support structure, the supply side ("technology push") instead of the demand ("market pull") side of the process. The recent contribution by Brush *et al.* (2001) takes the perspective of the entrepreneur, who needs to build up the resource base for his/her company — the physical, human, technological, social, financial and organisational resources. If and when such a resource base is necessary for a start-up company, then an incubator should play a role in building-up company resources (Van der Sijde 2002b). In this chapter we discuss the virtual incubator of the Comunidad de Madrid in Spain.

Virtual incubation in its strictest sense can (or should) be defined as an integrated support system without a physical infrastructure or physical personal contacts that provides services to clients via information and communication technologies, ICT (Van Tilburg *et al.* 2002). Van Tilburg also distinguishes three types of virtual incubators, "the stand-alone virtual incubator," in which all functions and tasks are provided virtually, "the virtual incubator as a network" partner, a combination of a virtual stand-alone incubator with (one or more) physical incubators, and "the virtual service" an incubator that provides virtual services. Nowak & Grantham (2000) propose virtual incubator that works in conjunction with physical incubators when needed, using distributed resources in which the private sector plays a lead role, while the university and the public sector provide support. Further, Nowak and Grantham discuss formalised management control systems, and the national and international focus. The virtual incubator in Madrid is

New Technology-Based Firms in the New Millennium, Volume III
© 2004 Published by Elsevier Ltd.
ISBN: 0-08-044402-4

primarily a public sector initiative in which other parties, as we describe below, play a supporting role.

In this contribution we present a case study of the virtual incubator in the Comunidad de Madrid as an example of the virtual incubation process. Of course, there are other examples which we describe in a previous publication (Van Tilburg *et al.* 2002).

The Virtual Spin-Off Incubator "VVE" of the Comunidad De Madrid (CM)

The Context

The region of Madrid has 12 universities, 50 (public and private) research institutes (PRI), twenty intermediary organisations, a large number of multinational companies, financial institutions and mass media businesses throughout the Madrid region. Approximately 50,000 people work in jobs that are closely related to scientific research and technological innovation; half of which belong to the public innovation system of the Comunidad de Madrid (CM). The innovation system of the CM includes not only universities, Public Research Institutes (PRI), but also the Chamber of Commerce of the region, and many other public administration departments and offices as well. A very significant part of these institutions are located in Madrid, as well as many firms (SMEs and large companies). All these are publicly recognised as the *Madrid Regional Innovation System* and are actively involved in the promotion and creation of (research) spin-offs. The objective of the regional government is to involve more of these institutions in a strategic scheme for spin-off promotion actions. For this reason "a toolkit" has been developed and is considered to be vital for the successful promotion and support of new spin-offs of the scientific research and technological innovation found in the Madrid region. The virtual incubator, Vivero Virtual de Empresas (VVE), is one of the instruments for the CM, used to reach these objectives.

The Virtual Spin-off Incubator — VVE

The virtual incubator of the Comunidad de Madrid, VVE, arose from the Regional Scientific Research and Technological Innovation Plan for the region of Madrid (III PRICIT; 1999–2003), which was launched by the CM. The VVE started its operations in November 1999 and has the following tasks key:

• Awareness building sessions (sensitisation).
• Courses for "would-be" entrepreneurs.
• Counselling and mentoring.
• Provision of information.
• Networking.

The VVE operates in the Internet environment at: www.madrimasd.org/vivero/default.asp. It is not a task for the virtual incubator to provide capital or physical incubation facilities.

Nevertheless, a close interaction with other providers of these services is established by the virtual incubator. The VVE is a top-down and bottom-up approach to spin-off creation; top-down because the virtual incubator acts as a promoter of the universities and research institutions and provides spin-off services and activities with an overall value for all institutions involved; and bottom-up, because the regional universities and research institutes have committed themselves to develop and to carry out spin-off programmes in their own institutions to promote entrepreneurial demand.

The PRIACES Project

In order to support the VVE, international expertise was made available via the PRIACES project.[1] The acronym PRIACES literally means (in Spanish) "Integrated Regional Policy to Support the Creation of Spin-off Companies." The activities of the PRIACES project in support of the VVE were:

- To promote and involve regional promoters of the (virtual) incubation and the creation of (research) spin-offs.
- To contribute to the learning process of the regional development promoters at the universities and at the research institutions, and the spin-off entrepreneurs via the exchange of experiences with other parties in the Consortium, international study visits, and regional training activities.
- To provide best practices and tools for international marketing and seed capital which might be used for the development of new support tools for the virtual incubator, VVE.
- Monitoring of the progress of the VVE, both internally and externally.

The PRIACES project provided a support tool for the further development of the VVE, and embed it internationally by adapting tools and models from elsewhere in Europe for use in the VVE in order to further support the creation and promotion of and for spin-off companies.

Operational Structure of the VVE

The operational structure of the VVE reflects an integrated approach to the promotion of spin-offs. There are a number of necessary activities that are important to procure the promotion process as depicted in the Figure 1, which are boldly highlighted. These activities give rise to tangible results or intermediate milestones in the process, which are contained in circles. Three different modalities of training actions support the promotion process. The overall process culminates with a number of services that are offered to entrepreneurs

---

[1] The main contractor of this EU-PAXIS project was the Comunidad de Madrid. The other members in the consortium were IUEE from Madrid, Spinno from Finland and Top Spin International from the Netherlands.

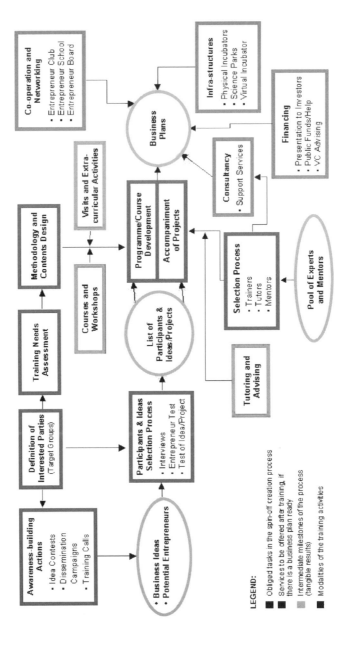

Figure 1: VVE spin-off promotion system.

upon business plan completion, which are highlighted in green (the lighter coloured boxes feeding into "business plan" on the right side of Figure 1). By no means is the operational structure of the VVE a close system. Activities, services, modalities and milestones can be added or withdrawn as new demands emerge or disappear during the putting into effect the different actions carried out by the VVE.

Phases of Support by the VVE

As a result of the operating structure, services and method, the VVE offers support during the company development process based on a procedure that is necessary to build a success-ful company. The first phase of this process is the dissemination and awareness-building phase, aimed mainly at institutions as well as "aspiring" or would-be entrepreneurs. The objective of this first phase consists in accelerating a paradigm change directed toward an entrepreneurial culture within institutions that lack such spirit. Likewise, during this phase potential business ideas/projects are identified. From the potential business idea/project, the entrepreneur undergoes a training programme that the VVE has standardised, endowing the entrepreneur with the necessary formative tools needed to develop the business idea into a business plan (BP). This BP is an essential tool to help the entrepreneur define the idea, and present to potential investors or venture capitalist (VC). The BP also serves as a reference for the future when the company is operating. During training, entrepreneurs are supported by experts (other entrepreneurs), which help them approach the business world from a realistic and real-life point of view. Once the BP is finished, the company then is incorporated legally and starts its business activities. During this stage spin-offs need a great deal of support from all kinds of institutions and organisations. The VVE provides them with a virtual platform to assess support for the entrepreneurs. The VVE also offers support to new firms during the period of consolidation and expansion. Throughout all of these phases, the actors in the process are a vital key in the establishment of successful companies. In this sense, all those involved participate in the process at different levels. Ad-ditionally, the services offered change, depending on the phase and users that are interacting at the time.

Key Players in the VVE

The main added value of the VVE is its capability to interconnect all key actors in the business development process. To this end, the VVE promotes the co-operation environment needed to encourage and facilitate necessary interactions, while promoting a culture of entrepreneurship. The VVE has five different types of key players:

- *"Aspiring" (would-be) entrepreneurs*: Those who decide to develop their company with the support of the VVE.
- *Institutions*: Universities and research institutes who are clients (who bring forward spin-off ideas and opportunities), suppliers (of would-be entrepreneurs) and also shareholders of the VVE.

- *Experts*: Trainers, businessmen, tutors and consultants that provide services that the VVE does not offer via its own services.
- *Companies*: Other companies looking for the opportunities the VVE provides.
- *Others*: All others who become familiar with the VVE and its services and want additional information.

Steps in the Spin-off Creation Process

The VVE sees the company incorporation process as a transitory phase for any entrepreneur. Each phase is an identified part of the process and once a given phase is underway, no more participants are admitted. However it is not necessary for a participant to undergo all the different phases of the incubation process. The process breaks down as follows:

- *Phase 0 is the dissemination and awareness phase*. There is a need for a culture change in the society for the results of the investigations of the universities and PRI to become valuable to the society through innovation.
- *Phase 1A is the support and assistance phase for entrepreneurs* who want to develop their business idea into a well thought BP.
- *Phase 1B is the advanced support phase for entrepreneurs*, lasting from BP Construction to the official launch of the commercial activities.
- *Phase 2 is the ratification of incorporated companies as "virtual companies"* on the Internet system of the VVE. Within the virtual incubator non physical support is provided to the newly incorporated businesses. This includes the value-added services required by spin-offs in their first few years of operation like financing, market research, internationalisation, human research management, search for partners to enhance the competitive position of spin-offs, capital restructuring and so on.

VVE Services

The VVE provides a number of support services that guide entrepreneurs through all stages of company creation and development, including the final phase, in which the company is fully operational. These services are also elaborated and implemented as solutions to the needs of support experts and institutions. In addition to the already cited services, the VVE offers the following:

- *Feasibility studies*: From detection of a business idea, a report is produced with a detailed analysis of the commercial and financial feasibility. Each business project is submitted by a member of the list of collaborators of the VVE.
- *Promoter training*: Training specifically designed for the promotional agents of the institutions. The training is targeted to a maximum of 20 participants. The goal of this programme is to disseminate the existing spin-off promotion mechanisms. Topics cover a large range of spin-off promotion issues including legal aspects, co-operative framework, financing, practical case studies, etc.
- *Expert support*: Services here include information unrelated to the training process, but which "aspiring" entrepreneurs with a business plan require. In such cases, experts

could be assigned to help them fine-tune their business plans and prepare them for the capital investment procedures.

- *Events*: The VVE promotes the entrepreneurial culture via summer courses, seminars and workshops and discusses issues with collaborators, key actors of the regional innovation system of the CM, and others regions of Spain, potential entrepreneurs, international experts or anyone interested in spin-off issues.
- *Virtual community*: The Internet site of the VVE enables experts to exchange opinions and experiences and to discuss the theory and practice of the entrepreneurial learning process.
- *Documentation centre*: The Internet VVE offers users a documentation centre and a virtual library where surveys, manuals, books, conferences held in the courses organised by the VVE, news, etc. can be found.
- *Fora*: The Internet side of the VVE is prepared for specialised fora broken down into subject matters. The identification of forum hosts and periodically renewing of the subject matters corresponds to the non-virtual aspect of the VVE.

VVE Products

The VVE designs and produces specific products for the promotion of the spin-off development process in the CM. These products are introduced to the Internet environment from where entrepreneurs, companies associated to the VVE, institutions and experts, can download them. Products are basically tools, simulations, literature, comparative research results and multimedia possibilities, e.g. videoconferences. The downloadable products of the platform include:

- Planning and financial control self-training programme, including an Excel document that the companies may apply in real-life situations.
- Business plan design self-training programme as an updated version of the currently available CD ROM application.
- Video-recorded conferences, seminars and international meetings.
- A business management simulator that enables team-play in a virtual competition.
- An investment analysis simulator for the beginning of the business activity.
- Creativity enhancing activities.
- A project planning and management programme that includes a MS-Project document that the companies may apply in real-life situations.
- Specific online training units liable to be used as support for the ongoing training of entrepreneurs and companies linked to the VVE.
- Written reports dealing with comparative good practices, surveys and an array of contributions from international experts.

VVE Technology

The Internet aspect of the VVE deploys all the designed capabilities of present day Internet technologies in order to perform efficiently its online services. This is a basic element of the virtual aspect of the VVE concept. For this reason one of the main tasks of PRIACES has been to design and develop a data communication platform capable of supporting as many online services, products and operations as possible with the maximum efficiency, while incorporating easy and compelling navigation aspects to the virtual side of the VVE. The deciding factor in the design of the architecture of this platform has been the quest for the necessary homogenisation of the architecture within the global environment featured in the Madrid portal where the Internet side of the VVE resides. The Internet VVE appeared for the first time in February 1999 in the form of a number of static web-information pages. Thanks to PRIACES the Internet side of the VVE has undergone major and important changes. These changes have given the virtual spin-off support (VSS) system a new approach and functionality designed to handle and manage all those interested in creating spin-offs and in promoting them. Since the beginning of the year 2002, the VSS system gives new business spin-offs and research and higher education institutions an "added-value site" that complements the various physical issues that each university and PRI wishes to establish within its campus.

Virtual and Non-Virtual Elements of VVE

There are "virtual" and non-virtual aspects of the VVE concept. The virtual aspect relates to the Internet spin-off support system and the non-virtual make up of the VVE organisation. Table 1 gives an overview of the cross-reference services/activities between the virtual and non-virtual parts of the VVE concept. In some instances services/activities overlap and no clear-cut distinction amid the two is obtained. In other cases, a differentiation between the two is clearly illustrated.

Results of the VVE Activities

Until the beginning of 2002, VVE services have assisted the creation of 26 new companies that are at an early phase of development and 27 new companies, which are at an advanced stage of formation (EC 2002).

Conclusions For and Recommendations to the VVE in Madrid

Among one of the PRIACES project tasks involved the building-up of an international expert panel able to analyse, assess and give feedback on methodological approaches, procedures, actions, results, etc of the project and the VVE concept. The following conclusions and recommendations emerged during a final workshop meeting held in Madrid with members of the regional innovation system:

(1) *A strong commitment of regional partners*: The input and participation of the VVE's clients (entrepreneurs, universities, promoters, organisations, etc) should be mobilized very intensively into the process to make sure that they truly get added value from the system and services provided by the VVE. The processes need to be adequately flexible in order to fit the needs of the users and really contribute to the promotion of

Table 1: Services/activities offered by the VVE.

Services/Activities Offered by VVE	VVE Concept/Model		
	Virtual	Both	Non Virtual
Services for entrepreneurs			
Awareness-building sessions			X
Entrepreneur training priority	X[a]		X
Tutoring and mentoring		X	
Networking priority		X	
Information dissemination		X	
Seed capital investment funds			X
Internationalisation and marketing schemes			X
Potential idea/project analysis priority			X
Feasibility studies priority			X
Individual strategic plan development			X
Virtual community	X		
Preparation and development of the business plan priority			
Intellectual property right assistance			
Other services			
Dissemination of the VVE concept		X	
Awareness-building and consensus sessions			X
Benchmarking priority		X	
Institutional networking priority		X	
Promoter training priority	X[b]		X
Expert support (tutors, mentors, trainers)		X	
Progress assessment priority		X	
Entrepreneurial growth, activities, actions, start-ups, follow-up, etc.		X	
Sponsoring and editing publications	X		
Conferences, forums		X	
Celebration of events			X
Documentation centre	X		
Orientation, guidance, research			X
Promotion, culture, innovation and network related to process of incubation			

Note: The added elements (in italic) as well as the priorities are provided by the experts.
[a] Although currently inoperative the VVE system is prepared for e-learning delivery.
[b] Although inoperative the Internet spin-off support system is prepared for e-learning delivery.

new innovations and entrepreneurship. Likewise more support for the VVE from the top-level organisations seems to be necessary.

(2) *Start where the action is*: The main problem seems to be that the VVE builds up a very comprehensive support architecture that is currently not yet receiving the necessary interest from institutions in the region of Madrid. It is true that successful spin-off examples should be established first in order to legitimise the VVE. But the current level of interest and awareness in the region of Madrid is still not very big, and is perhaps even decreasing, especially because the same key-people/institutions are always involved. It might be therefore a good idea to just focus on a limited number of active key-players that have shown a high interest in spin-off creation and take those elements of the spin-off support policy that they are most interested in. Just five or six concrete spin-off processes could be completed with them, and the experience of this process could then flow back into the methodology. Thus, it might not be necessary to readily offer all the services at the same time.

(3) *Clear objectives*: The objectives of the incubator should be expressed very clearly.

(4) *The main tasks are provision of content, networking, information and training*: The most important tasks that can be provided by a virtual incubator are provided by the members of the expert panel in Table 2.

(5) *Let services develop naturally*: The VVE services are concerned with the coaching of an incubator. A service becomes naturally virtual if the information technologies give better delivery and if the service can be willingly shared.

(6) *Focus on priority services*: The list of VVE services looks long and adequate. However, it needs to be born in mind that the services, activities and tools need to be constantly reviewed. Choose to fulfil now the most important ones (suggestions indicated as priorities in Table 4) and add/reduce as more information on the VVE clients' needs is accumulated. The idea should be to adopt a model of the project development with a fine tuning or evolutionary approach.

(7) *Combine virtual and non-virtual structure*: The combination of both a virtual and non-virtual structure should be highly recommended. In fact, it is not feasible to nurture entrepreneurship exclusively in a virtual mode. Moreover, the virtual component, legitimated by the fact that there is a vast number of research and higher education institutions to which a regional policy (ie. PRIACES) is directed, provides a powerful instrument of intervention within the entire community. Clearly, the VVE plays a horizontal and an integrative role within the spin-off promotion policy of the CM. At the same time it brilliantly leaves freedom of manoeuvre to each institution involved. However, an "Achilles Heel" is also present where successful co-operative arrangements are difficult everywhere and the CM is no exception. Regional institutions seldom co-operate and there is a need for a higher level of involvement and commitment. The "don't wait, just do it" approach is a bold move. It may as well be the only possible strategy. Undoubtedly, in the long run, the success of the VVE will be determined by the degree of participation of the regional institutions. On the other hand, one may expect that the VVE's own performance will be catalytic by contributing decisively in attracting general interest to the model.

(8) *Model to be copied*: One of the members of the expert panel pointed out that the PRIACES and the VVE model in particular are so attractive to him and to his

Table 2: Most important activities per task of a virtual incubator.

Major Tasks of an Incubator	Activities to be Accomplished by the Virtual Incubator VVE		
	Content	**Networking**	**Information and Training**
1. Human resources	Recruitment; brokerage; facilitate matching profiles	Networks; provide key-contacts	Provide information about managing HR; know-how on team building and management training
2. Social resources	Partnering	Networking; contacts	Training on negotiation skills, entrepreneurship, financing, how to find clients, etc.
3. Financial resources	Seed money; own funds; fund raising; reduce the risk for potential investors	Advice on financing; networks; introduction to different financing tools and organisations	Information and training about financial management and dangers (cash burn), comparison of different resources
4. Physical resources	Space at low costs, facilitate access to laboratories, image	Clustering, advice on available channels	Information and comparison of different resources
5. Technology resources	Advice on technological issues and potential partners, access to labs, due diligence	Links to researchers and research projects; partner search	Information and comparison of different resources; data base; web searching
6. Organisational resources	Legal assistance, mentoring, coaching and tutoring; book keeping; business planning; recruitment	Consulting networks	Information and training about organisational management

organisation that the possibility of a replication of this model in his own region (with necessary adjustments due to their own peculiarities), in a "softer" version, would be more than welcome. The dissemination of this model would be indeed living proof of the originality, the scope and the effective added value of PRIACES.

Concluding Remarks

The VVE is an example of a virtual incubator; time will tell whether it can be considered a best practice example. The VVE carries out the functions an incubator is supposed to do in the sense that it contributes and provides resources to (starting) companies interest on building a business platform (Klofsten 1998) in order to graduate from an incubator. Thus, the virtual incubator is a useful instrument in the "spin out" process.

References

Brush, C. G., Greene, P. G., Hart, M. M., & Haller, H. S. (2001). *From initial idea to unique advantage: The entrepreneurial challenge of constructing a resource base.* The Academy of Management Executive (February).

Centre for Strategy & Evaluation Sevices (2002). *Benchmarking of business incubators.* Luxemburg: European Commission.

EU (2002). Europe's cities — centres of innovation culture. *Innovation & Technology Transfer* (February).

Klofsten, M. (1998). *The business platform.* Luxemburg: TII.

Nowak, M. J., & Grant, C. E. (2000). The virtual incubator: Managing human capital in the software industry. *Research Policy, 29*, 125–134.

Van der Sijde, P. C. (2002a). Total issue, Functions and Tasks of an (Academic) Incubator. Paper prepared for the expert-meeting of the EU-PAXIS project PRIACES. Madrid (14 September).

Van der Sijde, P. C. (2002b). Introduction to special issue on "Developing strategies for effective entrepreneurial incubators". *International Journal of Entrepreneurship & Innovation, 3*(4), 233–236.

Van Tilburg, J. J. et al. (2002). Virtual incubation of research spin-offs. *International Journal of Entrepreneurship & Innovation, 3*(4), 285–293.

Chapter 7

The Formation of High-Tech University Spinouts Through Joint Ventures

Ajay Vohora, Mike Wright and Andy Lockett

Introduction

The spinning-out of university-based scientific inventions into separate companies represents a potentially important, but as yet under-developed, option to create wealth from the commercialisation of research (Shane 2001; Siegel *et al.* 2001). Traditionally, the commercialisation of university intellectual property has occurred through licensing. University spinout companies have become increasingly popular as a result of the problems of licensing and the desire to maximize the returns to intellectual property. For purposes of this study, a university spin out (USO) is a new company founded by employees of the university around a core technological innovation which had initially been developed at the university (Birley 2001). Exploitation of academics' inventions, outside what has historically been a non-commercial environment, raises new entrepreneurial challenges beyond those faced by new high tech ventures in general.

In a review of the literature, Shane (2002) suggests that there are two different models of entrepreneurial company interaction with universities. First, university faculty (or some-times other entrepreneurs) identify a university discovery that they seek to commercialise by starting a new company or growing a small company. Second, an external entrepreneur seeks assistance from the university to further develop his or her company. This paper examines an additional option, which involves the university working with existing outside firms to create joint venture spinout companies. We term these companies joint-venture spinouts (JVSOs).

While universities are capable of generating Intellectual Property (IP), they typically lack the resources and capabilities to successfully commercialise the IP through USOs. Furthermore, it is clear that academic entrepreneurs involved in creating USOs generally have little commercial awareness and/or experience of how to go about the process (Franklin *et al.* 2001; Vohora *et al.* 2002). One potential way to overcome these problems is to collaborate with an industrial partner in an Equity Joint Venture (EJV). An EJV is formed whenever two or more sponsors (parents) bring assets to an independent legal entity and are paid for some or all of their contribution from the profit earned by the

New Technology-Based Firms in the New Millennium, Volume III
© 2004 Published by Elsevier Ltd.
ISBN: 0-08-044402-4

entity (Hennart 1988). The formation of the EJV ensures that residual gains are divided amongst its parents on a predetermined basis, as outlined by its constitution. A JVSO is a new venture in which technology is licensed into a new company that is jointly owned by the university and the industrial partner. The academic scientist is commonly given an equity stake in the JVSO as a reward for discovering the new technology, and as an incentive to participate in the development of the technology into a marketable product. Although this paper focuses on the advantages of using JVSOs to overcome resource constraints we acknowledge that collaboration is not a panacea for firms with limited resources.

In this paper we compare the creation, and development, of JVSOs and USOs. The approach taken is to study the evolution of four high-tech university spinout companies, two of which are venture capital backed and two that were formed as a result of a joint venture between the university and an industrial partner. The first main section develops propositions relating to the relative attributes of JVSOs and USOs in relation to key stages in the development of the spin-outs company. The second section outlines the methodology used in this study, while in the third section findings from case studies are presented. In the final section the implications of the findings are discussed and conclusions drawn.

Development of Propositions

Spin-out companies can be identified as meeting the need to resolve four key challenges in their development. These challenges are: (1) opportunity recognition; (2) entrepreneurial commitment by a venture champion; (3) attaining credibility in the business environment; and (4) achieving sustainability through the ability of these new venture to become established firms within their respective markets. In this section we develop propositions relating to the relative attributes of JVSOs and USOs in meeting these challenges.

Opportunity Recognition

Entrepreneurial opportunities bring into existence new goods, services, raw materials and organising methods that allow outputs to be sold at more that their cost of production. These opportunities are recognised as different people have access to different levels of information, and often because they already possess prior knowledge (Shane 2000).

Three major dimensions of prior knowledge are important to the process of entrepreneurial discovery: prior knowledge of markets, prior knowledge of how to serve markets, and prior knowledge of customer problems. Within universities, the academic's pre-eminence in a research field may be important in providing the basis for an opportunity to be recognised. However, previous research suggests that academic entrepreneurs involved in creating USOs may not be the best people to champion the venture. One reason is that they may not have the necessary commercial awareness and/or experience to go about the process (Franklin *et al.* 2001; Vohora *et al.* 2002). Furthermore, only recently have universities begun to provide incentives for the scientists to think and behave entrepreneurially. Working in a non-entrepreneurial culture has the effect of reinforcing the non-commercial mindset of academics. This point is supported by the common use of surrogate entrepreneurs in USOs (Lockett *et al.* 2002).

Such factors suggest that universities might develop links with industry if they are to be successful at commercialising technology through USOs. The JVSO model of intellectual property commercialisation employs cooperative strategies (Shan *et al.* 1994) to achieve this. A university academic's new information about a technological discovery might be complementary to an industry partner's prior information about how markets operate, leading the discovery of the technological opportunity to require prior information about these markets. Important prior information about such markets might include information about supplier relationships, sales techniques, or capital equipment requirements that differ across markets (Von Hippel 1988). This prior information enables him or her to discover an opportunity in which to use the new technology. The above arguments suggest the following proposition:

Proposition 1. Cooperative links with potential industrial partners facilitate opportunity recognition as a precursor to the formation of a spin-out company.

Entrepreneurial Commitment

For USOs, once an entrepreneurial opportunity has been identified, there is an imperative to deal with the uncertainty surrounding the technology in its primitive state and the application of that technology to serve particular customer needs. The uncertainty arises in part due to a lack of information and the lack of an obvious market in which to exploit the technology (Miller & Friesen 1984). This absence of information creates decision uncertainty and decision complexity (Busenitz Lowell & Barney 1997). There is a need for an individual to be committed to resolving this uncertainty and intense complexity through championing the venture beyond start-up to commercial operation. Entrepreneurial commitment is necessary for a potential venture to be taken forward. Furthermore, the commitment of the academic may be particularly important to ensure a continued flow of innovations to enable the venture's product portfolio to develop.

However, this does not necessarily make the scientist the best candidate for the role of venture champion. First, developing the technology into a marketable product or service requires capabilities derived from prior industry and entrepreneurial experience, which the academic scientist may not sufficiently possess. Second, the academic scientist may be unable or reluctant to commit to the venture through either a lack of personal motivation or an institutional culture, which discourages commercial behaviour. Leaving an academic post and committing to the venture full time would mean going against accepted conventions.

In some cases, a surrogate entrepreneur (Franklin *et al.* 2001) may provide the solution. This occurs by enabling the academic to remain in his or her university role whilst contributing some effort to developing the technology into a product, whilst the surrogate entrepreneur manages day-to-day management of the business and coordinates transactions with suppliers and customers. Universities can have difficulty in luring people to champion new ventures in the absence of any funds and the ability of the university to identify suitable people to assume the role of a surrogate entrepreneur (Franklin *et al.* 2001). Furthermore, universities perceive risks in the surrogate's potential for differing objectives (to the academic and university) and cannot readily marshal the necessary resources to meet their strong demands for remuneration, in terms of both salary and equity. However, in the case of a JVSO, the industrial partner is potentially a rich source of such surrogate entrepreneurs and can supply managerial and marketing expertise to fulfil this requirement. Alternatively, the support of the industrial partner may make it easier to attract outside surrogate entrepreneurs who will be committed to the venture. The above arguments suggest the following proposition:

> **Proposition 2.** JVSOs are more likely than USOs to find a committed entrepreneur with the necessary commercial expertise to champion the venture.

Venture Credibility

Legitimacy, the institutional support of a powerful external actor, is often a critical ingredient for new venture success because of the liabilities of newness. A new USO has no track record, and hence there is a justifiable lack of confidence on the part of customers, distributors, and suppliers that the venture will survive and therefore little reason to provide patronage (Starr & Macmillan 1990). There may also be doubts internally by the academic or the university about the credibility of the new venture. Thus the venture faces a credibility crisis, and has to somehow create the impression of viability and legitimacy before it will receive support in acquiring resources.

Without this initial legitimacy and credibility, new high tech ventures in general, will not be able to overcome sceptical customer perceptions, gain access to product and financial markets and successfully achieve the transition from a "concept" to a "business" engaged in transactions in the marketplace.

Emerging high tech firms in particular tend to rely on alliances as sources of needed resources (Coombs & Deeds 2000). Through cooperative strategies, universities can partner with established firms in industry to pool resources and, build credibility and co-opt legitimacy (Starr & Macmillan 1990) in order to start the new venture. A key success factor in most entrepreneurs is building "network exchange structures with outsiders that are identified as critical resource suppliers, ones that can stabilize the new firm as a player in targeted markets" (Dubini & Aldrich 1991). The formation of a joint venture between the university and an established firm can be a mechanism for gaining access to resources, exchanging information, exercising influence, developing interorganisational commitments, avoiding conflicts and establishing legitimacy (Pfeffer & Salancik 1978). Based upon a networked form of governance, the JVSO model should lead to easier access to resources, with which to initiate and develop the venture as all the necessary ingredients are supplied by both parents (i.e. the university and the established industry partner). The above arguments suggest the following proposition:

Proposition 3. JVSOs are more likely to establish credibility much faster than USOs through a greater ability to gain access to resources and capabilities.

Venture Sustainability

As with other new ventures, the success of a USO relates to its ability to access, acquire and develop requisite organisational resources and capabilities (Eisenhardt & Martin 2000). For a USO, the challenges for growth are particularly high because intangible inputs need to be translated into tangible market returns by weak management teams that are largely untested. The venture's development is thus highly turbulent (Cooper *et al.* 1994) because the weak resource and capability position from which the USO starts from must constantly be restructured to cope with the challenges of growth. Routines and procedures need to be continuously redefined to cope with the need to adapt to market demands, whilst striving to develop further new products to secure future revenue streams (Vohora *et al.* 2002).

Unless the USO's inexperienced entrepreneurial team can acquire or develop the skills to manage the turbulence created by these challenges of growth (Shepherd *et al.* 2000), the venture will not have the infrastructure or capacity to continue to transfer a technological competitive advantage into a market competitive advantage. Over time, inability of the new venture to cope with emergent conditions will become exposed and hence it will fail to become established as a sustainable rent generating business (West & DeCastro 2001).

In contrast, Shoonhoven Bird *et al.* (1990) argue that new ventures founded by industry experienced individuals will gain advantage in the creation of favourable organisational outcomes, including the speed with which first products are developed and reach the market. A greater ability to access resources, capabilities and expertise leads to comparatively faster professionalisation of the new venture that transforms a start-up, consisting of a few founders with an idea, into a modern corporation (Hellmann & Puri 2001).

In the case of USOs, the notion of professionalisation can be extended beyond human capital resources to the accumulation of capabilities related to product development, manufacturing, marketing, sales and distribution, ensure that the venture can become established in the marketplace without delay. However, developing these organisational processes, routines and capabilities from scratch is costly and time consuming. Consequently, the ability to professionalise and therefore attain speed to market is the fundamental handicap which prevents USOs from engaging in sufficient revenue generating transactions. Hence, many USOs begin to stagnate and fail to develop as sustainable businesses (Vohora *et al.* 2002). It is at this juncture that a JVSO becomes a more effective vehicle for commercialising university intellectual property.

An industrial partner may be able to provide existing factors of production (FOP) resources and capabilities to the spinout. The industrial partner will already have an existing customer base and identifiable brand that provides a ready made route to market for the spin-out activity. The business and managerial expertise provided by the industrial partner is of significant importance to the professionalisation and therefore success of the JVSO. By not adopting cooperative strategies in creating non-joint venture USOs, universities experience greater difficulties in identifying and attracting credible managerial expertise to join the venture full time. The above arguments suggest the following proposition:

> **Proposition 4.** JVSOs become sustainable more quickly than USOs through the ability of both parents to professionalise the venture.

Data and Methods

Given the lack of prior research in this area, an inductive approach that provides a rich understanding of how spinouts evolve from research activities to commercial organizations is required (Van de Ven 1992). This study employs a multiple case study research design to examine four high-tech university spinout companies, two of which are USOs and two of which are JVSOs, in further detail. This method allows for close correspondence between theory and data, a process whereby the emergent theory is grounded in the data (Eisenhardt 1989).

Data were collected in a total of 24 in-depth face-to-face and telephone interviews with the management teams of the four spinouts, the joint venture partners and representatives from the universities during the period April 2001 to April 2002. The cases were drawn from universities ranked among the top ten research elite universities in the United Kingdom as measured by research income earned, and who are actively pursuing a programme of university technology transfer, through both licensing and spinout mechanisms. In performing this study we followed procedures commonly recommended for conducting case study research (Eisenhardt 1989; Yin 1994). The data were augmented by information provided by the spin-out company as well as publicly available information gathered from on-line data sources and press releases. For each of the cases, structured interviews were carried out with the head of the Technology Transfer Office (or equivalent), a number of business development managers (BDMs), managers at the joint venture partner, and the members of the USOs.

Interviews ranged from one to two hours in length. The interviews were openly recorded and directly transcribed. By using a number of key actors from each university, USO, and JV partner, we ensured that we elicited views on the universities' role in the spinout process to cross check our interpretation of events. Less formal telephone interviews were used to clarify and enrich previous points raised. Triangulation was also aided by the collection of archival data (Yin 1994), at the level of the university, and the USO. To avoid confirmatory biases, one of the authors was kept at a distance from the field observations and focused on conceptualisation and analysis of the material and interpretations developed by the other researchers (Doz 1996). Table 1describes the two USO and two JVSO cases used in this paper.

Responses from the interviews and other data were developed in a case study database, which included the use of table shells to record data (Miles & Huberman 1984). These table outlines ensured that the data collection was focused on the research questions and verified the same information was being collected for all cases and aided data analysis.

Table 1: The companies.

Company	Opportunity	Equity Distribution Amongst Founding Shareholders	Start-up Investment	Date Founded	Current Status
Nano-Technology Co.	To design, manufacture and distribute micro-sensors for detecting hazardous chemical substances	Academic entrepreneur: 22% Surrogate entrepreneur: 22% University: 21% Venture capitalist: 35%	Academic entrepreneur: Consultancy time Surrogate entrepreneur: Prepared to work pro-bono until USO makes a profit University: $50,000 seed funding; Licenses to patents in return for equity stake in lieu of royalties Venture capitalist: $420,000 seed investment	April 2001	Unprofitable Yet to ship first product to market Still located in university labs Unable to raise further funding from investors 2 engineers laid off to conserve cash
3G Wireless Co.	To design low power consumption microprocessors and license the IP to manufacturers worldwide in the mobile telephone market	Academic entrepreneur: 37% Surrogate entrepreneur: 33% University: 15% Venture capitalist: 15%	Academic entrepreneur: Consultancy time Surrogate entrepreneur: Prepared to work pro-bono until USO makes a profit University: Assigned patents in exchange for equity stake. Use of lab space for proof of concept and design work Venture capitalist: $300,000 seed investment	May 2001	Unprofitable Yet to sign up first customer Still located in university labs Looking for commercial partners to co-develop technology Cannot afford to hire new engineers Unable to raise further finance

Table 1: (Continued)

Company	Opportunity	Equity Distribution Amongst Founding Shareholders	Start-up Investment	Date Founded	Current Status
Automotive Co.	To design, manufacture and sell components for automotive engines to improve engine efficiency and reduce fuel consumption	Academic entrepreneur: 28% Management team: Share of 13% University: 17% Auto-Assembly Co. (Industry Partner): 42%	Academic entrepreneur: Consultancy time University: Licenses to patents in return for equity stake in lieu of royalties. Use of labs for testing new product performance Industry Partner: Facilities for product development, manufacture and repair, sales and marketing, customer and supplier contacts, $240,000 financial investment to support product success	October 2001	Profitable Awarded a contract to supply industry's leading manufacturer Relocated to industrial science park Working with industrial partner on developing new products for same market Looking to hire 5 new employees
Telecom Co.	To design, and manufacture specialised fiber optic components for use in telecommunication networks to be sold worldwide to device manufactures	Academic entrepreneurs: 21% shared between 3 scientists Management team: Share of 27% University: 12% Broadband Co. (Industry Partner): 40%	Academic entrepreneurs: Consultancy time University: Licenses to patents in return for equity stake in lieu of royalties. Access to equipment, labs and technicians Industry Partner: Facilities for product development, manufacture and repair, sales and marketing, customer and supplier contacts, $425,000 financial investment to support product success	April 2001	Seeking to be profitable in 1 year Awarded several new contracts by U.S. and Canadian customers Successfully raised $6 million first round venture capital Relocated to university science park Second new product released in Spring 2002 20 new employees New IP licensed from university Sponsoring more university research

Cross-case analysis (Eisenhardt 1989) was used to develop common and differential factors. Conceptual insights were in turn drawn out and refined during an iterative process as the case studies progressed.

Results

Opportunity Recognition

In the case of the two USOs, the drive for commercialisation was more technology led than in the two JVSOs where a clear customer need had been identified which the technology could fulfil. For example, regarding the USOs in the early stages of both Nano-Technology Co. and 3G Wireless Co., these ventures had numerous ideas for applications of their respective technologies in a number of different markets. It was more of a challenge for these USOs to focus upon who their customers were and what sort of business they should become in order to best serve these customers. In contrast, the JVSOs were customer oriented even before these ventures were formed. In the case of Automotive Co., even whilst the technology was in the university laboratory, the marketing director of the joint venture parent Auto-Assembly Co. knew precisely how much revenue could be derived from each potential customer and also how much money the product could save these customers. This presented a compelling case for commercial exploitation.

Likewise, Broadband Co., the JVSO parent of Telecom Co. had spotted a changing trend in their own market which made the technology created from research they had sponsored very valuable to their own success. This recognition stemmed from greater market intelligence than the scientists in the university had access to.

Unlike these two JVSOs, the academic and surrogate entrepreneurs in Nano-Technology Co. and 3G Wireless Co., could not sufficiently pin-point a precise market need by defining a value proposition for customers. The surrogate entrepreneurs in both USOs each had sales, product development and marketing experience in the relevant industries and were able to shape the potential of each technology in its early state in the university labs, into applications that were relevant to new or existing markets. However, and surprisingly, in both cases they believed that the technology was sufficiently compelling that a customer need could be created in new markets or that customer preferences in existing markets would change in their favour once their products were on offer. To date both of these so called "platform technologies" have only managed to find limited demand from a small niche of customers.

These results provide evidence that prior knowledge of the industry, its customers and suppliers, as well as manufacturing and sales techniques etc. are all necessary to recognise an opportunity. However, the results also suggest that there are benefits from being an established operator within a market in order to better recognise opportunities. These findings provide support for Proposition 1 that cooperative links with potential industrial partners facilitate opportunity recognition as a precursor to the formation of a spin-out company.

Entrepreneurial Commitment

In all cases examined, the academic entrepreneurs did not leave their university posts. As shareholders in the new ventures, they acted as technical advisors on a part time basis. In the case of the Nano-Technology Co. and 3G Wireless Co., a surrogate entrepreneur met the scientific academic and in recognising the market potential of the technology, agreed to start a USO with them in order to transfer the technology out of the university laboratory. The surrogate entrepreneur from Nano-Technology Co. describes how the new venture, the academic and the university all benefitted from this arrangement:

> He's going to use it in two ways, I mean, the academic will be research director of [Nano-Technology Co.] but he will retain his position within the university. He's literally got filing cabinets full of new ideas in his office, that are all in areas [Nano-Technology Co.] will be interested in funding. He gets to pursue his academic career as well as having this outlet for the technology that we may acquire to be commercialised.

In the JVSOs, the industry partner is also able to supply managerial and marketing expertise in order to bring a keen focus to the process of commercialisation. By their own admission, the academics all knew where their strengths lay and that they could best contribute the venture's success as technology experts at the forefront of scientific research. As an academic from Automotive Co. summed up:

> The major challenge, not just really for us but looking for all spinouts is actually having the people available to fulfil that area. Universities need to look at whether they can get some staff on board that could fill some of these interim roles in which they are paid on the basis of moving these ventures companies forward towards market.

Clearly the commitment of a full-time CEO becoming the venture champion is essential to starting the new venture, giving it direction and building momentum. Our case study evidence shows that an industrial partner is able to add more value in this area than a surrogate entrepreneur. By bringing industry knowledge, a commercial track record and networks of contacts, they are more able to resolve the uncertainty of how best to apply the technology in customer oriented applications and seek out profits. These findings provide support for Proposition 2 that JVSOs are more likely than USOs to find a committed entrepreneur with the necessary commercial expertise to champion the venture.

Credibility

For USOs, the credibility of the venture is the key issue in obtaining the necessary financial resources to allow the business to operate commercially. In turn, financial resources are essential to attracting and acquiring other resources such as premises and equipment. The issue of demonstrating credibility in acquiring financial resources becomes more significant for USOs when compared to JVSOs.

As Table 1 shows, the industry partners in both JVSOs contributed significant amounts of resource to the new ventures, including finance to develop products for market, as well as facilities in which the venture could operate in a commercial environment. This was because they had a better understanding of the market, and exactly what it would cost to commercially exploit the technology in order to gain a return on investment. In contrast, both USOs that received seed venture capital funding had not enjoyed the same access to resources and were still located in the university to the frustration of the surrogate entrepreneur from Nano-Technology Co.:

> From a commercial point of view, we really need to take the venture off campus or at least outside the university department. We're not doing ourselves any favours by still being here. To be honest, we don't fit in with the department's culture and we're not perceived to be a legitimate separate entity by outsiders.

In addition, in both JVSOs, the industry partner allocated significant managerial, marketing, and project management capabilities. For example, Broadband Co. the industrial parent of the Telecom Co. JVSO managed to entice industry experts in product development and manufacturing away from competitors, to come and work for the new venture. The chairman of a stock market listed company was also brought in on a part time basis to provide strategic advice and access to influential networks of people. The CEO of Automotive Co. clarifies:

> We've got a balanced team which is important going forward because we've got multiple inputs from all stakeholders. Whereas if you actually compare that with the sort of owner academic spinout, they don't want anyone else involved so they can only make a limited contribution that is confined to their own background and experiences.

Our results show that credibility for the new venture also derives from the university supplying intellectual property. The know-how supplied by academic scientists brings scientific credibility to the venture which the industrial partner cannot otherwise attain. In both Automotive Co. and Telecom Co. "strong academic backing" was an important signal to the market that the technology was proven and the product could be considered technologically superior to competitor's products. The CEO of Broadband Co. also revealed how the University is well placed with its facilities and know-how, to contribute to quickly resolving technical problems in the JVSO:

> Recently we've had a technical problem which we didn't have the resources within our own company to deal with. By taking it over to the University they contributed something like 15 man days within a week to solve the problem, whereas in the company with everything else going on it could have taken us two or three months to get there. That's why it's crucial to work in partnership with the university and reward them properly.

Similar examples in our research of "co-incubation" activities between JVSO parents show how product development can be done more rapidly in order to improve time to market, thus increasing the market value of the venture to both partners. In comparison, both USOs had still not reached a point where their technology is market ready. Interestingly, during the final interviews, the surrogate entrepreneur from Nano-Technology Co. revealed that they were in discussions with two companies, looking to form co-development partnerships and distribution licenses for the technology. Whether Nano-Technology is successful in this strategy or not, the example clearly shows the difficulties USOs experience in gaining access to market through a lack of resources, capabilities and credibility.

These results provide evidence in support of Proposition 3 that JVSOs are more likely to establish credibility much faster than USOs through a greater ability to gain access to resources and capabilities because their legitimacy is derived from the university, the track record of the industry partner that operates in the market as well as the reputation of the academics involved in the venture. This legitimacy enables the new venture to interact with the business environment, to access resources more easily and acquire and build capabilities to transform itself into a large complex organisation. Hence the risk to survival is comparatively lower for JVSOs.

The results also suggest that interim management in the form of a surrogate entrepreneur with business experience, together with seed finance may be insufficient to achieve this transformation, especially because of the problems in gaining access to customers. Leveraging the track record of an established market incumbent, in the case of Automotive Co., with Auto-Assembly Co., and Telecom Co., with Broadband Co., provides access to a clear route to market.

Finally, there is evidence for what appears to be a great deal more transparency in the relationships between the university, the industrial partner and the JVSO, in comparison to those between a USO and a venture capital firm. There appeared to be less information asymmetry between parties in the JVSOs than in the USOs. Although trust may be important in VC firm-USO relationships, as well as in JVSO relationships, the former may display features of the traditional principal-agent monitoring relationship between an investee and a venture capital firm, especially as the VC firm executives are less likely to have the time or the skills to become involved at the same detailed level (Gompers & Lerner 1999).

Sustainability

On examination of each of the new ventures, we found that JVSOs were better at integrating resources, finance and expertise in order to add value to the intellectual property that underpinned the existence of the venture. The results suggest that JVSOs are able to become established in the marketplace through a greater ability of both parent shareholders to professionalise the venture.

First, the industrial partner is able to provide physical resources such as space, facilities and capital equipment. Being located away from their university departments legitimised both Automotive Co. and Telecom Co. as business entities, whilst being close enough to access expertise and specialist laboratory equipment, accelerated product development. In contrast, both USOs felt commercially constrained in not having access to their own facilities and having to share resources with scientists in their university departments.

Second, the industrial partner is able to provide the JVSO with managerial resources allocated from its own personnel. Together with the academic scientists transferred into the venture from the university, a well-balanced team can be assembled quickly and with low transaction costs. Instead, both Nano-Technology Co. and 3G Wireless, had to rely upon pro-bono work from surrogate entrepreneurs, friends and part-time use of university researchers.

Third, organisational routines and procedures to manage the growth of the venture are transferred from the industrial partner. Establishing cross-functional teams, health and safety procedures, human resource management, project management and systems for ensuring quality control, management accounting and payroll are all examples recognised in the JVSO cases. In contrast, in both USO cases organisational routines, procedures and systems were either ad hoc, built from scratch or non-existent.

It was also clear in both USOs that the seed funding raised was only sufficient to support the development work necessary to transfer the technology discovered in the university labs into a working prototype, but insufficient to build a business. In the case of Nano-Technology Co., the surrogate entrepreneur lamented that they had achieved proof of concept in terms of the technology but failed to achieve "proof of market." Without financial resources to acquire capabilities for coordinating and controlling the growth of these USOs, the challenges of managing the complex and uncertain task of building a business became overwhelming for these entrepreneurs.

Fourth, the industrial partner is able to provide social resources such as access to networks of expertise, suppliers and customers with whom it has already established, business relationships, trust through previous transactions. A post-doctoral scientist who transferred from a university research group into Telecom Co. highlights the significance of a cumulative effect once all these advantages are present in the venture at the same time:

> The benefit is time to market. If it takes you six months longer to create the spinout, to secure funding and the rest of it, then it's twelve months late to market and costs eighteen months worth of lost revenue. There's a clarity real focus [in Telecom Co.] too, because we've got great management expertise, plus resources, plus money. It's great being able to say, we need that bit of kit, let's spend $15,000 on it and get the job done.

These findings emphasise that in order to professionalise these new high tech ventures, it is necessary to obtain new capabilities associated with recognising opportunities and threats, acquiring resources, and integrating them with existing resources and capabilities. JVSOs can have advantages over USOs in this process of professionalisation in relation to the ease and speed of access to resources and capabilities. Unless the venture is able to acquire a critical mass of capabilities it will not be able to create the level of sustainability necessary to maintain its original competitive advantage derived from its leading technology.

Discussion and Conclusions

This paper has provided evidence that JVSOs can offer a faster, more flexible, less risky and less costly business venturing route to commercialising university intellectual property in comparison to venture backed university start-ups. JVSOs can provide greater access to critical resources such as marketing, technology, raw materials, equipment, facilities, financial assets, managerial expertise and political influence (Kogut 1991).

JVSOs can allow universities and industrial partners to pool their resources and improve the competitive position of the new venture in a way that they could not do alone (Pearce 1982). This research highlights resource weaknesses and distinctive inadequacies which prevent or impede the creation of firm competitive advantage (West & DeCastro 2001). The study has shown that USOs in comparison to JVSOs faced problems in resolving four key challenges regarding their transformation from ideas into complex organisations: (1) opportunity recognition; (2) attaining entrepreneurial commitment by a venture champion; (3) attaining credibility in the business environment; and (4) achieving sustainability through the ability of these new venture to become established firms within their respective markets.

In examining the differences in performance between two different governance structures used to commercialise university owned intellectual property, the determinants of success and failure have also become more recognisable. First, given their prior knowledge of their industries and the superior market intelligence of the industrial partner, the JVSOs pursued opportunities, which better served markets and customer needs. Second, the industrial partner and the university were able commercially to cooperate in assembling a well balanced team with the necessary background to successfully exploit the technology. Third, the credibility acquired from both parent organisations enabled the JVSOs to access and acquire resources more readily and gain organisational momentum and access to markets. Finally, a more rapid process of professionalisation enabled the JVSO start-ups to be transformed into firms that had become established in their markets and sustainable.

The findings of this research offer support for the notion that JVSOs are at a comparative advantage in the creation of distinctive competencies by combining resource strengths offered by both parents and from the prevention of distinctive inadequacies through coordinating and integrating diverse processes throughout the organisation.

Cooperative strategies such as JVSOs provide a platform for organizational learning, giving new ventures access to knowledge of their parent firms (Grant 1996). It appears that USOs lack the resources and organisational routines to perform organisational learning. In comparison, JVSOs acquire this capability much earlier in their organisational evolution. Through a shared responsibility for commercialisation, mutual interdependence and problem solving, and observation of activities and outcomes, the JVSO partners "co-incubate" the new venture. This cooperation ensures that knowledge is created, transferred, assembled, integrated and exploited at a faster pace to create wealth through the new venture.

These results suggest that future researchers considering the commercialisation of university intellectual property should not limit themselves to the dichotomy of licensing and spinouts. JVSOs are an important organisational form with which to transform intangible assets owned by universities such as scientific know how and intellectual property, into wealth creating new ventures. JVSOs formed as a result of cooperative strategies between universities and industry increase the likelihood that these ventures will succeed, in comparison to venture backed USOs.

Further research needs to examine more closely the role of professionalisation by university technology transfer offices, surrogate entrepreneurs and venture capitalists in enabling spinout ventures to become sustainable, rent generating firms. Further research also needs to define measures for the propositions developed here in order to test their generalisability on a larger sample of USOs and JVSOs. Finally, while we have focused on the scope for JVSOs in university technology transfer, there is a need for further research that examines their potential downside. Industrial partners may, for example, have agendas that are not compatible with those of the entrepreneur, have shorter term horizons that are not compatible with the development of the technology, and may have considerably greater bargaining power over the distribution of gains at later stages in the process.

References

Birley, S. (2001). Universities, academics, and spinout companies. *International Journal of Entrepreneurship Education*.

Busenitz Lowell, W., & Barney, J. (1997). Differences between entrepreneurs and managers in large organizations: Biases and heuristics in strategic decision-making. *Journal of Business Venturing*, *12*(1), 9–30.

Coombs, J. E., & Deeds, D. L. (2000). International alliances as sources of capital: Evidence from the biotechnology industry. *Journal of High Technology Management Research*, *11*, 235–254.

Cooper, A. C., Gimeno-Gascon, F. J., & Woo, C. Y. (1994). Initial human and financial capital as predictors of new venture performance. *Journal of Business Venturing*, 9, 371–396.

Doz, Y. (1996). The evolution of cooperation in strategic alliances: Initial conditions or learning process. *Strategic Management Journal*, *17*(Summer), 55–83.

Dubini, P., & Aldrich, H. A. (1991). Personal and extended networks are central to the entrepreneurial process. *Journal of Business Venturing*, 6, 305–313.

Eisenhardt, K. M. (1989). Building theories from case study research. *Academy of Management Review*, *14*, 488–511.

Eisenhardt, K. M., & Martin, J. A. (2000). Dynamic capabilities: What are they? *Strategic Management Journal*, *21*, 1105–1121.

Gompers, P., & Lerner, J. (1999). *The venture capital cycle*. New York: Wiley.

Grant, R. (1996). Prospering in dynamically-competitive environments: Organisational capability as knowledge integration. *Organization Science*, 7, 375–388.

Hellmann & Puri (2001). Venture capital and the professionalization of start-up firms: Empirical evidence. *Research Paper No. 1661*, Graduate School of Business, Stanford University.

Hennart, J. F. (1988). A transactions costs theory of equity joint ventures. *Strategic Management Journal*, 9, 174–361.

Kogut, B. (1991). Joint ventures and the option to expand and acquire. *Management Science*, *37*, 1.

Lockett, A., Wright, M., & Franklin, S. (2002). Technology transfer and universities' spin-out strategies. *Small Business Economics* (forthcoming).

Miles, M., & Huberman, M. (1984). *Qualitative data analysis*. Beverly Hills, CA: Sage.

Miller, D., & Friesen, P. H. (1984). *Organisations: A quantum view*. Englewood Cliffs, NJ: Prentice-Hall.

Pfeffer, J., & Salancik, G. (1978). *The external control of organisations: A resource dependence perspective*. New York: Harper & Row.

Shan, W., Walker, G., & Kogut, B. (1994). Interfirm cooperation and start-up innovation in the biotechnology industry. *Strategic Management Journal*, *15*, 387–394.

Shane, S. (2000). Prior knowledge and the discovery of entrepreneurial opportunities. *Organisation Science*, *11*(4), 448–469.

Shane, S. (2001). Selling university technology: Patterns from MIT. *Management Science*.

Shane, S. (2002). University technology transfer to entrepreneurial companies. *Journal of Business Venturing, 17*, 1–6.

Shepherd, D. A., Douglas, E. J., & Shanley, M. (2000). New venture survival: Ignorance, external shocks, and risk reduction strategies. *Journal of Business Venturing, 15*, 393–410.

Shoonhoven Bird, C., Eisenhardt, K. M., & Lyman, K. (1990). Speeding products to market: Waiting time to first product introduction in new firms. *Administrative Science Quarterly, 35*, 177–207.

Siegel, D., Waldman, D., & Link, A. L. (2001). Improving the effectiveness of commercial knowledge transfers from universities to firms. NBER Working Paper.

Starr, J. A., & Macmillan, I. C. (1990). Resource cooptation via social contracting: Resource acquisition strategies for new ventures. *Strategic Management Journal, 11*, 79–92.

Vohora, A., Lockett, A., & Wright, M. (2002). Critical junctures in the growth of university high-tech spin-out companies. Working Paper, Nottingham University Business School.

Von Hippel, E. (1988). *The sources of innovation*. New York: Oxford University Press.

West, G. P., III, & DeCastro, J. (2001). The achilles heel of firm strategy: Resource weaknesses and distinctive inadequacies. *Journal of Management Studies*.

Yin, R. K. (1994). *Case study research*. Thousand Oaks, CA: Sage.

Chapter 8

The Netherlands Life Sciences Sector Biopartner: Stimulating Entrepreneurship in the Life Sciences

Haifen Hu, Wim During, Aard Groen and Nettie Buitelaar

Introduction

The life sciences are a promising industry from both economic and social points of view. As a science based industry, life sciences are acknowledged to be of great importance to job creation and innovation. In different parts of the world governments aim at capitalising on this potential by creating an environment in which universities need to market their knowledge to survive (the U.K./USA model), or by offering tax incentives and subsidies (the German/Canadian model) (Moret Ernst & Young 1998).

Traditionally the Netherlands has a strong science base in the life sciences, particularly in biomedical and agro life sciences, and an internationally oriented business climate. Although all the ingredients for a successful bioscience industry seem to be there, the number of life sciences researchers starting new companies trails behind neighbouring countries such as the U.K. and Germany (Ministry of Economic Affairs 1999). The Dutch government has recognised the value of life sciences start-ups and the need for a stimulating policy. In 1999, the Dutch Ministry of Economic Affairs launched the "Action Plan Life Sciences 2000–2004" (Ministry of Economic Affairs 1999). BioPartner is the practical result of this Life Sciences Action Plan. BioPartner aims to stimulate entrepreneurship in the life sciences in the Netherlands by improving the entrepreneurial climate for bio-starters and bio-business and by removing obstacles to bio-entrepreneurship, such as lack of information, financing, facilities and incubators. This initiative is expected to result in the creation of 75 new life sciences start-ups in the period 2000–2004 and comprises the following five policy instruments: BioPartner Network, BioPartner First Stage Grant, BioPartner Centers, BioPartner Facilities Support and BioPartner Start-up Ventures (for more information on BioPartner, see Appendix).

One of the main tasks of BioPartner Network is to monitor the progress of the BioPartner programme and the status of the Dutch life sciences sector. To this end, BioPartner Network

New Technology-Based Firms in the New Millennium, Volume III
Copyright © 2004 by Elsevier Ltd.
ISBN: 0-08-044402-4

has performed a benchmark study, the so-called "nulmeting" in the beginning of 2001 (BioPartner 2001). Additionally, in each of the three following years BioPartner will conduct follow-up studies to observe the development of the emerging Dutch life sciences sector and to measure the success of the Action Plan.

The results of the "nulmeting" and the 2002 survey are reported on in this paper.

The purpose of this paper is to give insights into *new* life sciences ventures (founded since 1990) in the Netherlands. It reports on the number of companies and their origins, employment statistics, sector, business and cluster activities, characteristics of the sector, the companies, and their founders. These data provide benchmarks for bio-starters, policymakers and others by giving an overview of Dutch bio-entrepreneurial activities.

Methodology

The survey is a quantitative study of *new* life sciences ventures in the Netherlands. Various methods, including database analysis, secondary research and telephone interviews, were used to collect relevant data. The list of companies was compiled using data from BioPartner, Senter, the Regional Development Agencies, the Dutch biotechnology association (NIABA), Holland Biotechnology, Ernst & Young and others. In addition to the databases of these organisations, the personal contacts of BioPartner personnel and secondary research were used to compile the list. Questionnaires and telephone interviews complemented the available data and checked its correctness. The results of the "nulmeting" and the 2002 survey were collected in a specialised CRM programme, which allows easy analysis and comparison.

Scope of the Study

Life sciences companies are defined here as: "those companies that apply the possibilities of organisms, cell cultures, parts of cells or parts of organisms, in an innovative way for the purpose of industrial production. They may also supply related services, and hardware and software." Existing technological fields, including biotechnology, pharmacology, biology, chemistry, physics and informatics, are integrated into this definition. The purpose of life sciences is to utilise these technologies in order to contribute to new medicines, improved treatments of diseases, a cleaner industry and environment, improved enzymes and new foods, etc.

This study is aimed specifically at "*new* Dutch life sciences companies." The criteria for inclusion are the following:

- founded during or after 1990;
- registered at the Dutch Chamber of Commerce;
- independent — not subsidiaries of an existing company or a public research institute (existing companies or public research institutes may not exercise control over the company);

- involved in life sciences R&D activities, consequently using insights generated by multi-disciplinary sciences in which research is conducted on the building blocks and the life processes of plants, micro-organisms, animals and humans.

By limiting the scope of the survey to "*new* Dutch life sciences companies," only a select subset of the Dutch life sciences companies were analysed.

Structure of the Paper

The outcomes of the study are presented in two sections: Section A reports on the *new* life sciences companies, Section B reports on the founders of these companies.

Results

Section A. New Dutch Life Sciences Ventures in Review

The number of new life sciences companies in the Netherlands
Based on the criteria used here for *new* Dutch life sciences companies, the survey listed a total of 94 new companies by the end of 2001.

At the end of 2000 there were 79 *new* Dutch life sciences companies (ref.: 75 identified in the "nulmeting" (BioPartner 2001) plus four that had not been previously identified). In 2001, 18 new life sciences companies were established in the Netherlands. Three companies ceased their activities in the same year. Thus, the 94 *new* life sciences companies at the end of 2001, equates to a net growth rate of 19% for *new* life sciences companies throughout that year.

The year 2000 showed a record number of life sciences start-ups in the Netherlands. The number of start-ups reached an all-time high of 22 companies in that year. In 2001, with a reduced number of 18 start ups this performance was not matched (Figure 1).

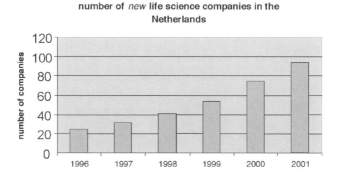

Figure 1: Number of new life sciences companies in the Netherlands since 1996.

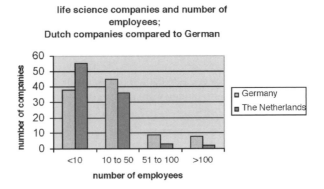

Figure 2: Percentage of life sciences companies with, respectively, <10, 10–50, 51–100 and >100 employees; comparison of Dutch companies to German companies in 2001. *Source:* German data: VBU.

Employment

Employment statistics are often used as indicators for sectoral growth. In February 2002, *new* Dutch life sciences companies provided employment for a total of 1,442 people. When the companies that were not identified in the previous report are taken into account, the net increase of employment in this sector was 29% in 2001.

Despite such growth, the majority of the *new* Dutch life sciences companies are still very small. Figure 2 compares the sizes of German (listed No. 1 in terms of number of companies in the E&Y European report of 2001) life sciences companies to the Dutch companies. As can be seen, the majority of the Dutch companies still have less than ten employees, whereas the largest category of the German companies employ between 10 and 50 people.

The majority of people active in life sciences work in Research & Development departments (63%). This is not surprising since R&D is the main activity of *new* life sciences companies (see Figure 3 and also Figure 6).

Sector

Most *new* Dutch life sciences firms are engaged in general biotechnology (mainly platform technologies or contract research services) or human health activities (49 and 40%, respectively). The Dutch life sciences industry is particularly strong in these two sectors, with special regard to platform technologies (18 companies), therapeutics (17 companies) and diagnostics (14 companies).

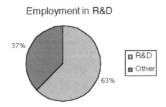

Figure 3: Percentage of employees in R&D in 2001.

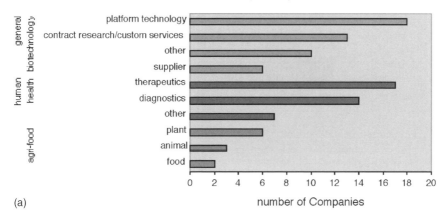

(a)

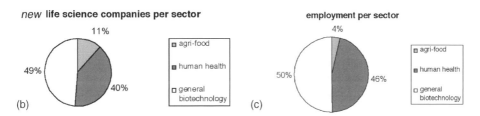

(b)   (c)

Figure 4: (a) New life sciences companies by sub-sector in 2001. (b) Percentage of agri-food, human health and general biotechnology companies in 2001. (c) Percentage of employees in agri-food, human health and general biotech in 2001.

The agri-food sector is trailing behind in terms of numbers of companies: only 11% of the companies are active in agri-food. Furthermore, the agri-food sector is the smallest contributor to employment: agri-food only accounts for 4% of total employment, whereas human health and general biotechnology are by far the two largest sectors (together accounting for 96% of total employment — see Figure 4c).

Origin
A remarkably high percentage (44%) of the *new* Dutch life sciences companies are spin-offs from existing organisations: 29 companies are spin-offs from universities or research institutes, while 15 are spin-offs from corporate firms. About 47% of the *new* Dutch life sciences companies, are independently established and the remaining 9% belong to the category "Other," which includes joint ventures and M&A (mergers and acquisitions). Our findings reveal that M&A activity has not significantly affected the Dutch life sciences sector to date. This is mainly due to the juvenile state of the sector. A good example, however, is the merger of IntroGene and U-BiSys in 2000, which combined to form the double listed (AEX and Nasdaq) company Crucell (Figure 5).

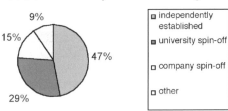

Figure 5: Percentage of independently established firms, spin-offs from universities and companies, and other types of company in 2001.

Business activities

Research and development is undoubtedly the main business activity for many *new* life sciences companies: 73% of the companies have designated R&D as their primary business activity. The other companies can be characterised as service companies (17%), supplier/distributors (6%) and production/manufacturing companies (4%) (Figure 6).

Life sciences clusters in the Netherlands

The clustering behaviour of companies is a well-documented phenomenon in various industries, particularly in high tech sectors. The main reasons for start-ups to be located within life sciences clusters are the presence of necessary infrastructure, availability of facilities, access to partners, short distance to knowledge institutes (During 1998), the presence of a talent pool (skilled students and researchers) and the proximity of lead users and potential clients (Figure 7).

The Dutch life sciences industry also has a tendency to congregate around centres of research excellence. In the Netherlands, geographic concentrations of life science activities can be found in several locations: Amsterdam, Groningen, Leiden, Utrecht, Nijmegen, Wageningen, Maastricht, Lelystad and Delft. The companies in these clusters represent 80% of the total employment by *new* life sciences companies in the Netherlands. Leiden is the most successful, both in terms of number of companies *and* employment (see Figure 8a and b). The success of Leiden can be partially explained by the presence of both large foreign and Dutch companies such as Centocor, Pharming, Mogen (all founded before 1990) and

Figure 6: Percentage of R&D, services, production and supplying companies in the life sciences in 2001.

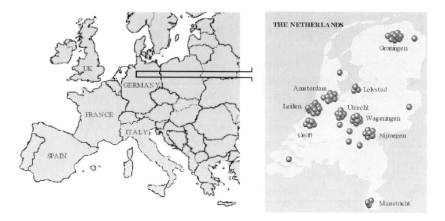

Figure 7: An overview of new Dutch life sciences ventures distributed throughout the Netherlands.

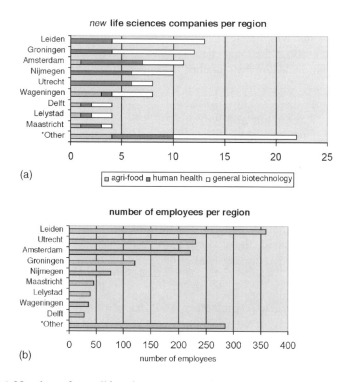

Figure 8: (a) Number of new life sciences companies per region in 2001. (b) Number of employees per region in 2001.

Crucell, which attract a lot of other business activities, as well as the early start of a bio-business park in the early eighties (making this the longest established such park in the Netherlands).

A critical factor in ensuring the further growth of the life sciences sector is the availability of infrastructure, such as laboratories and premises. In order to sustain the growth of the Dutch life sciences sector in the near- and long-term, six life sciences incubators (BioPartner Centers) will emerge in the next few years. They will be located in the following cities with life sciences research centres: Amsterdam, Groningen, Leiden, Maastricht, Utrecht and Wageningen. Each will offer residence to up to ten life sciences start-ups.

Besides these BioPartner Centers, there are several science/business parks located in the Netherlands that are fully or partially dedicated to life sciences: for example, the BioScience Park in Leiden, Amsterdam Science Park in Amsterdam, Zernike Science Park/BioMedCity in Groningen, Agro-Business Park in Wageningen and UBC (Universitair Bedrijven Centrum)/Mercator Business Park in Nijmegen.

Section B. The Entrepreneurs Behind the Start-Ups

The primary target groups of BioPartner are current and potential prospective entrepreneurs in the life sciences. Knowledge about the founders of existing life sciences companies will give BioPartner insight in its target groups and help identifying the life scientists with plans or the potential to start a life sciences company.

This section reports on the main characteristics of the founders of the *new* Dutch life sciences companies including their number, their gender, age and education.

Teams

Analysis of this survey shows that most life sciences companies were set up by a team of entrepreneurs rather than "solo-entrepreneurs." This is consistent with the findings of existing entrepreneurship research on successful high-tech firms (Roberts 1996).

The average size of the starting team is 1.8 people. The majority of the start-ups were founded by more than one entrepreneur: 34% of the companies had two founders while 15% had three or more. Forty-one percent of the companies listed in this survey have one sole founder. The remaining companies (10%) are the result of M&A, or another cause.

These results of the survey indicate that the *new* life sciences companies are rarely managed by a single individual: the average size of the management team is 2.8 persons. Only 20% of the companies are managed by one person (Figure 9).

Gender and age

As noted in our previous report (BioPartner 2001), the percentage of female entrepreneurs in the Dutch life sciences sector is extremely low at 6%. The results of the 2002 survey indicate that the percentage of female life sciences entrepreneurs has dropped from 6 to 4%. The cause behind this decrease is that none of the companies started in 2001 were founded by females (Figure 10).

The majority of the entrepreneurs involved in life sciences are between 40 and 50 years old. The average age was 42 at the time of the company's formation. The relatively high

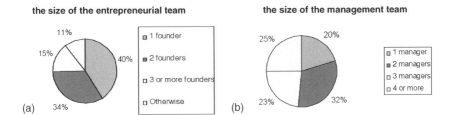

Figure 9: (a) Percentage of companies with 1, 2, 3 or more founders. (b) Percentage of companies with 1, 2, 3 or more persons in the management team.

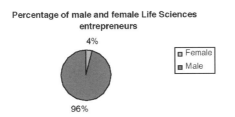

Figure 10: Life sciences entrepreneurs by gender in 2001.

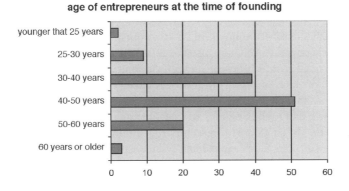

Figure 11: Life sciences entrepreneurs by age.

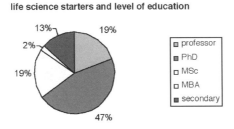

Figure 12: Life sciences entrepreneurs by educational background.

age of life sciences starters correlates with the educational background of the starters: life sciences starters are, in general, highly educated: about 87% had a Masters degree in some field of life sciences; 66% also held a Ph.D., while 19% were full professors (Figures 11 and 12).

The distribution of life sciences entrepreneurs by age can be found in Figure 11.

Discussion

The main purpose of BioPartner is to improve the entrepreneurial culture in the life sciences sector in the Netherlands. In order to achieve this, BioPartner needs to bring about a significant change in the attitudes of the life scientists to entrepreneurship. Since a change in culture is difficult to quantify, the number of start-ups is taken as a surrogate measure by which success can be measured. The BioPartner initiative is expected to result in the creation of 75 *new* life sciences start-ups in the period 2000–2004, equal to an average of 15 new companies a year. To put this number in perspective; in March 2001 a total number of 75 *new* life sciences ventures could be found in the Netherlands (BioPartner 2001). The number of start-ups in 2001 was 18 compared to 22 start-ups in 2000. We believe that this decrease is due to a combination of deteriorated economic conditions in 2001 and a natural annual fluctuation, and compared to the national net growth rate of 2.7% (for the year 2000; CBS, press release 2001), there is no need for concern.

This paper shows that the life sciences sector lives up to the expectation that the sector has great potential to contribute enormously to job creation. Although the absolute numbers are still small, the net increase of employment is extraordinarily high. The figures also show that the Dutch life sciences sector is composed of relatively small and young companies, that tend to cluster around centres of research excellence and are primarily either independently established ventures or spin-offs. However, it is expected that as the Dutch life sciences sector consolidates in the mid- and long-term future, the M&A activity among Dutch firms and between Dutch and foreign entities will become more commonplace. This anticipated consolidation of the Dutch life sciences sector in the future should also lead to more alliances and joint ventures as companies expand and sell their products and services.

The founders of the life sciences companies can be characterised as highly educated males with an average age of 42, who tend to start companies on a team basis. The low percentage of female entrepreneurs in the life sciences is alarmingly small compared to the national average of female entrepreneurs in all industries, of 27% (van Tilburg & Hogendoorn 1997). However, a slight increase can be expected, as a relatively high percentage of female prospective entrepreneurs are involved in life sciences start-ups at the moment.

Conclusions

This paper is focussed on providing an insight into Dutch bio-entrepreneurial activities and an overview of the number, dynamics and characteristics of these companies and their founders. These data indicate that the Dutch life sciences sector has gained momentum since 1999, the year that the Action Plan Life Sciences was launched. Both the number of entrepreneurial life sciences companies and the employment numbers have increased since then, although most companies remain relatively new and small. At this point is hard to determine how much of this seemingly accelerated growth of the sector is the effect of BioPartner. Continuation of research and deepening of the annual surveys are necessary to come to well-grounded conclusions and to determine whether BioPartner's stimulation policy is adequate, or needs to be adjusted. In 2002, BioPartner Network began with two additional research projects: an international comparative benchmark study and in-depth interviews with founders and CEOs of life sciences companies. The data collected from these studies will be used to, besides the above-mentioned purposes, identifying success factors and developing good practices in starting a life sciences company. The results of these studies are expected to be available for publication in the near future.

References

During, W. (1998). Co-operation between technology-based firms in business and science parks: An exploratory study into the situation in the Netherlands. *New technology-based firms in the 1990s* (Vol. 5, Chapter 6). London.

Ministry of Economic Affairs (1999). *Life sciences action plan 2000–2004*. Den Haag.

Moret Ernst & Young (1998). *Strategies for accelerating technology commercialisation in life sciences. An international comparative analysis of seven regions.*

Roberts, E. B. (1996). *Entrepreneurs in high-technology: Lessons from MIT and beyond.* New York: Oxford University Press.

van Tilburg, J. J., & Hogendoorn, P. (1997). *The success of innovative entrepreneurship. The spin-offs of the entrepreneurial university* (in Dutch). Enschede.

Further Reading

http://www.cbs.nl/nl/publicaties/persberichten/2001/pb01n199.txt.
http://www.dechema.de/biotech/vbu/vbu.htm.

Appendix: About Biopartner

BioPartner is the practical result of the Action Plan Life Sciences (Ministry of Economic Affairs 1999), which was launched in 2000 by the Dutch Ministry of Economic Affairs. BioPartner aims to stimulate entrepreneurship in the life sciences in the Netherlands by improving the entrepreneurial climate for bio-starters and bio-business and by removing the obstacles identified for bio-entrepreneurs, such as lack of information, financing, facilities and lab space during the three phases of entrepreneurship ("seed," "start" and "solo"). This initiative is expected to result in the creation of 75 new life sciences start-ups in the period 2000–2004 and comprises the following five instruments: BioPartner Network, BioPartner First Stage Grant, BioPartner Centers, BioPartner Facilities Support and BioPartner Start-up Ventures.

BioPartner Network

BioPartner Network is the central contact point for entrepreneurs in life sciences. BioPartner Network focuses on facilitation and stimulation of entrepreneurship in the life sciences in the Netherlands. BioPartner Network assists potential entrepreneurs, e.g. by establishing and expanding networks, both national and international, providing courses on entrepreneurship, brokerage between starters, investors and other parties, promoting the Dutch life sciences discipline at home and abroad and so on. BioPartner Network has a board that consists of prominent life scientists, life science entrepreneurs and investors.

BioPartner First Stage Grant

This subsidy aims at stimulating researchers to translate their knowledge into a viable business plan. Project proposals must be submitted as a joint application of a researcher and the public research organisation where he/she is employed. An external and independent advisory board will judge the proposals.

BioPartner Centers

BioPartner Centers offer adequate housing for starting life science companies. The Centers comprise office space, laboratories, the required permits and the full infrastructure that a starting entrepreneur shares with other life science start-ups in the Center. The Centers will host at least ten life science start-ups. The Centers will be located in the vicinity of universities or research organisations active in life sciences.

BioPartner Facilities Support

This fund enables start-ups to use advanced equipment and other research facilities in collaboration with universities or research organisations. The fund provides loans for pre-financing of the facilities. The equipment must be innovative to the Netherlands and of crucial importance to the start-up. The application for this fund must be done by a university or research organisation together with the start-up. A part of the costs of acquisition must be returned within five years using money earned by contract research with the new facility.

BioPartner Start-up Ventures

BioPartner Start-up Ventures provides risk capital to life science start-ups. An important condition for investment is matching funding provided by private investors. The investment is provided in exchange for shares of the start-up company.

Part IV

Clusters

Chapter 9

Overcoming Learning Uncertainties in the Innovation Process: The Contribution of Clustering to Firms' Innovation Performance

Paul Benneworth and David Charles

Introduction

The promotion of innovation among SMEs has recently become a key interest of Governments at both national and European levels, particularly in those less favoured regions without strong traditions of innovation (Lagendijk 2000). There are a wide range of policies and tools which are being developed to promote innovation, and increasing emphasis is being placed on what have been termed the "softer" aspects of innovation promotion (Florida 1995; Larsson & Malmberg 1999). These softer elements refer to activities and organisations which seek to promote networking and encourage groups of firms to collaborate to improve their innovative performance.

Although these networking activities have been recognised as an important part of innovation promotion, more recently, their conceptual utility has been muddied by their entanglement with the emerging "clusters" debate (Bryson 2000). The "clusters" approach, as promoted by *inter alia* Porter (1990) but permeating more widely into policy and academic debates (Benneworth *et al.* 2003), is increasingly being used as a framework for traditional policy instruments as well as a way of thinking about interactions and collaborative innovation (Bergman *et al.* 2001; OECD 2000; Roelandt & den Hertog 1998).

This is problematic, because the clusters approach is not solely about innovation promotion; as Gordon & McCann argue (2000), there are other important elements in a cluster, including the agglomeration involving inter-firm traded and untraded linkages. Particular difficulty exists in addressing the considerable ambiguity which exists between the various elements of the cluster approach which tend to be used interchangeably, leading to a misapprehension of cluster dynamics. A particular example of this with which this chapter is concerned is between clusters and cluster policies, and whether cluster policies require the existence of clusters, or whether such policies alone can promote networking and collaboration (Benneworth 2002; Martin & Sunley 2001).

The second problem with the clusters approach is the question of the degree to which it represents a general theory, or is instead a theory of exceptions. Although cluster approaches have been developed from particular stories of high performing regions (Longhi 1999; Saxenian 2000), cluster theory is built as a set of assumptions and processes which are assumed to apply in a generic sense as well as in specific cases. In criticising the ambiguities of particular clusters, we do not wish to refute the argument that lessons can be learned from particular exemplar clusters, which can then be applied to less successful regions. Cluster advantage may be created in a variety of different ways, but in the absence of agglomeration, clustering is likely to bring benefits associated with changed sets of relationships between companies.

In this chapter, our interest is in whether the clustering approach adds value to the way we think about innovation support through networking and soft support. We therefore ask whether firms in less successful regions can replicate the clustering behaviour and successes of agglomerated firms in exemplar regions. This analysis provides the basis for a more general discussion about the value of the clusters idea in improving firm performance in less successful regions and beyond. Indeed we argue that looking at less successful regions helps to illuminate "taken-for-granted" features of clusters in core regions. We begin by examining in some more detail the genesis of the "clusters" idea, and why it has been seen as beneficial to innovation.

Clusters, Innovation and Firm Competitiveness

The key research question we ask in this chapter is "can a clustering promotion approach improve firms' innovation performance in a less successful region?" It is firstly necessary to unpack and differentiate between the various elements of the "cluster approach." The notion of clusters was developed as a policy tool rather than as a rigorous body of theory. This has given the idea diversity and robustness, and led to the emergence of a strong international community of interest. The work of Porter's Monitor Consultancy has been in the vanguard of those highlighting the practical ways that policy-makers can use a "clusters" approach to solve particular problems. The basis of the Porterian approach lies in identifying agglomerations of economic activity (the clusters) and assuming that they possess and reproduce competitive advantages (Porter 1998). The approach typically starts from those agglomerations, which are mapped, analysed, and from which policies are developed, to address barriers to success (Benneworth & Charles 2001).

The Porterian approach is not the only theory of clusters; it is a very *ex post* and macro-scale analysis of the key cluster processes. The argument here is that existing competitive advantage interplays with concentration and agglomeration to give a virtuous cycle of dynamic cluster growth. The agglomeration produces benefits, although it is not necessary to assume that cluster benefits depend on the presence of a "cluster" *qua* agglomeration. Benneworth *et al.* (2003) have segmented the clusters approach into five distinct elements, in Table 1.

In many cases of exemplar clusters, a key feature of their success involves high levels of government R&D investment in the region concerned, creating opportunities for

Table 1: The distinct conceptual elements of a "cluster approach."

Cluster (cf. Porter 1998)	An existing concentration of industrial activity, which is self-replicating and has competitive success built on agglomeration. However, it is more than a simple agglomeration as a result of the innovation arising from the co-operative interactions between firms.
Clustering (cf. Dosi 1987)	The general behaviour of firms who are collaborating in innovation. "Clustering" does not have to take place within an agglomeration — micro-clusters of c.10 firms can gain advantages from co-operative interaction without the existence of a macro-economic agglomeration.
Cluster activities (cf. Klein Woolthuis 1999)	The specific events in which clustering can take place, typically through the collaborative activities in which firms meet and co-operate. Effectively a subset of "clustering," these "events" can occur with free-standing organisations or networks, and are characterised by identifiable outcomes.
Cluster organisations (cf. Lagendijk 2000)	Formal organisations with a responsibility for organising cluster activities. These may be state-funded and have a number of different goals: removing barriers to collaboration, arranging meetings, collective purchasing, branding etc.
Cluster policy (cf. Gilsing 2001; Larousse 2000)	Policies by Government to support cluster development in one of three classes: • support for existing clusters; • support for businesses that already collaborate; or • establishing new collaborations between non-co-operating businesses.

Source: Benneworth *et al.* (2003).

state-funded collaboration to appear as autonomous commercial collaboration. It is hard to see how clustering can be induced in other regions in the absence of such state subsidies and existing vibrant innovation systems. This is a specific example of a more general feature of clusters approaches, in that they say very little about the mechanisms whereby advantage is produced, nor deal with the unique and irreproducible specificities of the exemplar situation. We define this process of advantage creation as "clustering," which tends to be theorised in a rather abstract way, assuming that co-location necessarily leads to co-operation which necessarily leads to better innovation performance, which removes agency and specificity from the model. To reinsert agency into the idea of clustering, it is necessary to look at the way the benefits arise in existing agglomerations in exemplar clusters.

What Can the "Clusters" Idea Offer to Less Successful Regions?

Central to understanding the benefits which the "cluster approach" can provide to less favoured regions is in understanding what precisely happens in exemplar situations that leads to their improved performance. In exceptional regions (those with identifiable clusters) "clustering" is the means by which localised agglomeration is translated into competitive advantage. Understanding clustering requires looking at individual micro-scale behaviour (Dosi 1988). Micro-scale research into how advantage is produced suggests that clustering is exemplified in such activities as direct collaboration, indirect interaction (e.g. through staff) and by use of shared territorial assets such as universities or research laboratories. However, "clustering" is more than an emergent property of particular agglomerations; it describes particular sets of behaviour which encourage innovation, and these types of behaviour can be found in regions which do not have "clusters." If these types of behaviour can be found in less favoured regions, and they improve firms' innovation performance, this would allow clustering to have salience in peripheral regions.

One approach would be to look at the behaviours by which individuals in existing clusters interact (e.g. Saxenian 2000), giving the sense that the cluster is in some way a community of practice (Wenger 1998). Individuals are participating in communities, which transcend firm boundaries, which encourage co-operation and creativity and provides a coherence to the growth of that community (Benneworth 2002). Cornford *et al.* (2001) offer the idea of a "knowledge pool" as a way of dealing with this issue. Individuals have far more capacity to innovate than they can exhibit in any particular situation. If individuals can identify innovative projects to which they can contribute, they can make use of these currently dormant capacities. Thus the "cluster-as-a-knowledge-pool" is a rich environment in which opportunities can act as condensation nucleii for dormant assets to come into play, and by participating in new projects, individuals create new capacities, which may revert to dormancy at the project's end.

A more firm-based approach begins from the way that shared assets in a cluster are used by firms to assist with innovation. These shared assets complement the firm's own assets, and combine to allow the firm to "punch above its' weight." Again, following Dosi, it is necessary to specify that these club goods are tangibly produced, and not to rely on abstracted concepts such as "untraded interdependency" (Storper 1995). There are four main areas in which such club goods can substitute for innovation assets, which together constitute the basis of "clusters-as-collective-assets" (cf. Kaufmann & Tödtling 2000; North *et al.* 2001), by substituting for:

- A lack of people with the time, the skills and the inclination to innovate effectively;
- A poor capacity to approach people with the capacity to assist, which hinders the development of innovation networks and leads to taking bad advice;
- A lack of influence over strategic partners to maximise the benefit from innovation (Gomes-Cassares 1997);
- A lack of size to shape markets, regulations and culture, making the firm extremely reactive and vulnerable to exogenous change.

Two definitions represent different versions of how clusters produce competitive advantage. In the first version, the cluster is a knowledge pool of creative and innovative individuals, whilst in the latter, it is the generalisable territorial assets which firms can substitute for other absent resources (Maskell & Malmberg 1999). Both approaches have important methodological similarities in not being bounded by artificial/synthetic definitions such as SIC or NACE codes (cf. Feser & Bergman 2000); the clusters are self-defined, either by participation in the knowledge pool, or drawing on the assets. Bathelt (2001) charts how the Boston high technology complex consists of six separate sectors, which nonetheless share common knowledge pools and collective innovation assets, with a key element of the cluster's success being mutual stabilisation and reinforcement between sectors.

Where these approaches differ is in how they argue competitive advantage is produced. The knowledge pool approach is concerned with the way that individuals enact relationships and perform innovation. The collective assets perspective conversely begins with the firm as the unit of analysis, and looks at the way that the firm takes cluster assets, and in using them to solve its business problems, regenerates and recreates those general assets. In both cases, firms draw on assets with less effort than if they were to build those capacities for themselves. These approaches are complementary rather than contradictory, and real clusters are likely to contain elements of both.

In this chapter, we are not concerned with the existence of a "cluster" but rather in the ways in which innovation assets — knowledge pools and collective assets — are created. Such assets can be created anywhere where there is knowledge spill–over, and need not be confined to regions with agglomerations; assets, not the agglomerations are of central concern (Lawson 1999; Maskell & Malmberg 1999). Our research question is "can a clustering promotion approach improve firms' innovation performance in a less successful region?" We answer this question by identifying and analysing firms' behaviours, which are creating spill-over benefits and contributing to knowledge pool assets on which other firms can draw.

Classifying Firms by Innovation Performance: The Sophistication Approach

In order to examine the relationship between clustering behaviour and innovation, it is necessary to have a model of the innovation process. Although the vernacular sense of innovation involves a broad spectrum of creativity, inventiveness and even play, innovation has become a popular policy goal because of its potential to contribute to economic development. Innovation is regarded as a vital element of ensuring that companies respond as effectively as possible to changing market conditions, structures and demands. An innovation in this sense is any new product, process or technique which improves the competitive position and long-term viability of the company and the economic prospects of its stakeholders (Alic 1997).

Although innovation has long been described in terms of a linear model, conceptual and practical problems with the model have led to a series of refinements, of which the most widely accepted version is Kline & Rosenberg's (1986) chain-link model. In this model (cf. Figure 1), the stages do not flow automatically from one to another, and ideas and technologies are not smoothly translated to products. The situation is much messier, with innovators encountering problems, making compromises, experimenting and reverting, to bring products to market.

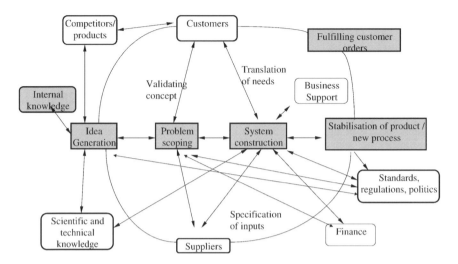

Figure 1: The firm-based innovation process within the context of a knowledge/innovation network.

This chain link model makes greatest sense when viewed from the perspective of an innovating firm; the firm feeds in external outputs at each stage of the process and attempts to materialise the idea. As the "idea" progresses through the firm, it becomes more tangible, becoming a project, a prototype, then a product. If the idea cannot be moved forward, then the firm attempts to access new ideas and resources to find other ways to advance. They may revert to an earlier stage to pursue alternative avenues; if a prototype proves unmanufacturable, then additional design and prototype work may be necessary before the designs can be finalised.

This model has a number of important implications, not least that the resulting product is not obvious at the outset, and the nature of the product is shaped by the interactions between those active in the project. Although at the end of a project it is possible to look back and to trace a single evolutionary path for the innovative product, a better description of the process is as a messy series of steps, blind alleys, and compromises. This "messiness" means that innovation management assumes an important role, as ideas are not passively shepherded through a series of gateways. Instead, decisions have to be taken about the direction of the development project, and control exerted to maintain sight of, then achieve, that goal.

This model also highlights the importance of interactions in the innovation process. In any innovation project, it is possible to identify a range of actors involved, including suppliers, universities, consultants, and inanimate agents such as markets, culture and regulations. One of the more commonly studied interactions is between users and producers, through which a common understanding of product parameters are shaped, and this increased user-producer interaction is regularly cited as one reason for improved innovation performance in clusters (Cawson *et al.* 1995; Porter 1998). However, using the idea of knowledge overspill, clustering suggests that other interactions are important, not least whereby dormant capacities and redundant assets are brought back into use. The

Table 2: The key features of interviewed firms by sophistication level.

	Practise	Performance	Reflexivity
Novice	Absence of innovation management systems. Innovation is implicit and limited to very few individuals in the firm. Weak and poorly handled linkages with external partners and contractors.	Innovation performance is out of control, with very rapid growth leading to potential overtrading and jeopardising the basis of the company. Cash may be generated but very little kudos or market leverage is consolidated from this.	Unawareness that the company is innovating. Unawareness that the company looks bad in comparison with its peers. Underestimating the effort involved in systematising the approach to innovation.
Inexperienced	Basic/generic quality systems, but still highly reliant on key individuals. Some strategic new product development planning. No managing of the drivers of R&D projects; customer-oriented firms over-reliant on firms, R&D teams too divorced from applications and marketing work.	Vulnerability to external shocks; too much volatility to really consolidate seamlessly. Good sales not always converted into profitability because of the R&D overhead. Firm begins to win outside recognition as a good company, and one that others can work with.	Aware of needs for systems but unaware of nuances of different approaches. Unsophisticated consumers of novel ideas outside main area of organisational competence, especially management thinking.

Table 2: (*Continued*)

	Practise	Performance	Reflexivity
Experienced	Sensitive management and systems; functional differentiation of R&D teams. Delegation of responsibility alongside work elements in development programmes. Broad staff involvement in strategic planning. Good understanding of positions in corporate, supply, technological hierarchies.	Problems in maximising the opportunities to be exploited by their technological and organisational asset base, gives steady state growth. Hiccups (e.g. flooding, market movement) mitigated by systems and strategies (disaster recovery). Seen as a sophisticated client for other firms' own innovation activities.	Individuals managed on a personal basis to get the best out of them; awareness that management needs grass roots knowledge. Good at understanding suppliers innovation processes and shaping them for their best advantage. Belief that own systems arriving at best practise.
Expert	Use of technology roadmaps in business planning so NPD totally focused on business goals. NPD process totally under control, world class performance. Corporate structures effectively allocate resources to NPD.	Firm is recognised leader in innovation. Firm is seen as a good partner to work with. Other firms benefit from their association with the firm. Sustainable and profitable growth.	Willingness to accept decisions can be totally wrong or wrong for current context. Adoption of continuous improvement to ensure own systems remain up to date and to prevent complacency. Able to mentor suppliers.

Source: Charles & Benneworth (1998, 2001).

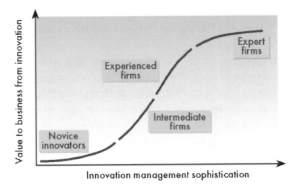

Figure 2: The relationship between performance and sophistication level in improving companies. *Source:* Benneworth & Charles (2001).

final product will only embody some elements created during the innovation process, but innovation involves experimentation, redundancy and failure, which create a wide range of dormant and residual assets that can later be re-used.

To analyse the relationship between clustering and innovation management, we use a synthetic variable, "sophistication level," which measures how firms manage the innovation process. In this approach (cf. Charles & Benneworth 2001a), we segment firms into four classes which we term "sophistication levels" as follows:

- Novice: an absence of formal systems and evaluation mechanisms, with a heavy reliance on luck to make the best of situations;
- Inexperienced: the presence of formal systems which sit uneasily with the culture and practises of the firm;
- Experienced: the presence of formal systems which have been modified to a degree to reflect the needs of the firm;
- Expert: the boundaries of the firm have disintegrated with respect to their innovation collaborators, but innovation always takes place in the best interests of the firm.

Some further detail on the characteristics of firms at each of the four levels are given in Table 2. Sophistication level is composed of three elements, practise, performance and reflexivity. Experience has shown that firms in one sophistication level demonstrate similar characteristics for each of the elements, except in those cases where firms are improving their sophistication level. In those cases, there may be an interval between firms improving reflexivity and practise to observing material improvements in performance. This relationship is shown in Figure 2.

Methodology, the Region and the Sample

This chapter presents a set of findings from a research project undertaken in the North East of England, the poorest of the British regions, with a per capita GDP rate of around 80%

of the national average, and with a population of 2.6 m and a workforce of 1 m (Charles & Benneworth 2001b). The region has lagged more successful regions since its decline began in the 1930s, since when, industrial policy has focused on managing industrial decline, and attracting new sectors through the use of incentives. This policy created volatile novel sectors heavily dependent on inward investment, with very low levels of innovation in both the domestic and foreign owned sectors (Charles & Benneworth 1999, 2001b; Williams & Charles 1986). The region fits precisely with the broader context of the research question sketched out in the introduction, because there is neither strong agglomeration nor high levels of R&D in the region.

The Interview Sample

In this research, because our focus was on how innovating firms enacted clustering behaviour rather than analysing a specific cluster, we interviewed innovating companies across a range of industrial sectors. A total of 38 companies were interviewed, with selection based upon three innovation indicators. *Ex post*, these proxies for innovative behaviour were robust, as all the firms interviewed *were* innovating, and many of the firms fell into more than one category as follows:

- Involvement with ReCET (a university technology transfer organisation);
- A (prestige or cash) award, including Millennium Award, Design Council Award, Queen's Award (for innovation) or a Smart award; or
- Involvement with a "club" organisation, Regional Services for Clustering or RTC North.

In order to contextualise the chapter, aggregate company details are presented below. Around 80% of the companies were locally owned (or managed in the case of plcs), and 75% were SMEs (not part of a larger parent company employing less than 200 workers); all those non-local firms were part of multi-site operations, but had some considerable local responsibility for their own new product development. Table 3 sets out a analysis of the firms on the basis of the industrial sector under which they would traditionally be classified.

In this chapter, we have looked at the interaction between firm and collective assets to analyse the contribution of "clustering" to innovation in peripheral regions and its utility as a policy tool for innovation promotion. The analysis segments the firms according to their sophistication level as a means of differentiating systemic variations in clustering behaviour between sophistication levels. This enables us to offer a more nuanced analysis of the levers policy-makers have available to influence and promote clustering-led innovation. In this analysis, we focused on three distinct ways in which firms' innovation interacted with "clustering":

- How clustering behaviour varied with sophistication level;
- How firms' ability to benefit from clustering behaviour varied with sophistication level; and
- How Firm innovation contributed to the stabilising collective innovation assets.

Table 3: Firm sample by industrial sector.

Industrial Sector	Firms
Electronics	11
Instrumentation	7
Utilities	5
Mechanical engineering	5
Capital goods	4
Construction	4
Food, drink & tobacco	1
Software	1
Total	38

Source: Authors' interviews.

Clustering Factors: The Innovation Process and Sophistication

The first step in this analysis is to segment the firms according to their sophistication level. The process for this segmentation is described in Benneworth & Charles (2001); the firms were classified following Table 2 according to the description which best fitted their circumstances; those whose level was the same in each of the three areas of practise, process and reflexivity were also categorised as "confident." As Table 4 shows, the classification process was successful, with 34 of the 38 firms falling unambiguously into one of the

Table 4: A preliminary classification of the sample according to innovation sophistication maturity level.

Maturity Level	Total	Confident	Locally-owned/Run	SME
Novice	13	11	13 (100%)	13 (100%)
Inexperienced	12	12	12 (100%)	10 (77%)
Experienced	9	8	5 (56%)	5 (56%)
Expert	4	3	1 (25%)	0 (0%)[a]
Total	38	34	31 (82%)	28 (74%)

Source: Authors' interviews and calculations.
[a] This should not be given to imply that locally-owned SMEs cannot be expert innovators. In the 1998 project, three of the four expert firms were precisely that, although one has since grown to no longer be an SME (Charles *et al.* 1998).

four levels.[1] There are clear differences in the way in which firms at different levels of sophistication "cluster" with other members of the knowledge pool (which also includes business support services, public sector research organisation/universities).

Novice Firms: Reactive, Historical and Non-contributory

The thirteen novice firms were best distinguished by the fact that they were extremely weak in controlling clustering activity to ensure that it contributed to innovation. Six of the novice innovators had formed themselves into what they termed a "cluster." This was an organisation which had won some ERDF funding to develop a collective sales and marketing effort, to jointly seek out contracts as a cluster organisation (called "Pegasus"), but which had not proven commercially successful. Although the cluster organisation persevered, and some firms had limited success, none of the firms grew beyond doing what they had done in the R&D centre, albeit as consultants rather than as employees. This vignette exemplifies four features about the way that novice firms enacted clustering behaviour.

The first was that the firms were highly reactive. The majority of the firms were formed to capitalise on ideas and contacts that the founders developed in other companies, but had difficulties in using their relationships to generate material contributions to their innovation and corporate growth in a timely way. Two of the other seven firms only used their inter-personal contacts when they were running out of work. Another two firms had been given some work by other local firms as a favour, but that favour had backfired because the novice firms had not been able to develop the products they had been asked to supply, and in both cases had to waste huge amounts of time ensuring they did not disappoint their contacts.

The second feature was that the firms did not contribute to generalising their highly contingent and unstable interpersonal contacts and specific behaviours into more general territorial learning competencies. Only two of the 13 firms were recruiting, training and developing staff with new high level skills (i.e. adding to the knowledge pool). Only one firm used public-sector grants to develop innovation capability rather than to buy a service or license a technology. Indeed, in the majority of the cases, it was uncertain whether innovation was creating anything which would outlive the retirement of the founders, or indeed that these firms and the cluster organisation represented anything more than a set of large-firm employees who had become consultants.

The third feature was that the firms more generally had a very limited understanding of innovation. Although all 13 were continually innovating in response to customer demands (often for one-off products), there was no management of the underpinning knowledge, technologies and ideas. Consequently, when one-off (and often very innovative) projects ended, little was black-boxed into modules which could be reused more cheaply on future

---

[1] Three of the remaining four were in such flux that the interviews tended to liberally mix statement, aspiration and perception hindering allocating them to a single category. The fourth was a firm which was bordering on "expert." The expert class is unique in a number of ways; it is a transient position, with backsliding a real possibility, and expert firms benefit their partners in a range of ways. Therefore, that fourth firm's process, practice and self-awareness all had elements of the expert situation, but it was clear that the firm itself was not an expert innovator.

projects or which could form the basis of a higher volume product. One Pegasus firm had developed a project pricing and decision-making tool-kit which could have been sold more widely by other consultants, but instead, they only used it internally. If any of these novice firms ceased trading, their assets would vanish, and nothing of broader territorial value would remain.

The final feature was that their business collaborators tended to be very cavalier in their treatment of these firms, not out of malice, but because these firms were not particularly important to their customers' technological portfolios. This meant that the novice firms could not develop a coherent technological trajectory, by building internal linkages and economies of scale between distinct knowledge sets. The formal cluster organisation was continually disrupted by changes in the industry into which they sold and their consequent inability to develop long-term profitable clients. Although the seven other firms all had key partners, none of those seven firms were themselves a key partner to anyone. This raises questions about the benefits which the novice firms were able to derive from "clustering."

Novice firms were weak innovators, but did participate in clustering activity, although it did not always contribute effectively to what they hoped to achieve. The best examples of successful "clustering" were in those cases where novice firms were able to draw on assets created by others. Novice firms' dealings with universities worked most smoothly when the relationship was limited to a single dimension. In practice this means that already funding had to be in place before those firms could have serious technological discussions with universities. More generally, novice firms were very weak at being able to create something on which others could draw.

Inexperienced Firms: Building New Relationships

In contrast to the novice firms, many of which could be regarded as at the end of the developmental line in particular industrial trajectories, the inexperienced cluster firms at least had a viable future. That future is not necessarily of high growth, but they at least have a degree of certainty and stability in contrast to the contingency and instability of the novice firms. The twelve firms interviewed all pursued viable developments. Although not necessarily creating territorial learning competencies, they contributed to a preservation of the existing knowledge pool and collective innovation assets.

The most popular approach to maintaining and reproducing themselves as businesses was due to the creation of new assets through the innovation process, which had as an incidental consequence the creation of more assets than functionally required by particular innovation projects (which contribute to dormant knowledge pool capacities). This could be through the purchase of intellectual property, the hiring of staff, and reselling old ideas in the form of new projects. In each case, the total sum of knowledge pool assets was greater at the end of each innovation sequence than at the start: new staff were taken on, sales increased and product portfolios diversified. The strategic use of public sector grants was important in levering in competitive advantage and producing innovation assets. The most obvious form this took was staff development; firms used teaching company schemes, traineeships and student placements (which we later refer collectively to as "learningships") as a means of pump-priming a sustainable staff asset that would outlive the short-term grant.

The second feature was that the firms were part of a sequential developmental trajectory, which was blossoming outwards, in contrast to the moribund contributions made by novice firms. A number of the firms were spin-offs from parent companies, but distinct from novice firms. Their competitive niches were an extension (not a replication) of the parent company. An inexperienced domestic electricals controls company, for example, hired new staff and moved into the locomotive controls market. This contrasts with a novice firm who also moved into the locomotive sector, but at the expense of existing competencies; that change was a sideways shift rather than a development.

The third main feature was that collaborative relationships were part of a more differentiated set of external relationships and interactions. The firms comprising Pegasus were unwilling to think about strategic relationships with firms outside the organisation. Inexperienced firms were more catholic in their choice of partners and partnerships, although the majority did not formally differentiate between types of relationship. One firm invited strategic partners from a wide range of sectors to review new product ideas, although the individual review meetings were not always productive. However, the majority of firms at least had some control over relationships, were not totally reliant on the largesse of others and were more systematic and strategic in choosing partners.

Experienced Firms: Taking Command of the Environment

What distinguished experienced firms was the degree to which their successful innovation contributed to clustering by supporting the development of shared goods whilst at the same time benefiting them. This is best exemplified in the way in which experienced firms managed their "clustering" relationships to maximise their benefits. One of the experienced firms used a number of the less experienced firms for design work; the net effect was that the experienced firm had smooth growth, although these sub-contractors remained heavily dependent on the largesse of the other firm. All the firms could be said to be "clustering," but it clearly benefited the experienced firm far more than the less experienced firms. It is possible to highlight three further substantive area of distinction.

The first was the balance of power in the inter-company relationships, with experienced firms being able to retain areas of control to shape the innovation projects to their evolving needs (and build internal knowledge linkages and economies of scale). Only two of the eight firms interviewed supplied finished products, but all eight firms had a strong degree of control of their internal development processes. None were in sub-contractor relationships — all eight had an expertise in their technological area and sold "innovation" as integral to their products. Five of the six companies involved in intermediate goods had formal development systems to ensure their partners could not disrupt their innovation processes, even when they were part of a larger group, or were spin-offs with close linkages to the previous incubator firm.

The second distinction was in the relationship of the firm to general territorial innovation assets. Novice firms made no contribution, and inexperienced firms were limited to contributing new assets at the end of the innovation process, particularly to the knowledge pool. The experienced firms also contributed to the knowledge pool, albeit in a more systematic manner, the example above of sub-contracting design work being a good example of this. Moreover, three of the eight firms worked with their PCB suppliers, and tried to get them involved co-operatively in their development programmes (i.e. to boost their suppliers' innovation sophistication).

The third distinction was the contribution made to territorial shared innovation goods, the business support and the public-sector technological infrastructure. Experienced firms were distinct in being proactive in engaging with the development of public sector business support. Two of the firms were engaged with the Regional Technology Centre (RTC North), one on a long-term programme, and the other through board membership. This contrasts to the less experienced firms whose involvement was limited to consuming the existing services offered by RTC rather than shaping the development of new services. Two firms had also been involved in setting up Manufacturing Challenge, a business networking organisation aimed at improving the performance of manufacturing in the North East as a whole (cf. Benneworth 2002).

Experienced firms had specific ways of dealing with the region's universities. Less experienced firms dealt functionally with universities, by buying consultancy services from universities, or participating in research projects/networks when offered money. Although experienced firms did use consultancy services, they also tended to shape the way universities provided those services. By encouraging university departments to be more accessible, they in turn made the services more useable for less experienced firms, a specific case of what Fontes & Coombs (2001) term "densifying the techno-economic network."

Furthermore, experienced firms were aware of the need to balance the opportunities and limitations of different types of relationships with researchers, segregating long-term (speculative) relationships (in areas of domain knowledge) from the purchasing of innovation services such testing or rapid prototyping. Because these more experienced firms were involved interactively with the development of domain knowledges (such as materials science or computer control algorithms), university research staff were also able to use these firms as assets in their own academic innovation programmes, as research partners and users, sources of prestige, and sounding boards (cf. Benneworth 2002). Thus, as well as drawing on regional knowledge assets from the public sector base, experienced firms were able to contribute to the development of elements of the science base, and leverage in mutually beneficial external research funds.

Expert Firms: An Obligatory Point of Passage for Clustering Behaviour?

This final set of four firms were the expert firms; these were a quite distinct class of enterprise because they were instrumental in controlling and configuring the environment within which the clustering took place. In the sub-heading, we allude to the idea that the expert firms are "an obligatory point of passage," that all the firms "clustering" are doing so in the regional socio-technical environment configured by these expert firms. Whilst experienced firms helped innovation support providers refine their existing services, it was the expert firms who engaged at an institutional level to shape the governance frameworks in which that support was provided.

It is clear that part of those benefits arose from the imperfect ability of the firms to monopolise the technologies they and their staff had developed. Those technologies were appropriable to some degree by the staff working there and were sufficiently valuable that third parties encouraged those left redundant to set up their own businesses replicating what was done in the expert firm. Although deliberate "intrapreneurship" is a well-documented feature, this appears to be a less intentional form of new business formation (Charles 1992; Moncada-Paterno-Castello *et al.* 1999). The expert firm has created assets, which are not allowed to lie dormant awaiting reactivation, but are spontaneously given a commercial life, independent of the parent organisation. This makes these firms central to the evolution and dynamism of the cluster.

The second area where they actively build local learning competencies is in the way that they condition the economic environment by working with local partner companies. Each of the four firms had their own way of managing this process. Given the small size of the sample of expert firms, it is difficult to distinguish common threads, but it is clear that each firm has developed a set of local linkages, which diffuse innovation capacity through other firms in the region.

- The first had set up a spin-off firm to supply its sub-contract electronics needs, and that "spin off" rapidly grew to be larger than the parent (and which diversified into much broader markets than its parent).
- The second firm had a small R&D fund reserve, which it distributed amongst local innovative micro-businesses, so that if the firm had a question it knew someone with a moral obligation to help them answer the question.
- The third firm was itself a spin-off of a local innovative SME with a number of subsidiary firms, all needing very skilled technicians and engineers, and so accepted some movement of its highly-trained staff as part of a strong local labour market (on which other less experienced firms also drew).
- The fourth firm had a deliberate policy of working with local suppliers to improve their systems and quality (and to contribute to their own quality).

The final area where the expert firms contribute to cluster building was in configuring the institutional environment to "densify" the regional innovation system. The relationship between high quality innovating firms and "upstream" research assets has long been understood (Massey *et al.* 1992; Porter 1990; Segal 1985), even if conceptualised through the somewhat reductionist lens of the linear innovation model. We have argued that experienced firms boost their regional innovation system and induce universities to develop research centres offering services which can easily be bought but draw on and contribute to the development of a world-class research base. Expert firms also behaved in this way, but are more strategic in their involvement, often working with universities at an institutional rather than an interpersonal level to change their institutional attitude to providing innovation support services.

Thus, a number of the firms at different sophistication levels had long-term relationships with Newcastle University's Department of Electrical and Electronic Engineering (EEE); a number of these were based around graduate recruitment, joint research projects, student placements and *ad hoc* consulting. What distinguished expert firms' roles was the strategic way they had been involved in the inception and development of two of EEE's spin-off consulting activities, supporting the individuals in the university who were promoting the idea to their more sceptical colleagues. Less expert firms had cited these two centres as providing services they used in their own innovation activities, but were not in a position to contract directly from academics. That they were able to do this indicated a service which had been deliberatively designed and stabilised as a collaboration between the university and strategic expert partners.

Analysis of Clustering and Innovation Asset Substitution

How can we then use the sophistication approach to analyse the way that clustering by North East firms improves innovation? Firstly, all of the firms interviewed were involved in "clustering" activity, in the sense that they used negotiated linkages to access external assets, and increase the scope of what they could achieve from innovating by substituting for missing innovation assets. However, there were very different reasons for behaving this way, and the more experienced the firms were regarding innovation, the more closely their approach correlated with best practise. The novice firms in the "cluster organisation" were clustering without it being clear what they were benefiting from in so doing, whilst the more experienced firms used these external relationships more positively to fill gaps in their innovation assets.

It is also possible to distinguish between sophistication levels in terms of the stability and permanence of clustering behaviour. Although Pegasus seemed the most stable of the clustering vehicles, it was not particularly significant as a catalyst because it created only very limited complementary capacities on which the firms could draw whilst innovating. Pegasus was highly dependent on a set of inter-personal relationships which had built up over long periods of co-working at the R&D centre, and was heavily dependent on grants for its capacity to achieve anything. Conversely, more experienced firms clustered in ways which created more permanent and stable benefits, and reduced reliance on inter-personal relationships to achieve more systemic and selective contacts. The policy implication is that it is not the clustering behaviour *per se* which is important, but the capacity in the firm to extract value from clustering.

Secondly, there were significant differences in the success that the firms had in drawing on the "collective assets" to which the clustering provided access. The better the firm was at innovation, the better the firm was at making use of the territorial collective assets provided by clustering. The novice innovators had the greatest difficulty in taking advantage of benefits, which arose from their interpersonal connections to companies who were willing to assist them. Indeed, in those cases when interviewed firms were mutually collaborating, when one firm was a demonstrably more sophisticated innovator than the other, the better innovator yielded much greater benefits from the collaboration than the more inexperienced firm.

This is not particularly surprising given that as firms become better innovators, they become better users of innovation support services. We argue that in common with innovation support services, "clustering" substitutes for missing assets, increasing the innovation capacity of firms (Robson & Bennett 2000). The corollary of this is that for "clustering" to be beneficial, innovating firms must be good innovators, and have a need for the assets created through clustering. The key issue here is of the substitutability of assets. The Pegasus cluster had a capacity geared towards knowledge sharing, but offered very little to address fundamental innovation problems in terms of developing a product which they could sell at high volume to create cash flow and fund business growth. This suggests that policies to support "clustering" may not succeed where innovation performance is weak, especially if there is not complementarity between what the firms are missing and what the clustering activity is potentially offering.

Finally, firms at different sophistication levels make very different contributions to collective innovation assets. Novice firms made almost no contribution, whilst the contributions of inexperienced innovators were limited to the creation of knowledge pool assets, lacking the managerial and technological capacity to engage with and shape innovation support providers. Expert firms were of most value to clustering, by helping to create systemic and stable elements on which other innovators could draw.

Both experienced and expert innovators played a key role in intensifying the techno-economic network, by inducing universities to provide services in packaged ways then easily absorbed by less expert innovators. Similarly, they were also involved in shaping what other support providers did, again with more general benefits. In terms of understanding clustering in a particular territory, these experienced and expert firms are the key drivers of the activities by which advantage is produced. This implies that policies seeking to create clustering activity to improve innovation performance ought to encourage better innovators to work with innovation service providers to create innovation support assets which are more easily absorbed by less expert firms.

Concluding Discussion

In this chapter, the overarching point to make is that there is clearly a difference in the clustering behaviour of firms at different sophistication levels, with differential impacts on the creation of a shared knowledge pool. Innovation sophistication can be regarded as an explanatory variable for looking at a particular "cluster," and opening up the "black box" of agglomeration, in order to begin to explain the dynamics by which competitive advantage is produced. We have begun to outline a map of the kinds of behaviour which characterise clustering firms, although space precludes a more detailed treatment.

This research suggests that less experienced innovators have much more potential to improve the knowledge pool than to create collective territorial R&D assets. Clustering benefits for the less sophisticated firms had come about either through direct state funding for innovation in the firms, or subsidised "learningships." These "learningships" all involved a relatively junior person being funded to perform a complete iteration of the innovation process in the firm, with two equally important outputs, the finished product and the "finished innovator." The first policy conclusion is that less sophisticated innovators benefit most from policies which increase their involvement in the knowledge pool. This can be through placing an existing member of the knowledge pool in an innovating position or providing the means for an existing employee to perform within that knowledge community.

The research also suggests that more experienced firms have a role to play in the development of cluster services, in which knowledge assets are purposively created. Fontes & Coombs (2001) use the idea of densification of the techno-economic network to refer to the way in which experienced innovators in sparse innovation systems can induce other actors (such as universities) to provide them with the services they need, but at the same time increase the capacity of those actors to serve the innovation needs of others. Densification is a good way of describing the way in which sophisticated innovators create "shared assets," but participation in that densification was limited to sophisticated innovators. Another policy conclusion is that developing clustering services for innovation in the periphery requires the involvement of experienced innovators, but not through mechanisms whereby those experienced innovators are regarded as the exclusive market for those innovation services. That outcome (new services for experienced innovators) is qualitatively less valuable than densifying ways that allow less experienced firms to access such services.

It is important not to overstate the claims we make in this chapter; we have spoken to less than 40 firms in one region of one country, but the findings suggest that the sophistication approach may be of value in analysing clusters, by providing an analytic variable directly related to the benefits assumed to come from the cluster. There are also important policy implications should the sophistication approach be valid, because it provides a benchmark against which to evaluate cluster activities. Policies will be judged good if they increase the stock of cluster assets in the knowledge pool or collective learning assets. Moreover, this provides a way of targeting clustering activity at appropriate classes of firms, encouraging less sophisticated innovators to participate in knowledge pools, and using more sophisticated innovators to densify the techno-economic network and build linkages to knowledge and innovation service providers. The next obvious way to develop this research agenda is the development of a cluster typology based upon the sophistication and inter-relations of clustering firms and the RIS.

At the start of this chapter, we remarked that the clustering approach was in danger of muddying the way that networking approaches to innovation were being applied in peripheral regions. This research, instead, suggests that the clusters approach deepens our understanding of what it is that networking and soft support provide innovating firms. The great weakness of these networking analyses is their focus on behaviour rather than performance, and the clustering approach helps to sharpen what counts as performance, i.e. a tangible increase in the stock of innovation assets. This is useful for building innovation policy for less favoured regions, and we have already highlighted the two main policy implications. However, looking at the stocks and flows of these innovation assets in the future may provide a more general means of analysing clusters, and more generally, the innovative milieux.

References

Alic, J. A. (1997). Technological change, employment and sustainability. *Technological Forecasting and Social Change, 55*, 1–13.

Bathelt, H. (2001). Regional competence and economic recovery: Divergent growth paths in Boston's high technology economy. *Entrepreneurship and Regional Development, 13*, 287–314.

Benneworth, P. S. (2002). *Innovation and modernisation of a peripheral industrial region: The case of the North East of England.* Unpublished Ph.D. thesis, University of Newcastle, Newcastle.

Benneworth, P. S., & Charles, D. R. (2001). Towards a policy for high tech small firms in peripheral regions. Paper presented to 9th Annual High Technology Small Firms Conference (31st May to 1st June 2001). Manchester Business School, UK.

Benneworth, P. S., Danson, M., Raines, P., & Whittam, G. (2003). Confusing clusters? Making sense of the cluster approach in theory and practice. *European Planning Studies* (forthcoming).

Bergman, E. M., den Hertog, P., Charles, D. R., & Remoe, S. (Eds) (2001). *Innovative clusters: Drivers of national innovation systems.* Paris: OECD.

Bryson, J. (2000). Spreading the message: Management consultants and the shaping of economic geographies in time and space. In: J. R. Bryson, P. W. Daniels, N. Henry, & J. Pollard (Eds), *Knowledge, space, economy.* London: Routledge.

Cawson, A., Haddon, L., & Miles, I. (1995). *The shape of things to consume: Delivering information technology into the home.* Avebury, Aldershot.

Charles, D. R. (1992). *The corporate dimension of research and development location.* Unpublished Ph.D. thesis, University of Newcastle, Newcastle.

Charles, D., & Benneworth, P. (1999). Plant closure and institutional modernisation: Siemens micro-electronics in the North East. *Local Economy, 14*(3), 200–213.

Charles, D., & Benneworth, P. (2001a). *Electronics applications investigations.* Report for Regional Centre for Electronics Technologies (ReCET), CURDS, Newcastle.

Charles, D., & Benneworth, P. (2001b). Situating the north east in the European space economy. In: J. Tomaney, & N. Ward (Eds), *A region in transition: North East England at the millennium* (pp. 24–60). Ashgate, Aldershot.

Cornford, J., Charles, D., Wood, P., Robson, L., Chatterton, P., & Belt, V. (2001). *Culture cluster mapping and analysis study.* Newcastle: One NorthEast.

Dosi, G. (1988). The nature of the innovative process. In: G. Dosi, C. Freeman, R. Nelson, G. Silverberg, & L. Soete (Eds), *Technical change and economic theory.* London: Pinter.

Feser, E. J., & Bergman, E. M. (2000). National industrial cluster templates: A framework for applied regional cluster analysis. *Regional Studies, 34*(1), 1–20.

Florida, R. (1995). Towards the learning region. *Futures, 27*(5), 527–536.

Fontes, M., & Coombs, R. (2001). Contribution of new technology based firms to the strengthening of technological capabilities in intermediate economies. *Research Policy, 30*, 79–97.

Gilsing, V. (2001). Towards second-generation cluster policy: The case of the Netherlands. In: E. M. Bergman, P. den Hertog, D. R. Charles, & S. Remoe (Eds), *Innovative clusters: Drivers of national innovation systems.* Paris: OECD.

Gomes-Cassares, B. (1997). Alliance strategies of small firms. *Small Business Economics, 9,* 33–44.

Gordon, I. R., & McCann, P. (2000). Industrial clusters: Complexes, agglomeration and/or social networks? *Urban Studies, 37*(3), 513–532.

Kaufmann, A., & Tödtling, F. (2000). Systems of innovation in traditional industrial regions: The case of Styria in a comparative perspective. *Regional Studies, 34*(1), 29–40.

Kline, S. J., & Rosenberg, N. (1986). An overview of innovation. In: R. Landau, & N. Rosenberg (Eds), *The positive sum strategy.* Washington, DC: National Academy Press.

Lagendijk, A. (2000). Learning in non-core regions: Towards intelligent clusters; addressing business and regional needs. In: R. Rutten, S. Bakkers, K. Morgan, & F. Boekem (Eds), *Learning regions, theory, policy and practice.* London: Edward Elgar.

Larousse, J. (2000). The political evolution of cluster policy in Flanders. Paper presented to Do clusters matter in innovation policy? OECD Cluster Group Workshop (8–9 May). Utrecht, The Netherlands.

Larsson, S., & Malmberg, A. (1999). Innovations, competitiveness and local embeddedness: A study of machinery producers in Sweden. *Geografiska Annaler, 81*(B), 1–18.

Lawson, C. (1999). Towards a competence theory of the regions. *Cambridge Journal of Economics, 23*(1), 151–166.

Longhi, C. (1999). Networks, collective learning and technology development in innovative high-technology regions: The case of Sophia-Antipolis. *Regional Studies, 33*(4), 333–342.

Martin, R., & Sunley, P. (2001). Deconstructing clusters: Chaotic concept or policy panacea? Paper presented to Regionalising the Knowledge Economy. Regional Studies Association Annual Conference (23 November). London.

Maskell, P., & Malmberg, A. (1999). Localised learning and industrial competitiveness. *Cambridge Journal of Economics, 23*(1), 167–185.

Massey, D., Quintas, P., & Wield, D. (1992). *Hi-technology fantasies*. London: Routledge.

Moncada-Paterno-Castello, P., Tübke, A., Howells, J., & Carbone, M. (1999). The impact of corporate spin-offs on competitiveness and employment in the EU: A first study. *IPTS Technical Report Series*. Luxembourg: OOPEC.

North, D., Smallbone, D., & Vickers, I. (2001). Public sector support for innovating SMEs. *Small Business Economics, 16*(2), 303–317.

OECD (2000). Do clusters matter in innovation policy? Proceedings of OECD Cluster Group Workshop (8–9 May). Utrecht, The Netherlands; Paris: OECD.

Porter, M. E. (1990). *The competitive advantage of nations*. Basingstoke: Macmillan.

Porter, M. E. (1998). *On competition*. Boston: Harvard Business School.

Robson, P. J. A., & Bennett, R. J. (2000). SME growth: The relationship with business advice and external collaboration. *Small Business Economics, 15*, 193–208.

Roelandt, T., & den Hertog, P. (Eds) (1998). *Cluster analysis and cluster based policies in OECD countries*. Paris: OECD.

Saxenian, A.-L. (2000). Networks of immigrant entrepreneurs. In: C. M. Lee, W. F. Miller, M. G. Hancock, & H. S. Rowen (Eds), *The Silicon Valley edge: A habitat for innovation and entrepreneurship*. Stanford, CA: Stanford University Press.

Segal, N. (1985). *The Cambridge phenomenon — the growth of high technology industry in a university town*. Cambridge: Segal, Quince & Partners.

Storper, M. (1995). The resurgence of regional economies ten years later: The region as a nexus of untraded interdependencies. *European Urban and Regional Studies, 2*(3), 191–221.

Wenger, E. (1998). *Communities of practice: Learning, meaning and identity*. Cambridge: Cambridge University Press.

Williams, H., & Charles, D. R. (1986). Electronics in the north east of England: Growth or decline. *Northern Economic Review, 13*, 29–38.

Chapter 10

Close Encounters: Evidence of the Potential Benefits of Proximity to Local Industrial Clusters

Thelma Quince and Hugh Whittaker

Introduction

Localised clusters of high technology small businesses are of increasing interest to politicians and academics. Underlying this interest has been the growing awareness of the economic role of high technology small firms. In the U.S. and much of Europe, the last quarter of the 20th century saw the importance of small firms increase significantly, particularly in terms of employment (Acs & Audretsch 1993). Further, activities experiencing rapid growth have tended to be those dominated by small enterprises (SBA 1999). This has been a particular feature of activities based on new technologies such as biotechnology, computer software, R&D services and telecommunications (SBA 1999).

The second factor prompting this development has been the resurrection of interest in localised industrial clusters (Porter 1990). The dense local networks of traded interdependencies (input–output relationships) based on high levels of specialisation which figure in descriptions of Italian industrial districts (see Pyke *et al.* 1990) were noted characteristics of earlier localised concentrations of industry such as the hardware, gun and lock industries of Birmingham and the Black Country (Allen 1929). More recent interest in industrial clusters has broadened to include untraded interdependencies (Storper 1995) and, for high technology firms specifically, notions of innovative milieux (see Lawson *et al.* 1997).

The consensus view, of interest to policy makers, is of a highly innovative small high tech firm deeply embedded in a local high tech cluster, experiencing rapid growth and serving essentially a niche market, globally oriented, and collaborating with other local organisations in ways which enable it to benefit from technology or knowledge spillovers from the local research base and localised specialist services (Goss & Vozikis 1994; Jaffe *et al.* 1993; Keeble *et al.* 1997, 1998; SQW 2000). But how far does location matter to high tech small businesses? Are those embedded in high tech clusters more innovative? Are they more export oriented? Are the same characteristics likely to be found among small high

New Technology-Based Firms in the New Millennium, Volume III
Copyright © 2004 by Elsevier Ltd.
All rights of reproduction in any form reserved.
ISBN: 0-08-044402-4

tech firms associated with more traditional manufacturing local industrial clusters? Against the consensus view are the findings that small firms in remoter or more peripheral locations may attempt to compensate for locational disadvantages by being more innovative and seeking larger international markets (Vaessen & Keeble 1995). What firms actually achieve is likely to be influenced, not only by location, but also by the nature of the activity undertaken, entrepreneurial objectives and ambitions, and more importantly, by some interaction between all of these.

Many studies addressing these issues have tended to focus on individual types of location such as Oxford or Cambridge (Lawson *et al.* 1997), West Midlands (Freel 2000), or specific industries such as opto-electronics (Hassink & Wood 1998). This paper reports on a more comparative approach. It draws upon a survey of high tech small businesses in the U.K. By defining each participating business's local area, examining the prevalence of their activity in that area, and assessing whether the local area was noted as an industrial cluster, participants were grouped according to their involvement in a local industrial cluster. The outline of this paper is as follows. Characteristics of the businesses studied are described in the next section, while methods used to classify the businesses according their potential involvement in local industrial clusters are outlined subsequently. The following sections describe the results derived from comparing businesses with differing levels of potential involvement in local industrial clusters in respect of indicators of niche markets, limitations encountered, collaboration, and finally, performance in terms of growth and innovation.

The Study

This work forms part of an ongoing comparative investigation of small high tech firms in the U.K. and Japan, and follows a similar survey conducted in 1998 (Whittaker 1999). The criteria for inclusion were employment size (less than 250), independent status and high tech activity (based on modifications to Butchart's (1987) definition of high tech activities (Hecker 1999). The sample comprised suitable respondents to the 1998 survey, together with an equal number of additional businesses. The latter were selected from the Dun and Bradstreet database. A postal questionnaire was administered to 781 firms in December 2000. The overall response rate was 34.1%.

The questionnaire covered:

- Descriptive characteristics of businesses and CEOs: (activity, age, size, ownership structure, educational qualifications and prior employment in the research base[1]).
- Markets, competition, and collaboration: (customer dependence, competitors, subcontracting, competitive advantages, limitations: collaboration and exporting).
- Performance: (employment and turnover growth, innovation, and R&D expenditure).
- Employment practices and policies: (qualifications of workforce, training and incentive schemes and levels of support for different HRM policies).
- Attitudinal variables: (CEO's personal objectives and approaches towards risk).

---

[1] HEIs, research institutes and hospitals.

Table A1 (Appendix) summarises the main characteristics of the businesses and their CEOs. The study focused on independent businesses[2] from which 65% were completely new start-ups. Almost 60% were manufacturing businesses. Service sector firms were further divided into two broad activity groups: computer services and telecommunications (CS&T) (19% of all firms), and research and development and technical services (RD&T), accounting for the remaining 23%. Two activities accounted for a quarter of the participants; "instruments for measuring, checking and other purposes" (SIC 33.2) and "software consultancy and supply" (SIC 72.2). A number of important high tech activities such as biotechnology, aerospace and the manufacture of computers, were weakly represented.

The median age of businesses was 16 years. There was a relatively even distribution of firms in terms of age, with almost as many businesses established before 1980 as since 1990. Responses were skewed towards smaller businesses since three quarters of respondents employed less than 50 workers, and less than 20% had a turnover of £5 million and over. The majority of participants had founded their businesses and had done so by collaborating with others (Roberts 1991; Whittaker 1999). Businesses owned 100% by the participant, together with those in which the only other minority shareholder was a spouse, were classified as "real or quasi sole proprietor," and accounted for 29% of all businesses.[3] The remaining collaboratively owned firms (71%) were roughly equally divided between those in which ownership was shared exclusively with other internal owners, and those in which there was some external ownership. Significant[4] sectoral differences were found in respect of age, size and marked differences regarding external ownership: manufacturing businesses tended to be older and larger, CS&T businesses were notably newer, while businesses engaged in RD&T were smaller and fewer had external owners.

The entrepreneurs were overwhelming male (92%) and "middle-aged," with a median age of 51, while over 70% were over 45. The significant age differences found between businesses in different sectors were reflected in differences in the participants' ages. Those managing manufacturing businesses were notably older, while those in CS&T were younger. The high tech entrepreneurs were highly educated and well qualified. Almost two thirds had a degree, and half of these also held a post graduate degree. The importance of the research base in providing a source of high tech entrepreneurs was demonstrated by the finding that in one in four cases, a member of the original founding team previously had worked full time in higher education or a research institution (including medical).

Hierarchical cluster analysis was used on sector, firm age, employment size, ownership and entrepreneur age variable to derive 5 robust "type" groups accounting for 95% of the businesses considered.[5] The characteristics of each type of business are summarised below.

---

[2] Single person businesses were not included.

[3] The legal requirements of incorporation frequently result in a spouse being designated as a director but having no real involvement in running the businesses.

[4] Unless stated otherwise tests of statistical significance used were non parametric tests: Mann Whitney for two group comparisons or Krushal Wallis for multigroup comparisons and "significant" was at the 5% level or better.

[5] Firms with no full-time employees were excluded.

Table 1: Locational distribution of businesses.

Locational Category	Percentage ($n = 221$)
In or adjacent to conurbation	38
In or adjacent to major town (pop $\geq$ 100,000)	44
In or adjacent to moderate town (pop $\geq$ 50,000 < 100,000)	6
In or adjacent "market town" (pop $\geq$ 10,000 < 50,000)	7
Rural (pop < 10,000)	5

Type 1   **Small Service Businesses: ($n = 49$)**
Mostly RD&T businesses with some CS&T but no manufacturing firms. Predominantly small, employing less than 20 workers, of varied age with CEOs mainly under 55. Virtually no external shared ownership.

Type 2   **New Activities Businesses: ($n = 26$)**
Mixed in terms of sector but with a relatively high proportion of CS&T firms. Predominantly new but large businesses (founded since 1990) with young CEOs. No proprietorships: almost all having external ownership.

Type 3   **New Manufacturers: ($n = 57$)**
Mostly manufacturers with a few CS&T but no RD&T firms, relatively new businesses, run by middle aged or younger entrepreneurs, employing less than 50 workers, mostly with shared ownership.

Type 4   **Older Manufacturers: ($n = 61$)**
Overwhelmingly manufacturers, predominantly older businesses (founded before 1980) medium to large size, run by older CEOs, with varying ownership structures.

Type 5   **Manager-run Businesses: ($n = 19$)**
A small group in which the participants were generally young with no ownership stake in the business, mostly older manufacturing firms.

Spatial and Industrial Cluster Analysis

The location of each business was examined and participants allocated to one of the 5 categories given in Table 1. Whether or not a business was regarded as "adjacent" to a major population centre was based on location within a 10 mile radius, but account was taken of accessibility by road (but not of possible congestion).[6] The majority of businesses (88%) were located either in, or adjacent to, major urban centres.

---

[6] With the exception of Greater London.

Table 2: Strength of involvement in local industrial clusters.

Criteria	Group	Strength of Involvement in Local Cluster	Percentage ($n = 215$)
(a) >0.45%	[I]	Local high tech cluster activity	14
(b) LQ > 1.25	[II]	Activity noted as associated with or supporting a local cluster	11
(a) <0.45% (b) LQ > 1.25	[III]	A locally over-represented activity associated with local cluster	10
(a) >0.45% (b) LQ < 1.25	[IV]	Local concentration of an activity associated with local cluster	14
(a) <0.45%	[V]	Activity not clearly associated with local cluster	52

To gain some idea of the local industrial cluster in which participant businesses could potentially be embedded, each individual business's "local area" and the corresponding Local Authority Districts (LADs) covering that area were identified. Borrowing methodology used in the Sainbury report (DTI 2000) employment data (1998),[7] at 3 digit SIC level, for the LADs in each business's "local area" were used to determine:

(a) the relative importance of the activity each business undertook in its own "local area" (a cut off of 0.45% of total employment in the local area was used); and
(b) the extent to which that activity was over-represented in that "local area" as compared to its national distribution (a location quotient of 1.25 cut off was used).

Information from the Sainbury Report (DTI 2000) on clusters was then used to classify the businesses in terms of the potential strength of their involvement in a local cluster on the basis given in Table 2.

The participating businesses were almost evenly split into those with some clear potential involvement in a local industrial cluster and those without such potential involvement. Businesses in Groups I and II were seen as potentially strongly embedded in a local industrial cluster.[8] Not only was there a significant concentration of activities broadly similar to their own within their local area, but also such activities were strongly related to identified industrial clusters. By comparison, those in Groups III and IV were potentially less strongly embedded since the activities they undertook were either less well represented or less concentrated in their local area. Businesses undertaking activities which appeared unrelated to any local industrial cluster were further differentiated on the basis on the information provided in Table 1 between those located in major population centres (which

---

[7] Excluding public sector employment, for Great Britain only, and for the year of the original study.

[8] It was not possible because of the relatively small numbers involved to further differentiate between high tech and non high tech industrial clusters.

Table 3: Cluster involvement categories.

Cluster Involvement	Percentage ($n = 215$)
Remote/isolated[a]	16
Urban related	35
Weak involvement	24
Strong involvement	24

[a] The majority of the business in this category were in "rural" locations. Not all of those locations were in peripheral regions but they were relatively inaccessible.

may benefit from the effects of agglomeration), and those in more remote locations. The resulting classification is given in Table 3.

Marked differences were found in the level of potential involvement in local clusters of businesses of different types (Table 4). Type 1 businesses, comprising largely of those undertaking RD&T and to a lesser extent CS&T, conformed to expectations since, outside of parts of the South East and Cambridge there are relatively few clusters based on scientific R&D alone. These activities, and those of technical testing, tended to be supportive activities. Interestingly, almost a quarter of these businesses were classified as relatively isolated.[9]

The small group of new activity businesses (Type 2) included those undertaking telecommunications and software consultancy, and were strongly related to local clusters and major urban centres. Types 3, 4 and 6 were dominated by manufacturing businesses, overall less involved in local clusters, but more likely to be related to major urban centres. This was particularly true for the older manufacturing businesses (Type 4).

Niche Markets and Competitive Advantage

CBR surveys have consistently suggested that SMEs operate in highly segmented and "niche" markets (Kitson & Wilkinson 2000). The potential indicators of "niche" markets studied include customer dependence, number and size of competitors and incidence and level of subcontracting relationships, together with perceptions of competitive advantage. Small high tech businesses, differing in their involvement in local clusters, are likely to vary in the degree and nature of the "niche" markets they operate in.

Although SMEs are traditionally seen not to be involved in exporting (Storey 1994), there is a relative lack of work (in the U.K. at least) on the spatial variations in small business export activity (Gorton 1999). High tech SMEs are considered to be more export oriented, particularly those in localised high tech clusters. Apart from exports and subcontracting, no information was collected on the spatial distribution of customers.

Overall the participants in the study appeared to operate in "niche" markets (Table A2 of Appendix). The majority displayed moderate levels of dependence on their principal

---

[9] Other analyses, not reported here, suggest that this type of businesses contained the highest proportion of potential "life-style" entrepreneurs.

Table 4: Cluster involvement of business types.

Cluster Involvement	Business Types				
	Type 1 (*n* = 48)	Type 2 (*n* = 26)	Type 3 (*n* = 56)	Type 4 (*n* = 59)	Type 5 (*n* = 17)
Remote/isolated	23	15	11	19	12
Urban related	15	31	41	49	47
Weak involvement	48	15	21	15	0
Strong involvement	15	39	27	17	41

customers. Although relatively few (11%) were dependent on a single large customer for more than 50% of their turnover, just over a quarter were dependent on one customer for more than 25%, and only 10% obtained less than 10% of revenue from their top three customers. While few respondents (5%) reported no serious competitors, half reported less than five, and almost three quarters reported less than ten. Participants were also asked how many of their serious competitors were located overseas. Some 36% reported no serious overseas competitors, but a similar number (35%) reported that the majority of their serious competitors were located overseas. Supporting the view of the greater export orientation of high tech SMEs, almost two thirds of the businesses were engaged in exporting, and for almost a quarter, exports accounted for over half of their turnover.

Fewer businesses strongly involved in local clusters had high level of dependence on a single customer (Table 5). There were also marked differences in the number, size and

Table 5: Niche market indicators.

	Involvement in Cluster Category			
	Remote/ Isolated	Urban Related	Weak Involvement	Strong Involvement
Dependence on 1 customer (*n* = 207) (% of turnover)[a]				
<25%	72	71	67	84
≥25%	28	29	33	16
Competitors (*n* = 207) (number of serious competitors)[a]				
<5	61	39	58	40
≥5 < 10	21	33	14	27
≥10	18	28	28	33
Larger competitors (*n* = 191) (% of all competitors)				
<50%	21	25	23	15
≥50 < 100%	36	26	19	25
100%	43	49	57	60

Table 5: (*Continued*)

	Involvement in Cluster Category			
	Remote/ Isolated	**Urban Related**	**Weak Involvement**	**Strong Involvement**
Overseas competitors ($n = 186$) (% of all competitors)				
None	41	34	48	30
$1 \leq 75\%$	15	40	21	30
$\geq 75\%$	44	26	32	40
Exports (incidence) ($n = 184$)				
Exporters	66	71	60	62
Exports (level) ($n = 120$ exporters only) (% of turnover)[a]				
<25%	26	48	61	41
$\geq 25 < 50\%$	21	15	14	26
$\geq 50\%$	53	37	25	33
Subcontracting for (incidence) ($n = 215$)[a]				
Subcontractor	43	36	17	46
Subcontracting for (level) ($n = 140$ subcontractors only) (% of turnover)				
<25%	40	37	61	54
$\geq 25 < 50\%$	10	22	12	14
$\geq 50\%$	50	41	28	32
Origin of subcontract orders ($n = 137$ subcontractors only)[a]				
Only local	10	15	19	32
Local and elsewhere U.K.	55	50	63	54
Some overseas	35	35	19	14
Subcontracting to (incidence) ($n = 215$)				
Client	37	29	17	25
Subcontracting to (level) ($n = 158$ clients only) (% of turnover)[a]				
<25%	54	40	65	59
$\geq 25 < 50\%$	32	31	26	31
$\geq 50\%$	14	28	9	10
Location of subcontract order placed ($n = 144$ clients only)				
Only local	30	33	24	29
Local & elsewhere U.K.	60	47	62	55
Some overseas	10	20	14	16

[a] Statistically significant.

location of competitors seen by businesses with differing levels of involvement in clusters. For example, isolated businesses saw significantly fewer serious competitors. There were also differences in the size of competitor seen, with entrepreneurs involved in local clusters regarding their businesses as confronted by proportionately more larger competitors. It has been argued that businesses in high tech clusters operate in global markets. Although not quite statistically significant, both businesses strongly involved local clusters and those in isolated locations faced more overseas competitors. For the latter, this was reinforced by significantly higher export intensity (proportion of turnover exported). This finding confirms other work of the CBR, which suggests that businesses in remote locations may attempt to compensate for locational and restricted market disadvantages (Vaessen & Keeble 1995). There was little variation in the incidence of exporting between firms with different involvement in local clusters.

Turning to subcontracting, almost two thirds of all participating businesses undertook subcontracting work for others. For those doing so, there was a bi-modal distribution; for half it counted for less than 10% of turnover, but for a fifth, it accounted for 75% or more. Further work, not reported here, suggests that high levels of subcontracting may not be conducive to innovation. Three quarters of the businesses put subcontracting work out to others, but for most, this counted for less than 25% of turnover. Participants were asked about the location of these links. As has been found in other studies (Whittaker 1997), there was a slight asymmetry: orders received were more geographically dispersed than orders placed.

Businesses weakly involved in local clusters were significantly less likely to be subcontractors, or to do so less intensively. The bi-modal distribution for all participants was strongly evident among isolated businesses. There was some evidence of the importance of a local cluster in generating demand for subcontract work. Businesses strongly involved in clusters were distinctive in receiving significantly more orders for subcontract work from "local" clients, and together with businesses weakly involved, tended to receive fewer orders from overseas, as compared to their counterparts with no involvement in clusters.

In respect of putting out work to subcontractors, again businesses only weakly involved in clusters tended to do so less, and less intensively. Isolated businesses were more likely to use subcontractors but urban related businesses used subcontractors more intensively. Although this latter group made slightly more use of overseas subcontractors it was interesting that there were no marked differences between firms with different levels of involvement in clusters regarding the relative importance of local subcontractors. There was no clear evidence that clusters constituted a better source of such subcontract supply links.

The competitive advantages seen by entrepreneurs may also indicate the nature of the markets they operate in. Accordingly participants were asked to indicate the importance of 11 possible competitive advantages on a five point Likert scale (Table A3). Advantages indicative of niche orientations, such as "Personal attention/responsiveness to client needs," "Quality of product/service" and "Established reputation" were the most highly rated. By contrast advantages associated with atomistic competition such as "Marketing and promotion," and "Price/cost advantages" tended to be eschewed. Three groups of advantages were created from factor analysis, which focused: (a) on the product/service itself ("product/service"); (b) on aspects of producing or delivering that product/service ("delivery"); and (c) marketing of the product/service ("marketing"). "Product/service" advantages received the highest level of support overall whereas "Marketing" advantages received the least.

There were no pronounced differences between businesses with differing levels of involvement in clusters in respect of the major groups of competitive advantages. There were some differences in respect of individual advantages. Entrepreneurs in isolated businesses tended to give greater levels of support to "technical expertise" and "personal attention to the client/customer," and lower levels of support for advantages more related to atomistic competition.[10] This lack of clear differences in competitive advantages, seen by firms differentially involved in clusters, was not unexpected given the nature of the sample firms, the majority of which were likely to operate in some form of niche market. What might be of greater interest is identifying differences in these niche markets. A clear distinction emerged between competitive strengths in technical expertise, reputation and quality on the one hand, and design, novelty and specialised nature on the other. It is argued elsewhere (Quince & Whittaker 2002) that the former may reflect niche markets where the technology is embodied in the person, whereas the latter may reflect niche markets where the technology is embodied product or service. Businesses involved in clusters gave significantly less support to advantages reflecting niche markets in which the technology was embodied in the product or service (mean score of 3.39 compared to 3.61), which suggests that the primary benefits of proximity may be in respect of tacit knowledge.

Supportiveness

The extent to which a local industrial cluster is supportive may be indicated indirectly by the constraints or limitations experienced by businesses. Participants evaluated the importance of 11 limitations on a 5 point Likert scale. Overall responses to these questions were muted: few were seen as "significant" or "crucial." "Increased competition," "Overall growth of demand," "Lack of marketing/sales skills" and "Access to new markets" were the most important limitations. At the other extreme, protection and acquisition of intellectual property rights, and the "Availability of manual/clerical skills" were the least important. Factor analysis produced three groups of limitations, in order of decreasing importance: demand, supply and technology.

There were large differences in the evaluation of limitations by businesses with differing levels of cluster involvement. The most pronounced differences were found in respect of demand limitations, both as a group and among individual limitations. "Increased competition" was a significantly less important limitation for isolated businesses and those strongly involved in a local cluster. Similarly the latter saw "Access to new markets" as less

---

[10] Combined mean for technical/scientific expertise and personal attention for isolated 4.2 compared to 4.02 for all other businesses. Combined mean for marketing and promotion and price/cost advantages for isolated 2.75 compared to 3.10 for all other businesses.

important, especially when compared to urban related businesses. Overall there were fewer pronounced differences in respect of supply limitations. "Lack of technical expertise" was a more important constraint for businesses involved in clusters, particularly those weakly involved. However interestingly "Cost and availability of finance" was regarded as more important by those strongly involved in clusters. Differences were also found in respect of technological limitations, where the most pronounced was found in "Acquisition of IPR," which was seen as a significantly more important constraint by businesses strongly involved in clusters.

Isolated businesses together with those strongly involved in clusters appeared to confront fewer demand constraints. For the former this may be a reflection of their greater involvement in export markets. For the latter this does not imply that the cluster alone represented demand. Rather, some aspects of the cluster may have given greater access to new markets. However the high tech businesses strongly involved in clusters considered lack of technical expertise, the acquisition of IPR and the cost and availability of finance, to have been more significant limitations on their actions than other businesses. How far did these evaluations reflect differing local environments as opposed to differences in firm behaviour and entrepreneurial ambitions? Pursuit of growth may increase awareness of constraints. Performance achieved and growth sought are discussed later (Table 6).

Opportunities for collaboration also indicate the supportiveness of a local industrial cluster. Research has pointed to the role collaborative arrangements play in the development of high technology SMEs, particularly in respect of innovation and foreign competition (Keeble *et al.* 1998; Klein Woolthuis 2000; Oliver & Blakeborough 1998). Just as founding tends to be a collaborative affair, so does developing a business.

For the study as a whole almost 60% of the businesses had entered into at least one such arrangement in the previous two years, and 60% of these had more than one agreement. Most commonly, collaboration was with suppliers closely followed by customers and other firms in the same line of business. By and large, these were with organisations elsewhere in the U.K., although (reflecting the subcontracting pattern mentioned above), local collaborations, tended to be with suppliers, while collaborations with overseas customers were more common than those with local customers. Collaboration with distributors also tended to be international. Just over one in five of the participants had collaborated with an organisation in the research base.

The reasons given for collaboration were multiple, the most common were related to expansion: to "expand range of products/services" (75%), "provide access to new markets" (56%) and "develop services/products for current customers." "Sharing research and development activity" was mentioned by just under 40%.

Slightly more isolated businesses, and those with strong cluster involvement had entered into collaboration. However isolated businesses were more likely to collaborate with only one type of organisation compared to other businesses. This is reflected in fewer isolated businesses collaborating with each of the different types of organisations identified in Table 7, with the exception of research based organisations. Isolated businesses were significantly less likely to have collaborated with suppliers and to a lesser extent with customers. Interestingly businesses located in major urban centres were the most likely to have collaborated with the research base, whereas those involved in clusters were slightly more likely to collaborate with firms in the same line of business.

Table 6: Importance of limitations.

	Involvement in Cluster Category			
	Remote/ Isolated	**Urban Related**	**Weak Involvement**	**Strong Involvement**
Limitation	Mean score	Mean score	Mean score	Mean score
Demand[a]	2.31	2.79	2.63	2.42
Increased competition[a]	2.31	2.96	2.88	2.80
Overall growth of demand in main product markets	2.29	2.59	2.61	2.42
Access to new markets[a]	2.14	2.61	2.39	1.90
Supply	2.13	2.20	2.30	2.25
Marketing/sales skills	2.11	2.51	2.53	2.56
Management skills	2.34	2.17	2.47	2.30
Availability and/or cost of finance	1.97	2.28	2.00	2.40
Lack of technological/ scientific[a] expertise	1.89	1.84	2.49	2.06
Availability of manual/ clerical skills	1.71	1.83	2.02	1.72
Technological[a]	1.56	1.78	1.49	1.82
Difficulties implementing new technology	2.06	2.11	1.78	2.08
Acquisition of IPR[a]	1.20	1.45	1.39	1.64
Protection of IPR	1.29	1.51	1.31	1.52

[a] Differences between all or some groups statistically significant at 5% level.

The nature of the questions used and the small numbers involved prevent any detailed analysis of the spatial orientation of these collaborative links. However some general comments can be made. Among isolated businesses collaboration with firms in the same line of business tended to be local, whereas those with customers included overseas links. For businesses located in or near major urban centres collaboration with suppliers tended to local and these businesses were also distinctive in having more collaboration with their local research base. Businesses weakly involved in clusters tended to collaborate with organisations located "elsewhere in the U.K." rather than local ones. The distinctive feature of the spatial orientation of the collaborative links of businesses strongly involved in clusters was the extent of links with overseas organisations. This was found for all types of organisation with the exception of those in the research base, and was particularly strong for suppliers and distributors.

There was little difference in the number of purposes for collaboration cited by businesses with differing levels of involvement in clusters, but there were differences in the relative

Table 7: Collaboration with other organisations.

Type of Organisation	Remote/ Isolated (%)	Urban Related (%)	Weak Involvement (%)	Strong Involvement (%)
All collaboration	69	56	53	64
More than one organisation	13	35	22	30
Suppliers	21**	65	52	55
Customers	38	50	56	55
Distributors	17	35	22	39
Firms in the same line	38	43	52	58
Research base	25	40	26	24
Number of purposes				
More than one	71	78	70	76
Share R&D	58	38	26	36
Expand range of products/services	67	80	78	73
Improve market/financial credibility	12	25	30	24
Meet current customer needs	25*	53	67	52
Spread costs	21	12	11	9
Provide access to new markets	54	58	52	55

*Significant at 5% level or better.
**Significant at 1% level or better.

importance of these purposes. The most notable differences were found in respect of isolated businesses which were significantly less likely to have undertaken collaboration to meet current customer needs or for the purpose of improving market or financial credibility. Sharing R&D was a more important reason for collaboration for these businesses than for others.

For the study as a whole, collaboration was associated with both recent and long term growth and innovation. As mentioned earlier differences in awareness of limitations may reflect differences in objectives and strategies towards growth as much as real environmental constraints. It is to the issue of performance that this paper now turns.

Performance and Objectives

Participants were asked about performance — growth of employee numbers and turnover in the past two years — and innovation and associated aspects such as R&D and applications for intellectual property rights protection. The overall results are presented in Table A2 (Appendix). Almost two thirds (62%) experienced some increase in turnover in the two years prior to the survey, but there was somewhat less growth in terms of employment, with half of participants increasing full time employee numbers. Large increases in both employment and turnover were reported by a small number of 19 businesses.

Assessing growth in small businesses is problematic: relatively small changes appear large in percentage terms because of small initial numbers. Growth in turnover and employment was assessed in relation to size to produce four categories: no growth or contraction, and low growth, average growth and high growth for size. Employment and turnover growth were combined to produce a composite measure of recent growth. Businesses can pursue different growth strategies, such as expansion of turnover but stable employment, particularly over a period of time as short as two years.[11] Almost 30% of the businesses had experienced no growth in either turnover or employment in the two years prior to the study whereas 22% recorded high balanced growth in turnover and employment. As a measure of *long-term* employment growth, size in relation to age was used, in those businesses in which the participant had a founding role (see table below). Just over a quarter (27%) were considered to have achieved high growth for their age.

	Employment Size Groups (Number of Employees)		
Age Groups	Smaller (<20)	Medium (≥20 < 50)	Larger (≥50)
Newer (since 1990)	2	3	3
Established (1980–1989)	1	2	3
Older (before 1980)	1	1	2

Scores: 1 = Low growth for age; 2 = Growth in line with age; 3 = Good growth for age. Only firms in which the participant had been involved in a founding role were included. Acquistions and MBO/MBIs were excluded.

Innovative activity was high. Participants were asked if their business had in the past two years undertaken innovation "new to their firm but not to their industry," (non novel) or "new to both their firm and their industry" (novel), in terms of product, process or logistics. Almost four in five had undertaken some innovation in the preceding 2 years. Two thirds of the innovators (56% of all businesses) reported introducing a "novel" innovation. The majority of businesses (61%) undertook R&D. However, just under 20% devoted more than 10% of their turnover to R&D. As other CBR studies (Wood 1997) have found, the level of spending on R&D was strongly related to both whether a firm undertook innovation or not, and the level of innovation undertaken. Almost two thirds of the non-innovators recorded no spending on R&D.

Much of the interest in high tech clusters arises from the assumed positive impact on performance of involvement in such clusters. As can be seen from Table 8, being part of cluster was associated with better growth over time, particularly when compared to isolated businesses. Although there was no relationship between cluster involvement and overall recent growth (composite growth), apart from fewer firms with involvement in

---

[11] Very few firms expanded employment while contracting turnover.

Table 8: Performance.

	Cluster Involvement Category			
	Remote/ Isolated (%)	**Urban Related (%)**	**Weak Involvement (%)**	**Strong Involvement (%)**
Long term growth (n = 147)[a]				
Low growth	48	29	22	18
Average growth	31	48	51	39
High growth	21	23	27	42
Recent growth (Composite growth) (n = 213)				
No growth	34	35	19	26
Low	14	24	27	22
Moderate/uneven	31	20	31	27
High balance	20	21	23	26
Employment growth (relative for size) (n = 199)				
No growth	50	53	43	48
Low	19	11	16	17
Average	22	24	20	10
High	9	11	20	25
Turnover growth (relative for size) (n = 202)[a]				
No growth	42	40	35	31
Low	10	26	18	17
Average	45	19	20	21
High	3	14	27	31
Innovation (incidence) (n = 212)				
Innovator	83	83	83	76
Innovation (level) (n = 172 innovators only)				
Non-novel	31	27	35	37
Novel	69	73	65	63
R&D (n = 215) (Spending as a percentage of turnover)				
None undertaken	40	34	40	40
<10%	34	49	54	31
≥10%	26	17	6	29

[a] Significant at 5% level or better.

clusters recording no growth, there were differences in respect of recent turnover growth. Businesses associated with clusters were more likely to have achieved high turnover growth, for their size, compared to those without such involvement (Table A4 of Appendix).

The relative lack of clear differences in innovative activity between businesses with differing potential levels of involvement in clusters (Table 8) may appear to fly in the face of evidence suggesting that that cluster based firms are more innovative. However innovative activity and novel innovation in particular was strongly related to activity: manufacturing businesses were significantly more likely than those undertaking RD&T to have introduced a novel innovation. The latter group included businesses undertaking contract R&D. This perhaps raises questions as to the appropriateness of the innovation measures used. Such businesses may well play a vital role in facilitating the innovative activity of others. Finally the sample specifically included very new businesses founded since 1998. Although there was relatively little difference in the incidence of R&D undertaken by businesses differing potential cluster involvement there were differences in the level of spending. As found in respect of exporting, there was some evidence of businesses potentially disadvantaged by location attempting to compensate for that disadvantage. Isolated businesses together with those strongly involved in clusters tended to undertake R&D more intensively.

Many studies of have suggested that SME's are not growth oriented (ACOST 1990). However, CBR surveys have consistently shown a majority aiming for at least "moderate" growth. In this study of high tech small businesses the overwhelming majority (90%) sought growth, with just marginally more seeking "moderate" rather than "substantial" growth (48% compared to 42%). Participants were also asked to describe their approach to risk. Two broad categories of approach were identified: a "closed" approach characterised by risk avoidance or aversion, and an "open" approach characterised by willingness or calculated willingness to assume risk. Slightly more entrepreneurs displayed an open approach (54% compared to 46%).

Strong associations were found between performance, both over time and in the recent past on the one hand, and future growth objectives, and, to a lesser extent, attitudes

Table 9: Growth objectives and approach towards risk.

	Cluster Involvement Category			
	Remote/ Isolated (%)	Urban Related (%)	Weak Involvement (%)	Strong Involvement (%)
Growth objective				
No growth	17	7	15	6
Moderate growth	40	61	40	43
Substantial growth	43	33	44	51
Approach toward risk				
Closed	58	46	42	39
Open	42	54	58	61
Venture Capital[a]				
Sought	20	14	14	33

[a] Significant at 1% level or better.

towards risk. (There could be various interpretations of this.) In over two thirds of businesses recording high balanced growth in the preceding two years entrepreneurs sought "substantial growth," compared to just under a quarter in businesses stagnating or contracting. Similarly entrepreneurs in over half (56%) of the businesses performing well over time were aiming for "substantial" growth, compared to just over a quarter of those in firms which had performed less well over time. Recent, and to a lesser extent long term, performance was positively associated with a more open approach towards risk: 72% of recent high growers displayed such an approach compared to 47% of non-growers. Seeking venture capital can also be seen to reflect a positive orientation towards growth. Only 39 participants had sought venture capital and those doing so were significantly more likely to aim for "substantial" growth and to have an open approach towards risk (Table 9).

Businesses strongly involved in clusters were more likely to seek "substantial" growth, particularly when compared to businesses located in major urban centres. By contrast, isolated businesses were less likely to adopt an open approach towards risk, while those strongly involved in local clusters were more likely to so do. These differences were not quite statistically significant. There was, however, a pronounced and significant tendency for more businesses strongly involved in clusters to have sought venture capital.

Conclusions

This comparison of small high technology firm with differing levels of potential involvement in local industrial clusters has illustrated differences in respect of the structure of their markets, innovative activity, growth and global orientation. Few differences were found to indicate that businesses strongly involved in local clusters tended to operate in niche or highly segmented markets to a greater extent than businesses in other locations. But the former were more aware of facing more overseas competitors of larger size. Similarly no difference was found between firms in different locations in respect of the incidence of either innovation or exporting, although those strongly involved in local clusters spent high proportions on R&D and exported more. These businesses were also more likely to have collaborated with overseas organisations.

The study provided some evidence that local industrial clusters were more supportive. Again it was not that businesses more involved in local clusters were more likely to act as subcontractors but that those doing so were more likely to be serving local clients. Further businesses located in industrial clusters saw fewer demand limitations, and were more likely to have experienced high levels of employment and turnover growth in the two years prior to the study. Their entrepreneurs were more strongly growth oriented and open to risk and more likely to have sought venture capital. These findings suggest that, rather than traded interdependencies, the most potent forces in local industrial clusters may be indirect influences raising aspirations, lessening perception of risk, and providing potentially greater access to venture capital. There is also some suggestion that businesses differing in their potential involvement in local clusters also encounter different types of niche markets. The results point to need to look more closely at exact nature of interorganisational linkages within clusters and more importantly how the mediating or facilitating of linkages occur.

These findings also support the view that what may be more important in understanding the dynamics of small high technology firms is not just the potential benefits to be derived from particular locations but also interactions between location, activity, entrepreneurial ability, objectives and ambitions. The isolated high tech small firms in this study appeared to attempt to compensate for locational disadvantages through exports and R&D activity (Vaessen & Keeble 1995), but unlike their counterparts supported by local industrial clusters, their performance was weaker. This may reflect the lessening dynamism of rural locations noted in other CBR studies (Keeble 2000). Perhaps a salient reminder to politicians might be that the economic expectations placed on small high tech firms must include, not only those within clusters, but also those in more disadvantaged locations.

References

ACOST (1990). *The enterprise challenge: Overcoming barriers to growth in small firms*. London: HMSO.
Acs, Z. T., & Audretsch, D. B. (1993). *Innovation and small firms*. Cambridge, MA: MIT Press.
Allen, G. C. (1929). *The industrial development of birmingham and the black country, 1860–1927*. London.
Butchart, R. L. (1987). A new U.K. definition of the high technology industries. *Economic Trends*. No. 400 (February).
DTI (1998). *Our competitive future: Building the knowledge driven economy*. London: HMSO.
DTI (2000). *Business clusters in the U.K.: A first assessment*. London: HMSO.
Freel, M. (2000). External linkages and product innovation in small manufacturing firms. *Entrepreneurship and Regional Development, 12*(3).
Gorton, M. (1999). Spatial variations in markets served by U.K. based small and medium sized enterprises. *Entrepreneurship and Regional Development, 11*(1).
Goss, E., & Vozikis, G. S. (1994). High tech manufacturing: Firm size, industry and population density. *Small Business Economics, 6*(4), 291–297.
Hassink, R., & Wood, M. (1998). Geographic clustering in the German opto-electronics industry: Its impact on R&D collaboration and innovation. *Entrepreneurship and Regional Development, 10*(4).
Hecker, D. (1999) High-technology employment: A broader view. *Monthly Labor Review* (June).
Jaffe, A. B., Trajtenberg, M., & Henderson, R. (1993). Geographic localisation of knowledge spillovers as evidenced by patent citation. *Quarterly Journal of Economics, 108*(3).
Keeble, D. (2000). North-South and urban-rural differences in SME performance and behaviour. In: A. Cosh, & A. Hughes (Eds), *British enterprise in transition: Growth, innovation and public policy in the small and medium sized enterprise sector 1994–1999*. ESRC Centre for Business Research, University of Cambridge.
Keeble, D., Lawson, C., Lawton-Smith, H., Moore, B., & Wilkinson, F. (1997). Internationalisation processes, networking and local embeddedness in technology intensive small firms. ESCR Centre for Business Research, Working Paper Series No. 53, University of Cambridge.
Keeble, D., Lawson, C., Lawton-Smith, H., Moore, B., & Wilkinson, F. (1998). Collective learning processes and inter-firm networking in innovative high tech regions. ESCR Centre for Business Research, Working Paper Series No. 86, University of Cambridge.
Kitson, M., & Wilkinson, F. (2000). Markets, competition and collaboration. In: A. Cosh, & A. Hughes (Eds), *British enterprise in transition: Growth, innovation and public policy in the small and medium sized enterprise sector 1994–1999*. ESRC Centre for Business Research, University of Cambridge.

Klein Woolthuis, R. J. A. K. (2000). *Sleeping with the enemy: Trust, dependence and contracts in inter-organisational relationship*. University of Twente, The Netherlands, FebroDruk.

Lawson, C., Moore, B., Keeble, D., Lawton-Smith, H., & Wilkinson, F. (1997). Inter-firm links between regionally clustered high technology SMEs: A comparison of Cambridge and Oxford innovation networks, ESCR Centre for Business Research, Working Paper Series No. 65, University of Cambridge.

Oliver, N., & Blakeborough, M. (1998). Innovation networks: The view from inside. In: J. Mitchie, & J. Grive Smith (Eds), *Globalisation, growth and governance*. Oxford: Oxford University Press.

Porter, M. (1990). *The competitive advantages of nations*. London: Macmillian.

Pyke, F., Becattini, G., & Segenberger, W. (1990). *Industrial districts and inter-firm co-operation in Italy*. Geneva: International Institute for Labour Studies.

Quince, T. A., & Whittaker, D. H. (2002). High technology small businesses and niche markets. ESCR Centre for Business Research, Working Paper Series No. 234, University of Cambridge.

Roberts, E. B. (1991). *Entrepreneurs in high technology: Lessons from MIT and beyond*. New York: Oxford University Press.

Segal, Quince, Wickstead Ltd. (2000). *The Cambridge phenomenon revisited*. Cambridge: SQW.

Storey, D. J. (1994). *Understanding the small business sector*. London: Routledge.

Storper, M. (1995). The resurgence of regional economies, ten years later: The region as a nexus of untraded interdepedencies. *European Urban and Regional Studies*, 2(1), 199–221.

U.S. Small Business Administration Office of Advocacy (SBA) (1999). *The facts about small business 1999*. U.S. Small Business Administration Office of Advocacy, Washington.

Vaessen, P., & Keeble, D. (1995). Growth-oriented SMEs in unfavourable regional environments. ESCR Centre for Business Research, Working Paper Series No. 6, University of Cambridge.

Whittaker, H. (1999). Entrepreneurs as co-operative capitalists: High tech CEOs in the U.K. ESCR centre for business research. Working Paper Series No. 125, University of Cambridge.

Wood, E. (1997). SME innovator types and their determinants. ESCR Centre for Business Research, Working Paper Series No. 72, University of Cambridge.

Appendix

Table A1: Characteristics of the businesses and of the participants.

	%
Type of business (*n* = 235)	
Completely new start	65
Spin out/off	16
MBO/MBI/acquisition	15
Other	4
Size employment (*n* = 232)	
<20 employees	37
≥20 employees to <50 employees	37
≥50 employees	26

Table A1: (*Continued*)

	%
Age (*n* = 233)	
Newer: founded since 1990	37
Founded between 1980 and 1989	28
Older: founded before 1980	35
Gender (*n* = 234)	
Male	92
Female	8
Qualifications (*n* = 210)	
Post graduate degree	32
First degree	32
Vocational/professional (*n* = 235)	53
Sectors (*n* = 236)	
Manufacturing	58
Services	42
Computer activities and telecommunications	19
R&D and technical services	23
Size Turnover (*n* = 215)	
<£1 million turnover	38
≥£1 million to <£5 million	45
≥£5 million turnover	17
Ownership	
Real and "quasi" sole proprietors	29
Shared (internal only)	36
Shared with external owners	34
Age of CEO (*n* = 234)	
Younger ≤45	30
Middle aged >45 to ≤55	39
Older >55	31
Experience of working in HEI/research/similar (*n* = 160)	25

Table A2: Niche market indicators.

Dependence	On One Customer for (n = 213) (%)	On Three Customers for (n = 218) (%)
<10% of turnover	31	10
≥10 to <50% of turnover	58	57
≥50% of turnover	11	33

Perception of Competitors

Number of Serious Competitors (n = 214) (%)		Proportion of Larger Competitors (n = 198) (%)	
<5	47	<50	21
≥5 to <10	25	≥50 to <100	25
≥10	28	100	54

Table A3: Competitive advantages.

Competitive Advantages (n = 222)	Mean Score
Product/service	3.95
Quality of product/service	4.34
Technological/scientific expertise	3.94
Specialised product/service	3.83
Design of product/service	3.65
Delivery	3.84
Personal attention/responsiveness to client needs	4.40
Speed of service	3.97
Price/cost advantages	3.19
Marketing	3.32
Established reputation	4.14
Being first in the market with new products/services	3.01
Marketing and promotion	2.87
Other	3.27
Range of products/services	3.27

Table A4: Performance.

Growth	Turnover (*n* = 208) (%)	Employment (*n* = 206) (%)
Contraction/no growth	37	49
Low growth for size	19	15
Average growth for size	24	20
High growth for size	20	16
	Percentage	
Composite growth (*n* = 219)		
Contraction/no growth (in either turnover or employment)	29	
Low/uneven growth (contraction/no growth in either turnover or employment)	22	
Moderate uneven growth	26	
High balance growth (average or high growth in both turnover and employment)	22	
Long term growth (*n* = 151)		
Low for age	29	
In line with age	44	
Good for age	27	
Innovation (*n* = 219)		
Incidence (*n* = 219)		
Innovator	81	
Level		
Low level (non novel innovators)	26	
High level (novel innovators)	56	
R&D Spending as a percentage of turnover (*n* = 221)		
None	39	
Less than 10%	43	
≥10%	18	

Chapter 11

On the Role of High-Technology Small Firms in Cluster Evolution

Olav R. Spilling and Jartrud Steinsli

Introduction

In his opening chapter of the book "Technology-based firms at the turn of the century", Metcalfe (1999) presents an analysis of the basic mechanisms of technological evolution in the capitalist economy. With his title "Restless capitalism, experimental economies", his article indicates essential aspects of the dynamics of capitalism; in which the system is continually evolving, and where a steady state of equilibrium does not exist. Change is something that happens all the time, and the system is open-ended in a way that future evolution cannot be predicted. Metcalfe characterises capitalism as a "developmental system *par excellence*" and by this follows in the tradition of Schumpeter (1934–1996), who characterised the capitalist economic system as a method of economic change and development. Schumpeter pointed to the entrepreneur as the main agent of change, who, by organising new business activities based on "new combinations," contributes to the disturbing of the "circular flow" in the current economy, and in this way represents a dynamic element that drives the system forward.

Formation of new firms provides significant contributions to evolutionary processes. At any time a large number of entrepreneurs introduce new businesses. Each entrance, and each step of development in an existing firm, may be regarded as a contribution to the evolution of the system as a whole. As knowledge may be regarded as the most important resource in the economy, and the process of learning as the most important process for development (Asheim 1996; Lundvall 1992), the evolution of capitalism may be regarded as based on processes of collective learning (Camagni & Capello 2000; Keeble & Wilkinson 2000). From this perspective, each new step may be more or less regarded as based on previous events and each step will further add new knowledge to the system. The system is continuously evolving and is restless (Metcalfe, 1999).

New Technology-Based Firms in the New Millennium, Volume III
Copyright © 2004 by Elsevier Ltd.
All rights of reproduction in any form reserved.
ISBN: 0-08-044402-4

Small Firms and the Evolutionary Approach

In recent years, a number of studies have been published in the field of cluster evolution (cf. for instance Garnsey 1998; Isaksen 2001; Keeble & Wilkinson 1999; Kuijper & van den Stappen 1999, 2000; OECD 1999, 2001; Wintjes & Cobbenhagen 2000). Many of these are particularly focused on the development in the high-technology area, and on the role of small firms (Audretsch 2001a, b; Dahlstrand 1999, 2000; Oakey 1999; Vatne & Taylor 2000).

The purpose of an evolutionary approach is to explain how changes occur over time, and what are the basic mechanisms of evolution. Based on Schumpeter's ideas of the entrepreneur as the agent of change (Schumpeter 1934–1996), the firms they are starting, mostly small ones, may be regarded as the main vehicles of implementing such change.

However, one can hardly speak about "the evolutionary approach" as if there was only one approach (Verspagen 2000), more it is a tradition, based on many different elements. But as pointed out by many different authors, there are some common elements that often are referred to such as variation, selection, retention and diffusion, or other concepts closely related to these (cf. Aldrich 1999; Edquist 1997). Edquist (1997: 6) follows up on this by identifying the following elements as essential parts of an evolutionary approach:

(1) The existence and reproduction of entities, like genotypes in biology or certain technologies and organisational forms in innovation systems.
(2) A mechanism for introducing novelty in the system, this may include significant random elements, but may also produce predictable novelties.
(3) A mechanism for selecting among entities present in the system, like "natural selection" or selection based on market mechanisms.

When analysing the role of small firms in this context, it is important to be aware that there will be problems with what is the most adequate unit of analysis. Applying concepts like "role" and "firm" may easily give the impression that these are static units, i.e. that the "role" of a firm is something stable during its life-time, and that the "firm" as a unit is also something constant. However, from the firm formation literature on population dynamics, it is well recognized that a firm is often a temporary unit, with a limited life expectancy, and during its life cycle a lot of significant changes may occur. Thus, Taylor (1999) has analysed the small firm as a temporary coalition. The resources that are organised in a company and its network may change significantly over time. When comparing the resources of the firm at different points in time it may reveal that the "firm" actually is two very different organisations, or coalitions, although the legal unit may be the singular.

Furthermore, the structure of the legal units involved in the development may change significantly. In her analysis of small and large firms in cluster evolution, Dahlstrand (1999, 2000) points to the fact that small firms on the one hand often are created through processes of spin-offs from larger firms, and later may be acquired by other firms if the business idea turns out to be successful. More generally, changes of legal unit may occur through different types of processes; for instance through mergers and acquisitions, or through other changes of the ownership.

In their analysis of characteristics of the development in Silicon Valley, Bahrami & Evans (1995) emphasize the importance of a high failure rate of new firms, and that processes of "recycling" of business ideas and other kind of business resources are continuously going on. Another aspect of this may be termed "re-starts," in which a new business, that started out based on an initial idea does not to work, and thus the business model will have to be "re-calibrated" for instance by introducing a new management team (p. 71). Furthermore Bahrami & Evans describe how different resources, in particular human capital, is "floating" around in the system and virtually continuously forming new constellations and new business units, and in this way contributing to a continuous process of re-cycling.

Thus, one way of regarding new, non-imitative small firm start-ups, is that their primary roles are to test out new business ideas. This approach was suggested in 1975 by Ramström (1975), and it is in line with the more recent understanding of the capitalist system as an experimental economy (Eliasson 2000; Metcalf 1999).

There are many possible outcomes for a new high-tech firm. In many cases the new firm will close after a short time when it turns out that the business idea is not viable. Nonetheless, the attempt of starting the new firm may be important to processes of collective learning as some experiences are developed (and the actors involved and other actors observing what is going on), may take advantage of these experiences in their future business strategies. And in line with what is referred to above (Bahrami & Evans 1995), such attempts may be the basis for the re-cycling of resources, and provide inputs for new processes of firm formation and development.

Of course, there may be more "positive" outcomes of a new start-up as well in the sense that the new business idea may turn out to be profitable and sustainable. In some cases, but not very many, the new business idea may reveal significant growth potential and can either be interesting as a candidate for acquisition by a large firm, or develop itself into a new and large independent firm. Other, and more likely, outcomes are that the firm will remain small, for instance based on serving a specialised market niche, or providing services to the local market.

However, in all cases it should be kept in mind that it is a temporary coalition with a limited life expectancy. Within firms, processes of reconfiguration of resources are going on continuously, and also at the firm level there will be processes of restructuring, implying that legal units may be reorganised. So, there is a great challenge to research in the field to develop adequate approaches for analysing these processes of evolution.

Methodology

The purpose of this paper is to contribute to the analysis of evolution of high-technology clusters by studying the role of high-technology small firms in two different Norwegian clusters in Oslo, the capital of Norway, and in Trondheim, which traditionally has been recognised as the "capital of technology" with its almost one hundred years' tradition of hosting the main technological university of the country.

As a part of this project, a survey was organised in Oslo and Trondheim, and the collected data provide an opportunity to look into some aspects of the role of small firms and their contributions to dynamic processes. However, when a survey is organised to provide insights into evolutionary mechanisms, there are several issues that should be addressed.

First, it may be commented that the best way of analysing cluster evolution, is to organise a longitudinal study, and follow a group of actors; partially actors such as individual entrepreneurs and members of entrepreneurial teams, partly formal institutions like firms, larger corporations and other bodies. As constellations of actors and firms may change continually (Bahrami & Evans 1995; Taylor 1999), it is important to be flexible regarding which units are included in the analysis. A cluster is an emerging field of business activity (Garnsey 1998), i.e. it is the "field" or cluster as a whole, or specific parts of the cluster, that should be the main focus of our interest, not the particular individual actors *per se*. As demonstrated through a case study of the development of the Internet technology in Oslo, there may be a whole complex of actors that are involved in the development — and there is a significant variation regarding which actors are involved in what ways at various stages of the development (Steinsli & Spilling 2002).

By its nature, a survey based on questionnaires, will be cross sectional and only provide information on specific actors at a certain point in time. However, as the alternative of a longitudinal study will require a lot of resources, one will often be confined to do some kind of cross-sectional analysis by collecting data from a sample of firms at specific points in time and try to reconstruct what has happened. When doing so, we have tried to draw upon previous experiences from the analyses of the Cambridge region (Keeble *et al.* 1999).

Second, we have the problem of choosing the appropriate unit of analysis: Is the most adequate level of analysis the individual entrepreneur or entrepreneurial teams; is it the firm, a group of firms, or, perhaps a whole agglomeration or cluster? Or should it be a more specific field of technology or knowledge? In our survey we have identified relevant populations of firms and contacted the managers of these firms. As a supplementary approach, we have also organised a case study in which we have reconstructed the development of a particular field of technology and identified all the individual actors in some kind of entrepreneurial position within this development. This is presented in a different paper derived from the project (Steinsli & Spilling 2002).

Third, we have the issue of what is high technology. Without going into too much detail, we will just comment that our approach is based on the much quoted article of Butchart (1987) and later modifications in which some service sectors have been included (Keeble & Wilkinson 2000, for further comments see Spilling & Steinsli 2003).

The relevant populations of high-technology firms in Oslo and Trondheim were identified through the databases of Statistics Norway and CreditInform by selecting all firms in the specified sectors with more than two, and less than a hundred employees in 1999. A sample

Table 1: Total population of firms, firms contacted and responses.

	Oslo	Trondheim	Total
Total population	288	145	433
Contacted	200	103	303
Not willing to participate	23	6	29
Received responses	81	36	117
Non-responses	96	61	157
Response rate	41%	35%	39%

of these firms was contacted by telephone, and those willing to participate in the survey, were sent the questionnaire, followed by one reminder. This gave a total of 117 acceptable responses, and a response rate of 39% for the two cities (see Table 1). Although we would have preferred a minimum response rate of 50%, the actual response rate may be regarded as fairly acceptable.

The Role of Small Firms in the Clusters of Oslo and Trondheim — The Static View

The two Norwegian cities of Oslo and Trondheim were selected for analyses because of their important roles in high-technology development in Norway.

Oslo, the capital of Norway, has a total population of about 900,000 people including the neighbouring municipalities. Of the total national employment in high-technology sectors of 102,000 workers, about 48% (48,800) is located in the Oslo region. The concentration of ICT firms is even higher, as about 61% (36,900) of the total employment in this sector is concentrated in Oslo.

Trondheim, on the other hand, is a much smaller city situated further north, with approximately 200,000 inhabitants when the neighbouring municipalities are included. The internationally renowned University of Science and Technology (NTNU) is located here, and the university and research community includes Scandinavia's largest independent research institute — the SINTEF group. In spite of this, the total employment in high-technology industries is only 6,400 workers, or about 6% of the total national employment. Within the ICT sector, Trondheim's share of the national employments is about 4.8%, while its share of the rest of the high-technology sectors is higher, i.e. 8.5%. When taken into consideration their population shares, the relative share of high-tech employment in the Oslo region is higher than that of Trondheim.

On average, smaller firms (with less than one hundred employees) account for about 50% of the total high-technology employment in Norway. There are significant differences between industrial sectors; larger firms dominate in sectors such as chemicals/pharmaceuticals and telecommunications, while smaller firms are much more prevalent in specialised retailing and wholesale activity, software development and specialised consultancy services with employment shares up to 60–80%.

While the size structure differs between the different sectors, it does not vary between regions. In the case of Oslo and Trondheim, the share of small firm employment is approximately the same (i.e. about 50%), and this implies that the sectoral determinants of the size structure is much more important than the influences of the region.

Firms covered by the survey were classified according to the main functions they perform. The classification is based on a value chain or production system perspective. The idea of the classification is that different companies tend to specialise in certain fields (Table 2).

The data shows that many of the firms have more than one role in the production system. In fact, among the 117 firms, only 52 firms (44%) reported one function, two functions were reported by 36 companies (31%) and the remaining 29 companies (25%) reported three or more functions. Thus, the majority of the small companies covered by the survey represent mostly what we may call multifunctional activities.

Table 2: Survey firms classified according to the functional roles ($N = 117$).

Standard products to end-user	19%
Specialised products to end-user	23%
Subcontractors	12%
R&D-services	15%
Software and system development	24%
Data and information services	17%
Consultants	57%
Sales	27%

Note: More than one answer possible, number of responses add up to more than 100%.

In this respect there is no significant differences between the companies in Trondheim and Oslo. However, when looking at the market orientation of the firms, some important differences are revealed. First, a higher proportion of the firms based in Oslo are exporters. In total 46% of the firms investigated produce goods or services for export. However, among exporters, only a smaller share of their sales are exported, and in the total sample, only 9% of the firms exported more than 50% of their sales. The share of non-exporting companies in Oslo was 47%, while the share of non-exporters in Trondheim was as much as 70%. Second, the importance of the local market is much more significant in the case of Oslo than Trondheim. For Oslo, 70% of the companies reported that their most important customers were in the local area, while only 28% of the Trondheim companies claimed the same.

These data indicate that the roles of smaller firms in the two clusters are much the same — they are accounting for approximately the same share of total activities and they are multifunctional. But in terms of market orientation, firms located in Oslo are more oriented towards the local market as well as the export market.

However, when it comes to data regarding how the relationships were organised inside the cluster, the differences are very significant. Firms in Oslo, in most cases, had their most important customers in the local area, while this is not so in the case of Trondheim.

The Role of Small Firms in Cluster Evolution — The Dynamic View

The majority of the firms included in the survey are of recent origin since about 60% were established during the 1990s, and around one third were established in 1996 or later. Mostly, the firms have been developed independently of other organisations (Table 3) while close to 60% of the companies report that their business idea has been developed independently. A smaller share of 29% or respondents reported that the idea had been developed in other companies or institutions. Ten percent of the cases were developed in collaboration with other firms or organisations.

As Table 4 shows, a significant share of the founders (75%) has backgrounds as previous employees of another firms, either as managers (33%) or in non-managerial position (42%). Among the remaining founders, only 12% were previously employed in the university and R&D sectors, while the rest were either unemployed or students.

Table 3: Development of the business idea.

Development of the Business Idea	Oslo	Trondheim	Total
Independent	57.5	54.3	56.5
In another company or institution	28.8	11.4	23.5
In collaboration with another company	10.0	31.4	16.5
Other	3.8	2.9	3.5
Total	100.0	100.0	100.0
N	80	35	115

It is important to notice that these data concern the role of the founders immediately before start-up. When asking about their more general background, the share of founders with R&D backgrounds increases to 26%, while the most significant group in this case turns out to have background in marketing (50%). In 29% of the cases, founders had a management background, while 28% had backgrounds in production. Since a team rather than an individual entrepreneur starts many of the companies, different backgrounds may be combined, and the percentages added up to significantly more than one hundred.

There are two points that may be made here. First, a fairly high share of the companies reported that their business idea was developed independently of existing organisations. Thus just a smaller share of start-ups may be regarded as spin-outs from existing organisations. Less than a quarter of the companies reported that they were based on existing firms or organisations.

One issue of special interest is the role of universities and R&D institutions as incubators of new firms, and to what extent one can talk about direct spin-outs from these kind of institutions. In the literature on cluster evolution, this route is regarded as a very important mechanism (Keeble & Wilkinson 2000; Segal Quince Wicksteed 1990, 2000). According to our data, the links to these institutions are rather weak. Among those reporting that their

Table 4: The role of the main founder before start-up.

Role of the Main Founder Before Start-Up	%
Unemployed	7.7
Student	3.4
University/R&D	12.0
Employed (not manager) in another company	41.9
Manager of another company	33.3
No information	6.0
Total	100.0
N	117

business idea was developed in another organisation, the vast majority specified that the idea had been developed in another firm (mostly larger companies), while in only a few cases was the incubator organisation a university or R&D institution. As a matter of fact, only eleven companies (9%) reported to have either been developed in a university, a science park or a research institute.

Based on these findings, it may be questioned what actually are the relationships between the academic institutions and the business community? Obviously, these findings do not indicate very impressive links. However, it is important to have a broader perspective on this. The total influence or contact patterns between universities and local firms may prove to be much more significant when one takes into account the total number of links between academic institutions and new firms. This may be discussed along two dimensions.

First, there may be indirect links. The founders may have their backgrounds in academic institutions, although they may not have directly served as incubator organisations. As reported from the survey, 12% of the companies reported that their founders were employed in an academic institution prior to start-up, and when asking more generally about the background of founders, as many as 26% reported the R&D sector as background for at least one of their founders.

Second, links to academic institutions may be even more indirect, since the commercialisation of academic knowledge may go through several stages of firm formation and processes of sequential entrepreneurship, as demonstrated in a case study of the development of Internet businesses in Oslo (Steinsli & Spilling 2002).

Third, other forms of links between academic institutions and the business community may be provided through former students that are employed, or by different forms of informal contacts. Thus, it is not easy to make a total assessment of the interface between the academic institutions and the business community.

However, this should not be an excuse for not analysing the opportunities for improving relationships and take more advantage of the potential for more spin-off firms (RITTS 2000).

To follow up on different aspects of evolutionary processes, the firms were asked to what extent they had been involved in, or contributed to, different types of processes. Data in Table 5 show that a diversity of processes had occurred. More than 20% of firms had been through a process of merger with another company, and a similar share had partially or fully acquired another company. As many as 32% of firms had contributed to the start-up of a new firm, while in 27% of cases, employees had left a firm to start a new business.

Table 5: Share of firms reporting different types of evolutionary processes.

Merged with another company	23.9
Acquired other companies (partially or fully)	22.2
Licensed *in* the right to other production	5.1
Developed and licensed *out* production rights	11.1
Contributed to a new start-up	31.6
Employees have left to start new business	26.5
N	117

The phenomenon of licensing out production rights does not seem to be of the same importance among these firms, since only 11% reported to have licensed *out* production rights, while 5% had licensed *in* similar rights.

Altogether, however, the data indicate that the pattern of development is fairly complex; evolution is diversified and is constituted by a number of processes in which restructuring is going on partly by forming new independent organisations, partly by spinning out new businesses from existing ones, or by a steady process of restructuring through mergers and acquisitions.

Small Firms and Innovation

According to the survey, there was a high level of innovation activity. In total, close to 60% of the companies reported to have organised R&D-activities internally during the previous year, and a significant share of the firms had also acquired R&D-services externally. When asked about innovation activities, virtually all firms reported to have performed some kind of innovation during the last three years. Only 7% did not report any kind of innovative activities as listed in Table 6. Most of the innovation activity was product oriented; 65% had developed new products or services, and 70% had improved products and services. In total, close to 80% of the companies had been involved in some kind of product related innovation. Significant shares had also been involved in other forms of innovation, among which application of new software or system solutions were the most frequent forms. Two thirds of the companies had applied new software or system solutions.

When analysing the importance of different actors in the innovation process, a rather typical picture is revealed in which the firms report that the most important contacts for the innovation process were organised along the value chain, with customers and suppliers as the most important partners (Figure 1). Interestingly, however, competitors are also of great importance to the innovation process, which implies that there are significant contacts and flow of information between firms although they are competing in the same markets.

Table 6: Share of firms with R&D-activities and innovation activities ($N = 117$).

R&D Activity Last Year	%
Own R&D activity	58
Acquired R&D-services	25
New products or services	65
Improved products or services	70
Process innovation	41
Innovation in marketing/sales	29
New markets	45
Applied new technology	44
Applied new software or system solutions	67

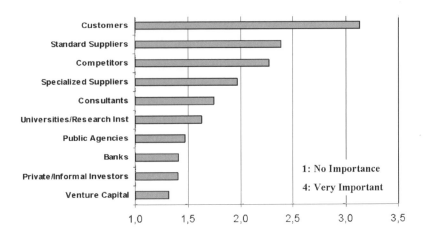

Figure 1: Importance of different actors in the innovation process.

On the other hand, other groups of actors are of less importance; consultants, universities and research institutes, public agencies, banks and other capital providers were not rated as important for the innovation process. However, it may be commented that these scores are based on the averages for all firms, and that there may be significant variations between different firms regarding which actors are important. We have two comments related to this.

First, comment may be made on why universities and R&D institutions are scoring so low in terms of contracts. As commented earlier in this paper, links between these institutions and the firms may be indirect, so many firms will not recognize them as important to the innovation process. However, a smaller share of the firms (i.e. around 15%), reported that these institutions were of high or very high importance. Thus, for some of the firms there were close interactions.

Second, comment is appropriate on the role of capital providers. It will be of no surprise that the banks in general are of low importance to the innovation process. More interesting is the finding that venture capital, and private investors, also score very low. Here, however, it may be kept in mind that such actors generally have small shares in start-ups. In fact, the data show that respectively only seven and 14% report high or very high importance of venture capital or private investors in the innovation process.

In summary, the general conclusion is that all types of actors present in the innovation system may be important to the innovation processes, but it varies as to what extent specific firms collaborate with different actors.

To shed more light on processes of innovation and the role of small firms we have organised a factor analysis of the importance of different actors in the innovation process. Three factors were identified, which provide a basis for grouping the firms according to what is their role in the innovation system (Table 7).

The three groups may be identified as: (1) the R&D based innovator; (2) the competition based innovator; and (3) the supplier based innovator, where the concepts applied indicate which parts of the innovation system is the most important to the group.

Table 7: Factors for identifying different types of innovators.

	Component		
	1	**2**	**3**
Standard suppliers			0.820
Specialized suppliers			0.768
Customers		0.603	0.316
Consultants		0.793	
Competitors		0.769	0.241
Universities/R&D	0.658		
Venture capital	0.803		
Private investors	0.771		
Banks	0.389		0.346
Public programs	0.650		

Note: Extraction method: Principal Component Analysis. Rotation Method: Varimax with Kaiser Normalization. For further details, see Spilling & Steinsli (2003, forthcoming).

In the case of the R&D-based innovators, the innovating firms have strong links to R&D institutions and obviously are working in close interaction with these institutions. Interestingly, this group also have close connections to the providers of risk capital, i.e. venture capital and private investors, and also public agencies providing financial resources. Probably, this group is oriented towards radical innovations, which require access to significant financial resources.

The second group, the competition oriented innovators, have strong links with customers, competitors and consultants, while other links are of less importance. The third group, the supplier based innovators, has their strongest links with their suppliers, and, the standard suppliers have a slightly higher factor score than the specialised suppliers.

In total, these groups of innovators cover different and supplementary parts of the innovation system, and it will be of interest to conduct more research to further develop the typology and develop more detailed insights regarding what characterises the different groups.

When comparing the data on innovation activities for firms located in Oslo and Trondheim, no significant differences were revealed, so it may be concluded that the processes of innovation in these two centres are very much the same. Approximately the same shares of firms are involved in the same types of innovation activities, and on average, the importance of different actors turns out to be very much the same.

Even when factor analyses are done separately for the firms from the two cities, virtually the same types of factors are identified, thus indicating that the typology suggested here is fairly robust. However, this should be further investigated by a larger and preferably more detailed dataset.

Although the process of innovation turns out to be virtually the same in the two cities, the data reveal significant differences between Trondheim and Oslo regarding to what extent actors of importance to the innovation processes are located in the city or not. Not surprisingly, the firms located in Oslo report that significantly higher shares of actors important to the innovation process are located in their region. Thus, firms in Oslo have the advantage of having a higher share of their important competitors, suppliers, customers and different service providers' locally compared to their colleagues in Trondheim. There is only one important exception to this, and that is the university and R&D institutions. Trondheim firms report closer links to the local institutions than is the case for Oslo. This indicates that generally the cluster of Oslo has better developed local relations, with the exception of the links to the university and R&D sector, which seems to be better developed in Trondheim.

The Clusters of Oslo and Trondheim Compared to Cambridge

In this section we comment on the situation of the clusters in Oslo and Trondheim compared with data from the Cambridge region. The growth and the development of the Cambridge region high-technology cluster has been given substantial attention during recent years (Garnsey 1998; Keeble, 1989; Keeble *et al.* 1999; Keeble & Wilkinson, 2000; Segal Quince Wicksteed 1985, 2000). A wide range of high-tech service and industries are located in Cambridge, and approximately 40,800 persons are employed in high-tech businesses. Historically, the cluster is focused on the University of Cambridge with its global reputation for research and scientific activity. This is reflected in the fact that the dominant sector is research and development, with 27% of the high-tech employment, while the second most important sector is electrical and electronic engineering, with 12.2% of total employment: Approximately 43% of this workforce is employed in companies with less than 100 employees. Engineering, chemicals and biotechnology have a high share of larger firms (Segal-Quince-Wicksteed 2000).

The data for the Cambridge region is based on a survey undertaken in 1996 by the ESRC Centre for Business Research (hereafter called the CBR survey) of 50 technology intensive SMEs, based on a stratified random sample designed to produce a representative balance of high-tech firms between manufacturing and services, but with inclusion of a somewhat higher proportion of larger SMEs than very small firms (Keeble *et al.* 1999).

In Cambridge 88% of high-technology SMEs were identified as spin-offs or new start-ups. In Oslo and Trondheim 80% of the high-technology firms were either spin-offs or new start-ups. Compared to Cambridge, the share of spin-outs and new start-ups are slightly less in the Norwegian study. However, the relatively high degree of new start-ups and spin-outs among the surveyed high-tech firms both in Cambridge and in Norway, imply a considerable diffusion of embodied knowledge from the "incubating" firms or institutions, through the set up of new firms, either by former employees or as spin-offs of existing firms.

In Cambridge 70% were founded by entrepreneurs formerly working for another company, while 25% of the chief founders were employed either by a university or a research institution prior to start-up, while in the Norwegian study 75% of the founders had previously been employed by another firm, while 12% had background from a university or a research institute. The role of Cambridge University as generator of new business ideas in the Cambridge region seems therefore more important than the similar roles of the universities (The University of Oslo and NTNU) for the development of the Oslo and Trondheim

Table 8: Comparison of data for the high-tech clusters of Cambridge and Oslo and Trondheim.

	Cambridge	Oslo and Trondheim
Spin-outs		
By another firm	12%	23.5%
As a spin-off	32%	56.5%
As an independent start-up	56%	16.5%
In collaboration with another company or institution		3.5%
Other		
Total	100%	100%
Previous employment of founder prior to start-up		
Manager/employed in another company	70%	75%
Employed by university/research laboratory	25%	12%
Self employed/unemployed/student	5%	11%
Total	100%	100%
External actors of importance to the innovation process[a]		
Customers	Important	Important
Standard suppliers	Some importance	Important
Competitors	Important	Important
Specialised suppliers	Some importance	Some importance
Consultants	Little importance	Little importance
University/research institutes	Important	Little importance
Public agencies	n.a.	None or little importance
Banks	n.a.	None or little importance
Private/informal investors	n.a.	None or little importance
Venture capital	n.a.	None or little importance

Table 8: (*Continued*)

	Cambridge	Oslo and Trondheim
Location of actors of importance in the innovation process	Main actors located in the rest of the U.K. Of actors of importance located in the Cambridge region the University and customers are most often mentioned as important.	In Oslo the main actors are located in the Oslo region, while in Trondheim the main actors of importance are located nationally.

[a] The questions asked in the Oslo/Trondheim study are not exactly comparable to the CBR study. The rating of the factors of importance in the Cambridge region is therefore based on our judgement.

regions. However, the difference is not as high as one might have expected, given the focus on Cambridge University as generator of the high-tech milieu in Cambridge. These results can, nonetheless, as discussed earlier in this paper, underestimate the importance of the universities, since many firms may have indirect links to a university, since first generation spin-outs from the university have themselves spawned new companies and so on. This seems also to have been the case in Cambridge. Even where the parent companies have not come from Cambridge University, the latter has constituted a basic reason for the organisation concerned to stay in the Cambridge region (Segal Quince Wicksteed 1985).

There are some differences with respect to what kind of external actors are of importance in the innovation process; however, in Cambridge as well as in Oslo and Trondheim, the customers are the most important external actors in the innovation process. Competitors are also emphasised as important in both surveys. In the Cambridge survey universities are also noted as important for the innovation process, while in the Norwegian study the universities are not regarded as an important actor in the innovation process (Table 8).

The two studies vary regarding location of the most important external sources for the innovation process. In Cambridge a higher proportion of firms or institutions rated as important for the innovation process, are located in the rest of U.K. or globally, than within the Cambridge region itself. Among the sources that were rated as important for innovation activity, such university resources were most likely to be located within the Cambridge region. But also in this case, universities located elsewhere in the U.K. were more often of importance than those located in Cambridge.

National and global innovation networks are thus notably more frequently rated as important than local networks in Cambridge for the innovation process. In the study of the Norwegian high-tech SMEs the results vary between Oslo and Trondheim. In Trondheim the national or international located actors are of larger importance for the high-tech businesses, than the regionally located actors (with the exception of university and R&D actors). For the high-tech businesses located in Oslo, the most important actors for their innovation process were located in the region.

Summary: The Role of Small Firms in Cluster Evolution

This paper has shed some light on the role of small high-technology firms in cluster evolution. Based on a static view, small firms on average account for about 50% of all high-technology employment in Norway. However, the sectoral differences are very significant. Based on a dynamic view, small firms show a diversified pattern, and contribute to evolution in different ways, partly through independent start-ups or spin-outs from other firms or organisations, partly through being parts of processes of restructuring, as for instance mergers and acquisitions. In this way small firms contribute to the development of the capitalist system, in addition to what a static view of the sector might imply, as described among others by Bahrami & Evans (1995) and Metcalfe (2000).

Small high-tech firms are highly innovative, and virtually all firms that participated in the survey were involved in some kind of innovative activity. Most of them were product oriented in their innovation, but significant shares were also involved in process innovation, market innovation or the implementing of new technology or systems solutions.

By means of factor analysis, a typology of innovating firm, or rather of innovation practices, has been suggested, by which the firms could be grouped according to which actors in the innovation system were of most importance to the innovation process. The three groups are the R&D-based innovators, the competition based innovators and the supplier based innovators.

A number of new research issues may be suggested based on the findings reported here, from among which we will very briefly make two comments.

First, more efforts should be given to the further development and testing out of the typology of innovating firms, partly by identifying what are the characteristics of the firms forming the different groups and partly by collecting more detailed data on firm characteristics and innovation practices. This may improve our knowledge on how innovation systems work.

Second, and more demanding, is the question of what is the role of smaller firms in cluster evolution compared to the role of larger firms. The present study is obviously restricted by methodological constraints. As in many other studies of small firms, there is no systematic comparison between the role of small firms and the larger ones. Without that kind of comparison, it is not possible to say what is the significance of smaller firms in relation to larger firms and to what extent do they supplement each other or take complementary roles. To what extent do they compete, and what are their distinctive roles in the innovation system and in other evolutionary processes?

References

Aldrich, H. (1999). *Organizations evolving*. Thousand Oaks, London and New Delhi: Sage.

Audretsch, D. B. (2001a). Research issues relating to structure, competition, and performance of small technology-based firms. *Small Business Economics, 16*, 37–51.

Audretsch, D. B. (2001b). The role of small firms in U.S. biotechnology clusters. *Small Business Economics, 17*, 3–15.

Bahrami, H., & Evans, S. (1995). Flexible re-cycling and high-technology entrepreneurship. *California Management Review, 37*(3), 62–89.

Butchart, R. L. (1987). A new U.K. definition of the high technology industries. *Economic Trends, 400*, 82–88.

Dahlstrand, Å. L. (1999). Industrial dynamics and ownership changes — incubation and acquisition of small technology-based firms. In: B. Johannisson, & H. Landström (Eds), *Images of entrepreneurship and small business — emergent Swedish contributions to academic research*. Lund: Studenlitteratur.

Dahlstrand, Å. L. (2000). Large firm acquisitions, spin-offs and links in the development of regional clusters of technology-intensive SMEs. In: D. Keeble, & F. Wilkinson (Eds), *High-technology clusters, networking and collective learning in Europe* (pp. 156–181). Ashgate: Aldershot.

Edquist, C. (1997). Systems of innovation approaches — their emergence and characteristics. In: C. Edquist (Ed.), *Systems of innovation. technologies, institutions and organizations* (pp. 1–35). London and Washington: Pinter.

Eliasson, G. (2000). *The role of knowledge in economic growth*. Stockholm: KTH.

Garnsey, E. (1998). *The genesis of the high technology milieu*. Cambridge: University of Cambridge.

Isaksen, A. (2001). Regional clusters between local and non-local relations. A comparative European study. Paper prepared for the IGU Conference on "Local Development: Issues of Competition, Collaboration and Territoriality", Turin (July).

Keeble, D. et al. (1999). Collective learning processes, networking and 'institutional thickness' in the Cambridge region. *Regional Studies*, *33*(4), 319–332.

Keeble, D., & Wilkinson, F. (1999). Collective learning and knowledge development in the evolution of regional clusters of high technology SMEs in Europe. *Regional Studies*, *33*(4), 295–303.

Keeble, D., & Wilkinson, F. (2000). High-technology SMEs, regional clustering and collective learning: An overview. In: D. Keeble, & F. Wilkinson (Eds), *High-technology clusters, networking and collective learning in Europe*. Ashgate: Aldershot.

Keeble, D., & Wilkinson, F. (Eds) (2000). *High-technology clusters, networking and collective learning in Europe*. Ashgate: Aldershot.

Kuijper, J., & van den Stappen, H. (1999). Clusters and clustering: Genesis, evolution and results. In: R. P. Oakey, W. During, & S. M. Mukhtar (Eds), *New technology-based firms in the 1990s* (pp. 105–121). London: Paul Chapman.

Metcalfe, J. S. (2000). Restless capitalism, experimental economies. In: W. During, R. P. Oakey, & M. Kipling (Eds), *New technology-based firms at the turn of the century* (pp. 4–16). Amsterdam: Pergamon.

Oakey, R. P. (1999). United Kingdom high-technology small firms in theory and practice: A review of recent trends. *International Small Business Journal*, *17*(2), 48–64.

OECD (Ed.) (1999). *Boosting innovation: The cluster approach*. Paris: OECD.

OECD (Ed.) (2001). *Innovative clusters. Drivers of national innovation systems*. Paris: OECD.

Ramström, D. (1975). De mindre företagen i ett dynamiskt perspektiv. (Small firms in a dynamic perspective.) In: D. Ramström (Ed.), *Små føretag stora problem* (*Small firms, big problems*) (pp. 156–169). Stockholm: Norstedts.

RITTS (2000). Summary report. Oslo business region 2000.

Schumpeter, J. A. (1934–1996). *The theory of economic development*. London: Transaction Books.

Segal-Quince-Wicksteed (1990/1985). *The Cambridge phenomenon. The growth of high technology industry in a university town*. Cambridge: Segal Quince Wicksteed Limited.

Segal-Quince-Wicksteed (2000). *The Cambridge phenomenon revisited*. Cambridge: Segal Quince Wicksteed Limited.

Spilling, O. R., & Steinsli, J. (2003). *Evolution of high technology clusters: Oslo and Trondheim in international comparison*. Report, Norwegian School of Management.

Steinsli, J., & Spilling, O. R. (2002). On cluster evolution and the role of small firms: The case of internet development in Norway. Paper prepared for the 12th Nordic Research Conference on Small Business, May, Kuopio, Finland.

Taylor, M. (1999). The small firm as a temporary coalition. *Entrepreneurship and Regional Development*, *11*(1), 1–19.

Vatne, E., & Taylor, M. (2000). Small firms, networked firms and innovation systems. In: E. Vatne, & M. Taylor (Eds), *The networked firm in a global world. Small firms in new environments* (introduction, pp. 1–16). Ashgate: Aldershot.

Verspagen, B. (2000). Economic growth and technological change. An evolutionary interpretation. Eindhoven Centre for Innovation Studies (ECIS) NS Maastricht Economic Research Institute of Innovation and Technology (MERIT).

Wintjes, R., & Cobbenhagen, J. (2000). Knowledge intensive industrial clustering around Océ, MERIT — University of Maastricht, Maastricht.

Part V

Networking

Chapter 12

The Competitiveness of New Technology-Based Firms: The Contribution of Trade Associations

Mario Cugini and Sarah Cooper

Introduction

The development of local, regional and national economies is underpinned by the activities of numerous economic and non-economic actors. Economic development organisations are increasingly focussed upon ways of enhancing the level of performance and competitiveness of businesses within their region, and offer a range of financial and non-financial forms of support and advice to organisations, targeted at enhancing development and growth. Regardless of what assistance is offered, the desire for enhanced competitiveness needs to exist within individual organisations. Decisions regarding the extent to which a firm adopts a competitive strategy and the exact shape which it takes are made at the level of the firm; those in the role of decision-maker need to be encouraged to see enhanced competitiveness and continuous improvements as central to their operations, be they service or product oriented.

The ultimate decision regarding the preferred approach to competitiveness rests with the firm; however, other agencies have a valuable role to play in influencing the decision-making process by communicating the potential benefits which can accrue and assisting in a variety of ways. Increased competitiveness at the firm-level benefits the firm. However, positive effects spread far beyond that micro unit of analysis. Other local, regional and national firms may improve their ability to compete, raising the level of industry capability. Regional benefits may flow from the positive perception of the area from the perspective of adoption and absorption of new, innovative behaviours. The region may attract inward investment by firms in the same or related industries wishing to be associated with an innovative location, and/or to benefit from the higher quality of collaboration (for example, as suppliers, customers, R&D partners) possible with innovative, local organisations. Increasing sectoral competitiveness enhances the collective benefits of adopting best industry practices. Innovative capacity may be influenced by the wealth and richness of firms, institutions offering research and technical support, and agencies such as Trade Associations (TAs) and Development Agencies providing more broadly-based business support.

New Technology-Based Firms in the New Millennium, Volume III
Copyright © 2004 by Elsevier Ltd.
All rights of reproduction in any form reserved.
ISBN: 0-08-044402-4

While economic development agencies are able to make general comments and offer broadly-based assistance to organisations, they are less well-placed to understand the sector/industry context and the resulting specific issues which determine the opportunities and threats which firms face, as well as the strengths and weaknesses which may influence a firm's ability to become competitive. Arguably, where an industry or TA exists it should be uniquely placed to understand the industry context and offer specialist support to assist firms through the minefield which lies between them enhanced competitiveness. This may be truest in the context of complex industries and dynamic markets, where the level of sophistication of know-how required to offer bespoke assistance is arguably greater than in less complex environments. TAs have the potential to play a role in regional and national innovation systems by contributing to the innovative milieu. This paper considers briefly the origins and activities of TAs, examines their possible contribution to improving member firm competitiveness, and presents the results of empirical research undertaken in the United Kingdom which focussed on 43 NTBFs that were TA members.

Trade Associations: Roles, Activities and Competitiveness

A TA is a "voluntary association of business firms organised on a geographical or industrial basis to promote and develop commercial and industrial opportunities within its sphere of operation, to voice publicly the views of members on matters of common interest, or in some cases, to exercise some measure of control over prices, output, and channels of distribution" (Encyclopaedia Britannica 2001). While literature on the origins of TAs is scarce, they are thought to have existed in ancient Egypt, China and Rome, although present-day TAs developed principally from 16th century English merchant guilds (Mack 1991). TAs are numerous, extremely varied in terms of their composition, scale and organisation, and undertake a wide range of tasks (Bennett 1998; May *et al.* 1998). Among the services which they offer to member firms, are representation of members' interests to government and undertaking coordination and regulatory tasks within the industry or sector. However, the representational function distinguishes them from other external business service entities (Boléat 1996). Their services can be divided into two broad categories: *inclusive* and *exclusive* (Bennett 1998). *Inclusive* activities, such as lobbying and self-regulation, are non-excludable and impact on the whole sector or industry represented, while *exclusive* activities, such as bulk purchasing and best practice fora, benefit only members. A firm may have a number of motivations for joining a TA, for example, to improve its profitability and/or strengthen the industry and general economic environment (Masten & Brown 1995). Boléat (1996) contends that TAs survive because they provide services to members more efficiently than individual members can undertake them due, for example, to economies of scale or their fundamental ability to do things better than lone members.

Achieving competitive advantage is increasingly viewed as a prime objective for individuals, organisations and nations (Xavier & Ramachander 2000). TAs are merely one of a number of actors within networks that may assist or hinder the competitiveness of firms (Nadvi 1999) and influence the wider competitiveness of an industry (Bennett 1998). Initially, TAs were criticised for "jumping on the competitiveness bandwagon" for political gain (Modic 1987). However, it is widely acknowledged that TAs have the potential to influence the competitiveness of their members (e.g. Bennett 1998; Boléat 1996; Nadvi 1999; Porter 1990; Vander Weyer 1996). In 1993 Michael Heseltine, then President of the Board of Trade, commented: "TAs in this country ought to be playing a much bigger role in promoting the international competitiveness of their member-firms," distinct from the provision of legal services, economic data, and basic lobbying which at that point was the principal function of most TAs (Heseltine 1993: 2). Heseltine encouraged TAs to be more proactive in assisting member competitiveness, and many responded by introducing new services (Boléat 1996) resulting in a shift in importance of TAs over the last decade. Bennett proposes (1998) that TAs are more important to, and by implication, add more value to small than to large firms. The potential of TAs to influence member-firm competitiveness depends in part upon the types of activities they undertake, and the level of participation and cooperation of member-firms. Previous studies have identified government lobbying as the most prevalent activity undertaken by TAs (Bennett 1998; Vander Weyer 1996). Porter (1990), however, argues that lobbying squanders a TA's most important potential benefit. He views TAs as important institutional mechanisms for investing in cluster-wide factor creation (i.e. assisting competitiveness), pointing to Italian and Japanese TAs which have led industry development, and nurtured internationally competitive industries. The British government shares Porter's enthusiasm regarding TAs as important industry change agents and potential contributors to member firm competitiveness: Proactive work to improve sector competitiveness was the dominant theme in the Department of Trade and Industry's *Model Trade Association* (1996).

Recent U.K. research suggests that TAs can play an important role in competitiveness (e.g. Bennett 1998). However, it remains unclear which drivers of competitiveness TAs can influence, how TAs might assess their contribution to member-firm competitiveness, and how member-firms view the services which they receive. Bennett (1998) believes that there are few services for which associations offer specific or unique selling points, describing their influence as "improving collective industry standards" (p. 258). Based on historical data of TAs in the U.K., and surveys undertaken by TA executives, he concludes that TAs do have the potential to influence the competitiveness of their members, although the whole sector tends to gain rather than individual businesses. Since few studies have provided empirical evidence of the perceptions of member-firms regarding the role of TAs on their competitiveness, Cugini (2001) questions the appropriateness of "playing down" TAs' influence on firm-level competitiveness without first examining the perceptions of member-firms.

"Competitiveness" and its Drivers

Several authors have attempted to define competitiveness (e.g. Heseltine 1994; Zahara 1999). However, competitiveness means different things to different people and organisations because a universal definition of competitiveness does not exist. Competitiveness may be seen as the ability to persuade customers to purchase your product or service over alternatives, or as the ability to endlessly advance product and process capabilities (Feurer & Chaharbaghi 1994). Heseltine defines competitiveness as "the ability to produce the right goods and services of the right quality, at the right price, and at the right time" (1994: 8),

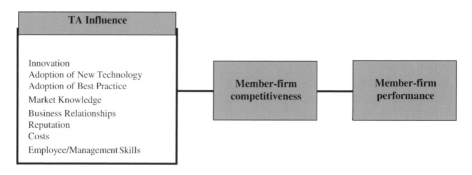

Figure 1: TA competitive assistance model (TACA). *Source:* Cugini (2001).

a satisfactory definition as the market determines what is "right." McFetridge (1995) uses the concept of profitability — "an unprofitable firm is uncompetitive" (p. 3), while others use competitiveness synonymously with achieving excellence (e.g. Xavier & Ramachander 2000; Zahara 1999). "To achieve competitive success, firms . . . must possess a competitive advantage in the form of either lower costs or differentiated products that command premium prices. To sustain advantage, firms must achieve more sophisticated competitive advantage over time, through providing higher-quality products and services or producing more efficiently" (Porter 1990: 10).

Innovation is central to competitiveness (Kay 1993) and competitiveness can only be maintained through continuous improvement of the products/services and capabilities of the organisation (e.g. Hamel & Prahalad 1990; Porter 1985). Apart from innovation, many drivers are suggested in the literature which are deemed important to building and sustaining competitiveness. Some factors are internal to the firm (e.g. employee skills) and generally more controllable, while others are external to the firm and firms generally cannot influence such factors alone (e.g. import tariffs). TAs are in a unique position in that they have the potential to influence both internal aspects of the firm and the external factors which determine member competitiveness. If a TA's mandate is to improve member competitiveness its activities must directly or indirectly influence the abilities of its members to produce more efficiently, or provide better quality products or services. Previous studies which have attempted to explore the impact that TAs have on member competitiveness (e.g. Bennett 1998; Nadvi 1999) have not defined what they mean by "competitiveness" and its causes.

This study focuses upon the drivers of competitiveness which TAs can influence, and how they influence them. After an analysis of the many organisational factors that influence competitiveness, common TA activities and services, and various TA case studies and reports (e.g. Nadvi 1999; NMI 2000), Cugini (2001) developed the TACA model, encapsulating eight firm-level drivers which TAs have the potential to influence (Figure 1).

The TACA model proposes that TAs may influence directly or indirectly the innovation, technology, operational practices, knowledge, reputation, costs and skills of members, which may subsequently lead to a change in member-firm competitiveness, and ultimately, a change in member-firm performance. In the following section each driver is discussed in the context of its relationship to competitiveness. Evidence will be included from the literature which points to the capacity of TAs to influence it and empirical evidence on the role that two Scottish electronics TAs have played in influencing member-firms' competitiveness. Both TAs operate in highly dynamic industries in which firms need to be globally competitive in order to survive.

The Research Context, Methodology and Findings

Context

Prior to a faltering of the economy towards the end of 2001, the previous 10 years had witnessed rapid, sustained growth in Scotland's electronics industry, with output increasing by 274% since 1990 (The Scottish Office 1999). Recent environmental changes have brought new pressures to the industry. Arguably, the greatest challenges faced by the industry concern the upgrading of the supply chain and skills shortages, while other issues include the impact of environmental legislation, growing competition and poor links with academia. As a young sector of the electronics industry, optoelectronics faces similar problems to those of the wider industry. However, in addition, awareness levels of opto-electronics across consumer and business-to-business markets are still relatively low, and the entrepreneurial skills pool limited. Against this background the challenges faced by the Scottish Optoelectronics Association (SOA) and Electronic Scotland (ES) are significant as they seek to assist the growth and development of their respective sectors/members.

ES is an independent TA, which represents the Scottish electronics industry. It was created in July 1999 following the merger of the two main Scottish electronics industry groups, with the dual purpose of defending the industry and providing services to assist members' competitiveness. It accepts as members all like-minded individuals and organisations interested in developing the industry. By the end of 2000 its membership was 90 firms, representing a penetration rate of 21%. Members range from large OEMs, such as Sun Microsystems and IBM, to small electronics design consultancies and component man-ufacturers. The SOA, launched in October 1994 to stimulate "economic growth in Scotland through the knowledge, manufacture and application of Optoelectronics" (SOA 2001), had 90 members at the start of 2001, comprising of companies, universities and research organisations. The SOA enjoyed a 100% penetration rate as all firms involved in the design or manufacture of optoelectronics technologies were members of the SOA. Its primary activities are focussed around bringing universities and optoelectronics firms together to collaborate and develop new technologies, as well as generating skilled engineers for the sector. The theme of collaboration extends throughout all of its activities and services.

Methodology

The following discussion is based upon data gathered from a number of sources including interviews with a senior officer from each TA, published material on each TA's activities and a postal survey conducted among high technology firm members of SOA and ES. The sample frame for the postal survey comprised only firms who were members of either of the

Table 1: Firm size profile of survey respondents.

Employees	Total	
	n	*%*
0 to 9	8	18.6
10 to 99	15	34.9
100 to 499	11	25.6
Over 499	9	20.9
Total	43	100.0

two TAs under investigation; members such as universities and governmental agencies were excluded given the research focus on firm competitiveness. Respondents completed a postal questionnaire and a small number of firms were contacted subsequently to elaborate on responses to particular questions. In total 161 questionnaires were sent out, 54 questionnaire responses were recorded (34% response rate) and once non-relevant firms were removed, 43 remained (a usable response rate of 27%). Non-relevant firms consisted of members who felt they were unable to comment fairly, due to only recently having joined the TA, or having never participated in any TA activities. Firms in the optoelectonics industry are relatively younger than those in other sectors of the electronics industry sector, reflected in the size profile of respondents (Table 1).

Research Findings

TAs and member firm competitiveness
This paper focuses on the perceptions of members in order to gain a deeper understanding of the role that TAs play in assisting member-firm competitiveness. Benefits which firms derive are heavily dependent upon their willingness to participate in TA activities and avail themselves of what is on offer. Prior to considering the individual activities of the TAs as they relate to drivers of competitiveness, objectives for membership and the perceptions of respondents regarding the wider contribution of TAs to competitiveness are explored.

Objectives for membership: The three prime objectives of TA membership were to network and develop contacts, access more information, and strengthen business relationships (Table 2). The objectives of reducing costs by gaining access to cheaper services and being part of a larger group to lobby government were not ranked as major motives of membership — notably the two main motives of membership argued by Boléat (1996).

Importance of TAs to competitiveness: Our findings suggests that smaller members perceived TAs in general to be more important to their competitiveness than larger firms did (Table 3). One TA executive commented that it is often difficult to keep close contact with people in large firms, whereas building and maintaining strong relationships is easier with smaller firms. The result is a better understanding of smaller firms' needs and interests and, consequently, the provision of services more closely tailored to the requirements of small

Table 2: Ranked mean objectives of membership.

Objective of Membership	Mean Ranking
Network & develop contacts	2.3
Access more information & advice	3.3
Strengthen relationships	3.7
Learn best practice	4.2
Collectively promote sector	4.2
Lobby government	4.9
Reduce costs	6.1

Where 1 = most important; 7 = least important.

firms. Another explanation may lie in the causality of competitiveness: smaller firms may be better able to determine specific events (e.g. participation in TA activities) which led to visible improvements in competitiveness and performance. In larger firms the causality of competitiveness may be more complex and difficult to attribute.

Effectiveness of TAs in assisting firm competitiveness: Over 41% of members felt that their TA was effective in assisting their competitiveness (Table 4). However, a sizeable proportion of members were "unsure." This again suggests that many firms have difficulty in assessing the impact of TA activities on their competitiveness.

Respondents assessed the effectiveness of their TA in assisting their competitiveness on a scale of 1–5 (5 = very effective). Micro- and small-sized firms perceived that their TA was more effective in assisting their competitiveness than did medium and larger-sized firms. Firms with fewer than 10 employees rated their TA at 3.87, compared with those firms with 500 employees or more who gave a rating of 2.22. This provides empirical evidence to support Bennett's (1998) assertion that TAs are more important to small firms in terms of assisting competitiveness. Again, this may be this is due to the fact that TA activities often do not provide highly discernable benefits to members' competitiveness, and those activities which are effective will tend to be felt more immediately by smaller firms.

Table 3: Perceived importance of TAs to members' competitiveness by firm size.

Importance of TA to Competitiveness	Number of Employees									
	Under 10		10 to 99		100 to 499		Over 499		Total	
	n	%	n	%	n	%	n	%	n	%
Important	5	63	10	67	3	27	1	11	19	44
Unsure	1	13	2	13	1	9	0	0	4	9
Not important	2	26	3	20	7	63	8	89	20	47
Total	8	100	15	100	11	100	9	100	43	100

Table 4: Perceived TA effectiveness in assisting members' competitiveness.

Perceived TA Effectiveness	Total	
	n	*%*
Very effective	3	7.0
Effective	15	34.9
Unsure	14	32.5
Not very effective	6	14.0
Not at all effective	5	11.6
Total	43	100

TAs and industry competitiveness: Nearly 75% of respondents considered that their TA was contributing to improving the competitiveness of their industry as a whole (Table 5), suggesting that members believed that their TA was assisting the competitiveness of the sector more than their own business. There is a danger that if firms perceive only sectoral improvements they may opt out of TA membership and "free-ride" the benefits, putting TAs under severe financial pressure and further, limiting their ability to help members in particular or the industry as a whole. Some larger firms expected their TA to act in the interests of the wider industry rather than primarily for their individual benefit, considering that a better business environment (e.g. educated labour pool) would enable them to prosper. In contrast, smaller firms tended to perceive their TA as vital in highlighting opportunities, transferring knowledge, and generally alleviating some of the burdens faced in running and growing a small business. Thus, larger firms appear to prefer TAs to undertake initiatives to assist factor creation in the longer-term, whereas smaller firms place greater value on services that help to address their short-term problems.

Drivers of competitiveness and the impact of the TA
Respondents commented upon elements within the TACA model (Figure 1) that evaluate TA contributions to the drivers of competitiveness (Table 6).
Innovation: "Firms create competitive advantage by perceiving or discovering new and better ways to compete in an industry and bringing them to market, which is ultimately an

Table 5: The role of the TA in improving industry competitiveness.

TAs Improve Industry Competitiveness	Total	
	n	*%*
Agree/strongly agree	22	74.4
Unsure	10	23.3
Disagree	1	2.3
Total	43	100

Table 6: Drivers of competitiveness and the role of TA involvement.

Involvement with the TA Led to	Level of Agreement of Members Regarding TA Involvement					
	Agree		Unsure		Disagree	
	n	*%*	*n*	*%*	*n*	*%*
Greater innovation	5	11.6	15	34.9	23	53.5
Adoption of new technology	7	16.3	16	37.2	20	46.5
Adoption of best practice	10	23.3	21	48.8	12	27.9
Greater market/technical knowledge	27	62.8	3	6.9	13	30.3
Improved business relationships	33	76.7	4	9.3	6	16.0
An improved business reputation	29	67.4	6	14.0	8	18.6
Cost savings	5	11.6	5	11.6	33	76.7
Improved employee/management skills	8	18.6	13	30.2	22	51.2

Sample size for each element: $n = 43$.

act of innovation" (Porter 1990: 45). Ultimately, the ability of a firm to gain and maintain competitive advantages depends significantly upon its ability to innovate. Firms may become more innovative by accessing external skills and know-how, as well as engaging in greater collaboration with universities and other research organisations (HMSO 1994). Firms which enjoy international competitive advantage have innovated in ways which anticipate the needs of both domestic and foreign markets (Porter 1990). Information plays a key role in the innovation process and the majority of TAs are active information providers. Boléat (1996) found that 67% of TAs provided market information to their members, and 88% provided general advice. Some TAs have assumed a research role, such as the British Leather Confederation which invested £1 million into technical innovation to achieve improvements in leather quality and performance (Boléat 1996). It is less likely in high-technology industries that members will give TAs significant responsibility for technical innovation, since technology is the primary source of competitive advantage. In such sectors TAs still have the potential to foster an environment more conducive to innovation and change through information provision.

Both TAs undertook activities which they considered would assist innovation within member firms, from launching websites to assisting design firms and manufacturers to identify common interests and form alliances to develop new technologies, and to showcase innovative actions of members at TA meetings. Despite such efforts, most members did not perceive their TA as assisting innovation (Table 6). Some members reported positive experiences, detailing research collaboration with other member-firms and universities which resulted in the development or the refinement of new products. Others commented on the value of TA meetings in providing opportunities to discuss how to make improvements to products with sub-contractors, and in raising awareness of newly emerging markets which helped with targeting R&D efforts. Indeed, 26% of respondents believed their involvement in the TA had led to their firm diversifying into new product markets.

Adoption of New Technology: The adoption of new technology will affect a firm's competitive advantage if it plays a considerable role in determining its relative cost position or differentiation (Porter 1985). New technology may enable a firm to reduce its costs or enable it to provide a higher quality product or service. Adopting new technology accelerates the speed of change enabling a firm to improve faster than competitors (Thompson 1996). Prior to adoption, information about new technologies needs to be communicated to firms in order for them to decide whether or not to adopt; professional associations may play an important role in imparting knowledge that is important for the diffusion of technology (Swan & Newell 1995). One of the study TAs was involved in formal technology transfer whereas activities of the other were relatively informal, based around information provision.

Again, a high proportion of members disagreed, or were unsure, as to whether involvement in their TA had led to the adoption of new technology (Table 6). One TA's members were anticipating the launch of an e-commerce portal, an initiative likely to improve the competitiveness of adopting firms by both reducing members' costs and/or allowing them to further differentiate themselves through more rapid response to market changes. The view of most of these members was that their TA has not yet played a significant role in the diffusion of new technology, although the attraction of more research-led, university members might increase opportunities to learn about new technologies.

Adoption of Best Practice: Best practice can be defined as the best organisational approach to tackling an activity or problem (Jarrar & Zairi 2000). The way activities are carried out in the value chain affects a firm's costs and its ability to produce a differentiated product/service (Porter 1985). Adopting best practice can enable a firm to improve several internal drivers of competitiveness, including quality, efficiency, and productivity (Slack *et al.* 1998). In order to implement a best practice, and consequently improve its competitiveness, a firm must first learn about it (Jarrar & Zairi 2000). The British government believes TAs have a major role to play in spreading best practices by organising competitiveness improvement programmes and running benchmarking clubs, developing a climate where members are willing to identify, share and adopt best practices (DTI 1996). Networking in conferences and training courses (Jarrar & Zairi 2000) are two other ways in whch best practice may be transferred indirectly, services widely provided by TAs (Bennett 1998).

Both TAs praised best practice at their meetings, however, the efforts of one were more informal than those of the other. One provided several best practice fora for members, focussing on aspects from the supply chain and e-commerce to the environment. It also collaborated with other associations to transfer best practice where common interest lay. Only 23% of members agreed that involvement with their TA had led to the adoption of best practice (Table 6). The formal approach of one TA proved more effective with 35% of its members reporting some adoption of best practice through membership, compared with only 10% of the other's members. In terms of process results, only 14% of members reported improvements in productivity, 16% in efficiency, and 11% in quality, as a result of participation. Nevertheless, these findings suggest that TAs can encourage process improvements through best practice initiatives.

Market and Technical Knowledge: Organisations can never know enough about their market place, and must attempt to gather information continuously on stakeholders such as customers, suppliers and distributors as well as the general macro-environment (e.g. Hitt *et al.* 1999; Kotler 1997; Porter 1980, 1985). Firms are in a race to gain "industry foresight" — a deeper understanding ". . . than competitors, of the trends and discontinuities — technological, demographic, regulatory, or lifestyle — that could be used to transform industry boundaries and create new competitive space" (Hamel & Prahalad 1994: 64). TAs provide a variety of information services to members that help link local producers with foreign markets, including data on markets, competitors and trade policies (Nadvi 1999). Furthermore, many TAs provide technical advice to their members, which may help them overcome problems (Boléat 1996). The speed at which firms acquire new knowledge and develop the skills necessary to apply it can be a key source of competitive advantage (Hitt *et al.* 1999). Thus, information and advice from TAs on areas such as legislation and environmental issues may provide members with the capability to seize opportunities and neutralise threats, thereby improving or sustaining competitiveness.

Both TAs disseminated market information and technical advice through a number of channels including telephone, e-mail, websites, membership meetings, visits to member companies and newsletters. They operated as information intermediaries; information and advice tended to be retrieved from expert sources with little market or technical knowledge developed internally. The availability of information via the Internet led to a view that TAs should add value by addressing longer-term strategic issues and avoiding the traditional information focus. The majority of members (63%) agreed that involvement with their TA had led to the acquisition of greater market/technical knowledge (Table 6). However, 60% of members felt that the provision of "market information" by their TA was effective in assisting their competitiveness while only 23% of members perceived its "technical advice" to be effective. These findings provide strong evidence that TAs can play an important role in highlighting market opportunities to members, which may result in subsequent diversification by member-firms to enhance their strategic competitiveness.

Relationships: Relationships involve organisations coming together to achieve shared or compatible ends. A firm's relationships adds value through the creation of organisational knowledge and the establishment of a co-operative ethic (Kay 1993). External relationships are vital for acquiring resources that are not accessible or are too costly to acquire alone. Relationships can provide greater access to resources, assist the development of resources, and foster organisational learning through information exchange (Håkansson & Snehota 1995). Successful relationships can also lead to improved competitiveness through lower transaction costs, greater efficiencies and synergies between actors (Thompson *et al.* 1991). TAs can frequently support the development of successful relationships (Lane & Bachman 1995). Through horizontal coordination amongst producers and vertical coordination of upstream and downstream linkages, TAs can ease members' internal constraints in relationship development (Doner & Schneider 1998). Lane (1997) argues that TAs can influence buyer-supplier relations, through actions such as disseminating and enforcing industry rules and regulations and providing information on members that allows them to be evaluated as potential suppliers or customers; resulting in greater mutual trust benefits both at firm and sectoral-levels of competitiveness.

Both TAs were actively involved in encouraging members to network and develop relationships both domestically and abroad, for example, by holding regular membership meetings and other events which allow members to network, and also through the provision of web-based membership directories which allow firms to learn more about each others' businesses. Over three quarters of members (76%) considered that their TA had assisted them to develop stronger business relationships (Table 6). Over 91% of members of one TA which provided special relationship building fora agreed that relationships had improved compared with 60% of the other. Evidence also suggests that TAs can assist members in developing international relationships through collaborative agreements with foreign TAs. For example, one R&D firm explained how its TA helped it to establish contact with an organisation in Taiwan which resulted in a visit to the major firm in Taiwan and the subsequent development of a strong sales relationship.

Reputation: Reputation is critical to firms in markets where customers are unable to determine easily product quality through search or their own experience. Building, maintaining and spreading reputation are vital tasks for any firm wishing to succeed in the market place, and firms often require the help of others to accomplish them (Kay 1993). Reputational capital is an intangible resource which can provide firms with a more sustainable competitive advantage as it is difficult to imitate (Collis & Montgomery 1995; Porter 1985). In globalised industries, such as electronics, an international reputation may be vital; firms must communicate with suppliers and customers worldwide, not only their ability to provide high quality, but also in order to assure quality (NEDC 1988). TAs can formulate local standards which may result in an international reputation for quality, boosting the international competitiveness of member-firms (Nadvi 1999). TA activities, such as joint marketing programmes and international networking, may help to build, spread and maintain reputation, by promoting members to potential customers at home and abroad.

Both TAs investigated in this research believed that companies gained reputational capital through their membership. Although neither plays a formal role in setting and regulating industry standards, both were actively involved in helping members spread their reputations; for example, one TA organised stands for its members at trade shows in the U.K. and abroad. As the TAs' international reputations grew members enjoyed positive "knock on" effects that helped them to establish stronger reputations as suppliers with global supply capabilities. In countries such as Japan and Taiwan TA membership is perceived as a status symbol and generates significant levels of trust: firms opting out of TA membership may experience difficulties partnering in some overseas countries. Over 67% of respondents agreed that TA participation had led to an improved business reputation (Table 6). The older of the two TAs appeared to have enhanced members' reputations more effectively (80%) compared with the younger TA (56%), suggesting that reputations are built over the long term. Members of one TA pointed to the benefits of its trade shows, which had increased awareness of their products throughout the world. Financial and logistical assistance from the TA had helped one member build its name internationally, resulting in it supplying firms in Silicon Valley. The "trust" gained through membership was confirmed by two members of one TA who commented that their TA's strong international reputation had helped gain access to customers and suppliers in Taiwan. Through membership screening and monitoring, combined with effective joint marketing campaigns, TAs can build strong reputations and establish trust internationally, which in turn, may "trickle down" to members and provide intangible competitive advantages.

Costs: Costs are central drivers of competitiveness. Porter (1985) identified a number of cost drivers which determine the cost behaviour of value activities. Through an analysis of

Porter's drivers, it appears that TA member-firms can potentially reduce costs via learning (e.g. best-practice), vertical linkages (e.g. alliances between members), and collaborating to influence institutional factors (e.g. lobbying government to reduce taxes). In addition, procurement generally has important strategic significance for firms (Porter 1985), as externally accessed inputs frequently constitute the largest part of a firm's total costs (Wilkinson *et al.* 2000). Examples of activities/services provided by TAs in an attempt to reduce members' costs are the provision of information, advice, bulk purchasing and joint marketing activities at discounted prices (Boléat 1996).

One of the survey TAs is proactive in developing cost saving initiatives while the other considers that its joint marketing activities (i.e. trade shows/missions) provide smaller members with cost-effective means to promote their products overseas and increase exports. Members generally disagreed that participation in the TAs has led to cost savings (Table 6). Surprisingly, the 83% of respondents from the TA which proactively arranged bulk purchase and other cost saving schemes, considered that they did not enjoy cost savings as a result of membership. Firms reporting cost savings were predominantly small firms who received financial assistance to attend overseas trade shows. While findings suggest that TAs have the potential to reduce members costs, members of both TAs indicated that cost reduction was not a prime goal of their membership (Table 2). Thus, procurement costs do not appear to be amongst the most important strategic issues faced by the industry.

Employee/Management Skills: The external labour market has a large impact on organisational competitiveness (Torrington & Hall 1998). Workforce skills are increasingly crucial differentiating factors in enabling firms to build and sustain competitiveness (Pfeffer 1995; Ulrich 1993). Porter (1990) highlighted that the availability of skilled labour can often determine the sophistication and sustainability of a firm's competitive advantages. The importance of highly skilled staff to technology-based firms is apparent from a number of studies (e.g. Cooper 1999). TAs often lobby government to provide financial assistance for training and education in the industries they represent (Boléat 1996). Several TAs promote their industry to school leavers in an attempt to widen the available labour pool, e.g. the National Microelectronics Institute (MNI) (2000). Some TAs offer excludable training services; for example, the U.K. Freight Transport Association (2000) provides over 200 training courses per year to members.

Both study TAs perceived the importance of the workforce as a driver of industry competitiveness and had attempted to address skill shortages in electronics with specific initiatives. One had partnered with universities to develop two specialist MSc-level training packages and attempted to make optoelectronics attractive as a career to the Scottish school child. The other was attempting to improve members' skills through initiatives such as discounted training courses, primarily in IT, and software programming and e-commerce master classes. It was also attempting to influence future skills provision by lobbying government to co-ordinate improvements in engineering education at graduate level.

A sizeable proportion of members considered that their TA did not assist them in improving their employee/managerial skills (Table 6). With extreme skills shortages in the industry, it is logical that both TAs promote the industry to young people. However, this activity is inclusive, in that the whole industry gains rather than sole members. Furthermore, benefits of promoting to school leavers will take several years to become visible to members, thus members' perceptions may be different in three or four years time. Both TAs are addressing skills issues with inclusive and exclusive initiatives. However, as with reputation, fostering improvement of skills can take a long time. Findings indicate that exclusive skills initiatives, such as training courses, are valued by small firms who lack the resources or expertise to conduct training themselves. Medium and large-sized firms appear more concerned with inclusive skills initiatives, which address wider labour market problems. Firms will not receive exclusive skills benefits through their membership as such initiatives tend to improve the whole industry's competitiveness. TAs need to provide both inclusive and exclusive skills initiatives in order to benefit both the sector and member-firms' competitiveness. While focussing solely on inclusive services will lead to some firms "free-riding" the benefits, the inclusion of exclusive services may encourage firms to join or renew membership.

Conclusions

For over a decade the British government has viewed TAs as having a role to play in influencing the competitiveness of firms in the sectors they serve. If they are to attract members, TAs have to offer an appropriate portfolio of support and services which benefit individual members through exclusive services benefits, and inclusive benefits which advance the standing of the whole sector. Achieving an attractive balance is vital as TAs can only assist individual members' competitiveness if they choose to participate. A comprehensive understanding of members' needs is essential to align TA services and activities and to encourage participation.

This research using the TACA model has enabled a deeper understanding of specifically *how* TAs can assist their members by contributing to firm-level competitiveness. TAs have been identified as playing an indirect role in members' competitiveness, acting predominantly as intermediaries or "stepping stones" and offering members the potential to benefit in certain areas of business activity. Examination of elements within the TACA model with respect to members of the SOA and ES revealed the generation of greater benefits in some areas than others. Both sets of members perceived that their TA had a relatively low influence on their innovation, adoption of new technology, best practice, skills and costs. However, elements of good TA practice were identified, such as involving universities in membership to transfer technology from academia to industry, and making effective use of IT to diffuse best practice and market knowledge. More importantly, evidence suggested that the TAs' greatest contribution to competitiveness involved the transfer of market knowledge, and assistance provided to strengthen members' business relationships and reputations. There are positive aspects to these results for TAs wishing to promote services to members and non-members.

Member diversity poses a challenge for TA executives (Bennett 2000) faced with meeting the needs of organisations differing, for example, in size, organisational culture, nationality of ownership, products/services, resources and membership objectives. A full discussion of barriers is beyond the scope of this paper, but quantitative and qualitative evidence supported Bennett's (1998) assumption that TAs have a greater influence on the competitiveness of small firms than large firms. Smaller members may place greater value on TAs' contribution to their competitiveness, due to the impact of TA activities being generally easier to assess. There is currently a lack of visibility to larger firms of benefits of TA activities. Resource constraints stimulated small firms to make greater use than larger firms of TA services and they tended to have a shorter-term strategic outlook than larger members, rarely valuing "inclusive" activities such as lobbying. Maintaining contact with personnel in small firms was easier for TAs than with personnel in large firms, which led to the provision of services more closely aligned to small firm needs. A high proportion of firms however were uncertain as to the effectiveness of their TA in assisting their competitiveness. This suggests that TAs need to emphasise the tangible gains that members can receive and have received through participation, and how their services allow members to enhance their relative cost position or ability to differentiate. The influence of TAs on competitiveness depends highly on member participation and cooperation. TAs must market themselves effectively to establish credibility as important contributors to members' competitiveness.

The government views TAs as promoters of best practice and there is some evidence that ES and SOA members have benefited from TA activities in this area. There is a need to create a climate where members are prepared to share best practice first, which can be encouraged through formal initiatives such as best practice clubs, and informally at membership meetings and other events. Using new technologies, such as the Internet, may facilitate the best practice transfer process for firms who are unable to attend meetings. Ensuring that services are consistent with the objectives of members and the strategic issues faced by the industry is likely to maximise participation rates.

These TAs have contributed to developing and strengthening members' supply chain relationships, and stimulating the formation of productive formal and informal strategic alliances. Encouraging members to collaborate and pool complementary resources can create powerful synergies, while collaboration with TAs overseas can assist members in developing relationships with firms abroad, thus strengthening international competitiveness. Evidence indicates that firms have gained reputational benefits through membership and participation in TA activities (Nadvi 1999). Both TAs had helped to build, spread and maintain members' reputation by undertaking joint marketing campaigns and assisting members to participate in trade shows, thereby increasing awareness levels in international markets.

TAs can make significant contributions to member-firms' competitiveness by providing services that somehow enable members to improve their cost positions or abilities to differentiate. However, they face many barriers in achieving this goal. TAs first need to develop a deep understanding of their members' objectives and the strategic issues that they face; only then can they develop effective services and activities that "fit" and help to fulfil the competitive requirements and aspirations of members. As industries evolve and the needs of members change, it is important that TA continue to evaluate what it is that their members want and establish whether member-firms are deriving the benefits from the services which the TAs think that they are gaining. In dynamic and fast changing high technology sectors the challenge faced by TAs is significant. Failure to monitor the environment of members will result in falling membership and a loss of a central opportunity to nurture and support sector competitiveness and development.

As nations become more dependent upon the economic contribution of NTBFs the role which different actors are able to play in influencing the competitiveness, and thus the survival and growth of firms, is important for policy makers. The findings of this research point to the important role which TAs are able to play in enhancing the competitiveness of technology-based firms. Findings suggest that some TAs have the potential to contribute to cluster development through enriching the innovative milieu, contributing another component to the institutional thickness exhibited by innovative regions. As such they may be viewed as key actors in local, regional and national economic development, with the potential to influence the attractiveness of a region and/or nation as a location for other start-up and expanding businesses.

References

Bennett, R. J. (1998). Business associations and their potential contribution to the competitiveness of SMEs. *Entrepreneurship and Regional Development, 10*, 243–260.

Bennett, R. J. (2000). Factors governing the effectiveness of business associations. *Proceedings of the Effectiveness of EU Business Associations Conference*, Brussels (18–22 September), 1–13.

Boléat, M. (1996). *Trade association strategy and management*. Association of British Insurers.

Collis, D. J., & Montgomery, C. A. (1995). Competing on resources. *Harvard Business Review, 75*(4), 118–128.

Cooper, S. Y. (1999). Sectoral differences in the location and operation of high technology small firms. In: R. P. Oakey, W. During, & S.-M. Mukhtar (Eds), *New technology-based firms in the 1990s* (pp. 170–184). Oxford: Elsevier.

Cugini, M. G. (2001). *The role of trade associations in assisting member-firm competitiveness*. Unpublished MA (Hons) thesis, School of Management, Heriot-Watt University, Edinburgh.

Doner, R. F., & Schneider, B. R. (1998). Business associations and economic development. Paper presented at the SSRC/IILS Conference, Geneva (November).

DTI (1996). *Model trade association*. Department of Trade and Industry, London.

Encyclopaedia Britannica (2001). Trade Association, http://www.britannica.com/eb/article?eu=75047&tocid=0.

Feurer, R., & Chaharbaghi, K. (1994). Defining competitiveness: A holistic approach. *Management Decision, 32*, 49–58.

FTA (2000). "Training", Freight Transport Association, http://www.fta.co.uk/training.

Håkansson, H., & Snehota, I. (1995). *Developing relationships in business networks*. London: Routledge.

Hamel, G., & Prahalad, C. K. (1990). The core competence of the corporation. *Harvard Business Review, 68*(3), 79–91.

Hamel, G., & Prahalad, C. K. (1994). Seeing the future first. *Fortune, 130*(5), 64–68.

Heseltine, M. (1993). Speech on Trade Associations at the CBI (17th June 1993) by the President of the Board of Trade, Press Release, Department of Trade and Industry, London.

Heseltine, M. (1994). Competitiveness: What, why and how? *European Business Journal, 6*(3), 8–15.

Hitt, M., Ireland, R. D., & Hoskisson, R. E. (1999). *Strategic management: Competitiveness and globalization*. Minneapolis/St. Paul: West Pub. Co.

HMSO (1994). *Competitiveness: Helping business to win*. Her Majesty's Stationary Office.

Jarrar, Y. F., & Zairi, M. (2000). Best practice transfer for future competitiveness: A study of best practices. *Total Quality Management*, S734–781.

Kay, J. (1993). *Foundations of corporate success*. Oxford: Oxford University Press.

Kotler, P. (1997). *Marketing management: Analysis, planning, implementation and control*. London: Prentice-Hall.

Lane, C. (1997). Trade associations and inter-firm relations in Britain and Germany. In: R. J. Bennett (Ed.), *Trade associations in Britain and Germany* (pp. 23–32). Anglo-German Foundation.

Lane, C., & Bachman, R. (1995). Risk, trust and power: The social construction of supplier relations in Britain and Germany. Working Paper 5, ESRC Centre for Business Research, Cambridge.

Mack, C. S. (1991). *The executive's handbook of trade and business associations*. Quorum Books.

Masten, J., & Brown, S. (1995). Problems facing business associations in emerging countries: The case of Ghana. *International Small Business Journal*, *13*(4), 91–96.

May, T. C., McHugh, J., & Taylor, T. (1998). Business representation in the U.K. since 1979: The case of Trade Associations. *Political Studies*, *XLVI*, 260–275.

McFetridge, D. G. (1995). *Competitiveness: Concepts and measures*. Occasional Paper Number 5. Ottawa, Industry Canada.

Modic, S. J. (1987). Just what is competitiveness? *Industry Week*, *233* (April 20), 7.

Nadvi, K. (1999). Facing the new competition: Business Associations in developing country industrial clusters. Discussion paper, International Institute for Labour Studies.

NEDC (1988). *Performance and competitive success: Strengthening competitiveness in U.K. electronics*. National Economic Development Council.

NMI (2000). *Annual Review 1999/2000*. National Microelectronics Institute.

Pfeffer, J. (1995). Producing sustainable competitive through the effective management of people. *The Academy of Management Executive*, *9*(1), 55–69.

Porter, M. E. (1980). *Competitive strategy*. New York: Free Press.

Porter, M. E. (1985). *Competitive advantage*. New York: Free Press.

Porter, M. E. (1990). *The competitive advantage of nations*. London: Macmillan.

Slack, N., Chambers, S., Harland, C., Harrison, A., & Johnston, R. (1998). *Operations management*. London: Pitman Publishing.

SOA (2001). *The Scottish Optoelectronics Association*, SOA, http://www.optoelectronics.org.uk.

Swan, J. A., & Newell, S. (1995). The role of professional associations in technology diffusion. *Organisation Studies*, *16*(5), 847–865.

The Scottish Office (1999). Statistical Bulletin: NO IND/1999/C1.9, www.scotland.gov.uk/library/documents-w8/eis-01.htm.

Thompson, G., Frances, J., Levaxix, R., & Mitchell, J. (1991). *Markets, hierarchies & networks*. London: Sage.

Torrington, D., & Hall, L. (1998). *Human resource management*. Prentice Hall Europe.

Ulrich, D. (1993). Profiling organisational competitiveness: Cultivating capabilities. *Human Resource Planning*, *16*(3), 1–16.

Vander Weyer, M. (1996). Trade bodies punch their weight. *Management Today* (June), 66–69.

Wilkinson, I. F., Mattsson, L. G., & Easton, G. (2000). International competitiveness and trade promotion policy from a network perspective. *Journal of World Business*, *35*(3), 275–299.

Xavier, M. J., & Ramachander, S. (2000). The pursuit of immortality: A new approach beyond the competitiveness paradigm. *Management Decision*, *38*(7), 480–488.

Zahara, S. A. (1999). The changing rules of global competitiveness in the 21st century. *The Academy of Management Executive*, *13*(1), 36–42.

Chapter 13

Organizational Influences Upon the Development of Professional Expertise in SMEs in the Netherlands [*]

Beatrice I.J.M. Van Der Heijden

Introduction

Small and medium-sized enterprises (SMEs), i.e. following the definition used by the European Union companies that employ fewer than 250 employees, form an increasingly important feature of the global economy, constituting over 99% of all enterprises in almost all OECD countries. In broad terms they contribute between 40 and 80% of employment in different countries and 30 and 70% of GDP (OECD 1997).

In the Netherlands no less than 99% of private enterprises consists of medium and small-scale businesses — a percentage representing more than 515,000 companies. Together they provide employment for 2.3 million people (60% of the Dutch labour force) and account for some 52% of the national income generated in the private sector (Royal Association MKB the Netherlands 2000).

Scientific interest in SMEs is fairly recent, and at international conferences, the topic is still getting marginal attention (Chanaron 1998). There is a considerable lack of empirical data concerning personnel functions in SMEs. In research and also in business management textbooks the areas of accounting, finance, production, and marketing all take precedence over personnel management (Hornsby & Kuratko 1990). Also in the debates about HRM that have characterized the 1980s and early 1990s (see, e.g. Storey 1992, 1994) SMEs have not figured prominently. The majority of SME-studies appears to concentrate on enterprises with 100 or more employees (Duberly & Walley 1995) and hardly considers the characteristics of smaller ones. Moreover, these studies are primarily based on the large-firm HRM blueprint for SME staff policy (Deshpande & Golhar 1994).

[*] The Publisher of the Journal of Enterprising Culture has granted permission to reprint the following article in this volume: Van der Heijden, B.I.J.M. (2002). Organizational influences upon the development of professional expertise in SME's. Journal of Enterprising Culture, Vol. 9(4), 387–406.

New Technology-Based Firms in the New Millennium, Volume III
© 2004 Published by Elsevier Ltd.
ISBN: 0-08-044402-4

Because of the fact that the major part of personnel management in smaller enterprises is determined by the general characteristics of these enterprises (accent on team spirit, informal working procedure and firm co-ordination by the employer), personnel management cannot be judged by existing models of HRM (Koch & De Kok 1999). Most of these normative models are (implicitly or explicitly) based on large enterprises and take the classical approach to strategy formulation (Legge 1995), which is not relevant for small enterprises. The assessment of personnel management in SMEs should take into account the specific characteristics, background and possibilities.

The effective management of human resources is emerging as a key variable in the success of SMEs. Because of the fact that the managerial characteristics and management style largely determine the characteristics of SMEs, it is very important to determine the relationship between differences in management and organizational characteristics and the amount of success of the company (During & Kerkhof 1998).

In order to meet with this need a study on the development of professional expertise in Dutch multinationals (Van der Heijden 1998) has been replicated in SMEs. Despite the fact that in literature smaller firms' competence development and acquisition is often described as improvised and short-term (Hendry *et al.* 1995; Storey 1994), the study of Ylinenpää (1997) has revealed that small manufacturing firms do invest significantly in developing and acquiring competencies. Investments consist of a mixture of different methods such as visiting expo's and trade fairs, courses, work rotation, study visits and delegation of work tasks.

Yet, the influence of career activities upon the development of professional expertise has to be investigated more thoroughly because of the previous lack of a profound operationalization of the concept (see also Van der Heijden 1998). Moreover, in most SMEs relatively few older workers are employed, implying that most companies lack an aging policy. Most managers "just" try to adjust job requirements according to the age of the individual employee (Koch & Van Straten 1997).

This contribution examines the relationship between four organizational predictor variables and the extent of professional expertise possessed by middle and higher level employees working in small and medium-sized enterprises. Firstly some theoretical outlines concerning the dependent variable professional expertise will be given, followed by some outlines and accompanying hypotheses concerning the organizational factors.

The Definition of Professional Expertise

For the measurement of professional expertise, a domain-independent and multi-dimensional instrument has been developed. This approach is found to be useful for several reasons. Firstly, a multi-dimensional operationalization permits the measurement and comparison of performance levels attained in different domains. Secondly, a multi-dimensional approach is useful in identifying those components most in need of being updated. Thirdly, the approach is useful in examining how specific factors in the working environment support or limit performance in particular expertise elements. Up to now little research has been done aimed at examining the relationships between specific influential factors and particular dimensions of professional expertise.

A multi-dimensional and domain-independent operationalization of the concept of professional expertise should in any case comprise the different types of knowledge that are inherent to a certain professional field. These different types of knowledge are declarative knowledge ("knowing that"), procedural knowledge ("knowing how") and conditional knowledge ("knowing when and where or under what conditions") (Alexander *et al.* 1991).

The intended dimension is termed the knowledge dimension and is closely related to the second dimension, called the meta-cognitive knowledge dimension ("knowing about knowing" or "knowing that one knows"). This second dimension, that has to do with self-insight or self-consciousness, is known by a wide variety of names: meta-cognitive knowledge, meta-knowledge, executive control knowledge, self-knowledge, regulative knowledge and meta-cognitive strategic knowledge, to mention but a few.

The third dimension, skill requirement, has to do with the skills an employee needs to perform the required professional tasks. Once the activities and responsibilities have been defined, it is clear which skills are necessary to perform a given job. A person can only be referred to as an expert if his or her overt behavior demonstrates the capacity to perform qualitatively well in a particular domain.

The three dimensions that have been discussed so far are fairly commonplace to earlier conceptualizations of the construct of expertise (Bereiter & Scardamalia 1993; Chi *et al.* 1988; Ericsson 1996; Ericsson & Smith 1991). Nevertheless, we do not share the opinion that expertise is fully explained by these three dimensions (Boerlijst *et al.* 1996; Van der Heijden & Boerlijst 1997).

Measurement of cognitive abilities and skills is not enough to fully cover the construct of professional development (Ericsson & Lehmann 1996; Trost 1993). Motivational aspects and self-insight, as well as social skills, social recognition and growth capacities are important interactors and moderators. That is to say, there is a compelling reason for the proposal of a broader type of measurement, in which cognitive abilities and overt skills play an explicit, but partial role. Thus, there is a need for alternative perspectives for the measurement of expertise.

Expertise can only exist by virtue of being respected by knowledgeable people in the organization. There are a lot of people who have a vast amount of knowledge and/or skills. However, not all of them can be considered highly skilled or experts, owing to a lack of social intelligence, communicative skills and so on. This fourth important aspect of professional expertise can be labeled the dimension of acquirement of social recognition.

A fifth dimension that has been added to previous conceptualization frameworks, is the dimension of growth and flexibility. People who are capable of acquiring more than one area of expertise within adjacent or radically different fields, or who are capable of acquiring a strategy to master a new area of expertise or expert performance in another territory can be termed "flexperts" (Van der Heijden 1996). These are people who are both flexible and in possession of expertise.

For the fact that the study that is reported in this article is a replication of a large-scale study in multinationals, earlier publications are referred to for more detailed information concerning the process of instrument development in order to prevent repetition (see Van der Heijden 1998, 2000).

Theoretical Outline and Hypotheses Concerning the Relationship Between Organizational Predictor Variables and Professional Expertise

This section presents a justification of the relevance of the predictor variables in the context of a person's professional development. In addition, the operationalization and scale reliability, if applicable, are given. For each predictor variable, the theoretical outline is concluded with accompanying hypotheses concerning both the occurrence of the organizational predictor in three different age groups, and with respect to the relationship between the predictor and scales of professional expertise.

As the majority of studies have been highly domain-specific (Bereiter & Scardamalia 1993; Chi *et al.* 1988; Ericsson & Smith 1991) and besides, the predictor variables that are proposed stem largely from our own research activities (Boerlijst *et al.* 1993), the hypotheses regarding the relationship between the predictor variables and professional expertise tend mostly to be framed with argumentation stemming from our own reasoning and insights obtained in earlier research (Boerlijst *et al.* 1993; Van der Heijden 1998).

Social Support from Immediate Supervisor

At the heart of the learning climate lies the relationship between the employee and his or her immediate supervisor. Good supervisory feedback and good communication between the two enhance the opportunity for advancement in the worker's capabilities. Social support from one's superior can generate a general feeling of satisfaction and faith in one's further career development. The work climate must be one in which mutual trust is prevalent, stemming from a lack of complaints, grumbling and negative attitudes.

In our previous studies on the relationship between age and individual career development (Boerlijst 1994; Boerlijst *et al.* 1993; Van der Heijden 1998) we have found that most supervisors fall short in devoting attention to the functions and functioning of their older workers. Particularly in the case of seniors, supervisors appear to be uncooperative and unhelpful with regard to professional development. In other words, the degree of social support from one's immediate supervisor is expected to decline when the employee gets older (Hypothesis 1).

As regards the influence upon professional expertise, it is expected that social support from one's immediate supervisor has a positive effect upon the extent of *knowledge* (Hypothesis 2), *meta-cognitive knowledge* (Hypothesis 3) *and skills* (Hypothesis 4).

Once a supervisor has provided his or her employee with information concerning strengths and weaknesses in job performance, meta-thinking can be initiated. Growing self-insight may be the result as well as perceivable improvements in knowledge and capabilities. No effect upon the amount of growth potential is expected. The items comprising the scale of social support received from the immediate supervisor relate to the job that is actually performed in the everyday practice of working life and not to future demands.

The strengths of the relationships, as formulated, are envisaged as being the same for each age group. In other words, no interaction effects are expected.

Social Support from Near Colleagues

Further expertise development is largely based on the progressive internalization of information gained through social interaction with colleagues. Exchange of information and positive feedback processes during work can make a major contribution to the formation of relevant knowledge and skills. In each working organization, one's peers must bear the responsibility of providing reliable information on current technical developments, for example by drawing one's attention to useful new journals or training courses. Colleagues must be willing to act as sounding boards for new ideas based on their own experiences. In this way, colleagues can be highly valuable sources of information and help.

In the case of the middle-aged employees, determination of possibilities for advancement in one's professional field seems to be a central theme (Schein 1978). Because of the fact that vertical progress is not within everyone's reach, owing to the increasing flattening of organizations, this gives rise to a great deal of competition between near colleagues. As a result, the individual's social network is subject to change in the course of life (see also Sarason *et al.* 1987).

Accordingly we expect a decrease in social support from near colleagues when employees enter the mid-career phase. The difference between the middle-aged and the seniors is envisaged as being minimal (Hypothesis 5).

As far as the relationship with professional expertise is concerned, the same pattern as that for social support from the immediate supervisor is expected. That is to say, positive relationships with the dimension of *knowledge* (Hypothesis 6), *meta-cognitive knowledge* (Hypothesis 7) and *skills* (Hypothesis 8) are expected.

Organizational Facilities

Organizational policies should be aimed at continuing expertise development. For an organization to be attractive to employees, it should provide learning opportunities and chances to grow in knowledge and skills and to improve their capabilities. As regards network participation, there is no doubt that an organizational environment that facilitates open communication and exchange among professionals, and between management and professionals, fosters the growth of professional expertise. First of all, this type of environment encourages the exchange of ideas, information and skills among employees in all parts of the organization. The interchange of ideas and skills facilitates qualitatively high performance. Moreover, in organizations with facilities for network participation, experts are acknowledged for their contribution to these networks and for their credibility as expert team members.

As for training and development programs, emphasis should increasingly be on enlarging employees' tasks by flexible "know-how" acquired from training and education in external expertise areas and in the domain of personal and social skills. These types of training and education make them less vulnerable in times of change. Training participation aimed at experience broadening, contributes to professional flexibility (Thijssen 1996).

The above-mentioned types of organizational facilities can really enhance employability. Even where companies *want* to keep the same employees long term and promise them a job for life, they can only afford this if they keep them growing and learning through elaborate training and retraining. In short, companies should keep them *employable*.

Consistent with the Human Capital Theory (Becker 1993; Schultz 1971) and what we have called the "pay-off period" of investments (Boerlijst *et al.* 1993) it is expected that the availability of organizational facilities will diminish with age (Hypothesis 9).

Regarding the relationship with professional expertise, depending on a thought-out structure, organizational facilities may increase both professional *knowledge* (Hypothesis 10), *meta-cognitive knowledge* (Hypothesis 11) and *skills* (Hypothesis 12). Furthermore, the *acknowledgement* (Hypothesis 13) one receives could increase, for example owing to the possibilities to participate in social networks as well as courses aimed at further personal development. Or because of the recognition from relevant other people of significance as being up-to-date in knowledge and skills. Finally, organizational facilities may increase one's *growth potential* (Hypothesis 14). No interaction effects for age are expected.

Attention from Immediate Supervisor for a Further Career Development

In our earlier studies (Boerlijst 1994; Boerlijst *et al.* 1993; Van der Heijden 1998) it was found that shortcomings on the part of supervisors seriously endanger the mobility and employability of most non-executives at middle and higher levels of functioning. It was hypothesized that these shortcomings cause pitfalls in the career of many highly positioned employees, limiting or hampering their (cognitive) development.

Supervisors need to stimulate employees to participate in training and development programs, to the exchange of information and to thinking about following career steps. A basic requirement for growth is the formulation of an individual development plan that is drawn up by the employee in co-operation with different organizational bodies (Stickland 1996).

Management perceptions regarding the possibilities for further education are often negative where the older employee is concerned (Thijssen 1996). Accordingly, older workers are less stimulated to participate in training and development programs (Boerlijst *et al.* 1993; Onstenk 1993; Plett *et al.* 1991).

The previously mentioned Human Capital Theory (Becker 1993; Schultz 1971) implies that investments in educational activities decrease with the age of the employee. Although the variable "*attention from immediate supervisor for a further career development*" comprehends more than stimulating educational activities, we expect that the shorter the pay-off period of the career investments becomes, the more the immediate supervisor gives up stimulating the older employee to take part in career activities (Hypothesis 15).

Regarding the relationship with professional expertise, depending on the focus, attention from one's immediate supervisor may increase both professional *knowledge* (Hypothesis 16), *meta-cognitive knowledge* (Hypothesis 17), and *skills* (Hypothesis 18). Furthermore, the *acknowledgement* (Hypothesis 19) one receives may increase, for example owing to growth in self-esteem if the employee is exposed to a working climate with considerable involvement on the part of the superior in his or her career development. Finally, we expect a positive relationship with one's *growth potential* (Hypothesis 20). The strengths of the relationships are expected to be the same throughout the career.

For sake of clarity, all hypotheses in this contribution are summarized in Table 1.

Table 1: Hypotheses on organizational predictor variables.

Hypothesis 1	The employee's age is negatively related to the degree of social support from his or her immediate supervisor.
Hypothesis 2	There is a positive relationship between social support from the immediate supervisor and the extent of professional knowledge. The strength of the relationship is the same in each age group.
Hypothesis 3	There is a positive relationship between social support from the immediate supervisor and the extent of meta-cognitive knowledge. The strength of the relationship is the same in each age group.
Hypothesis 4	There is a positive relationship between social support from the immediate supervisor and the number of professional skills. The strength of the relationship is the same in each age group.
Hypothesis 5	Starters receive greater social support from their near colleagues, compared with the middle-aged and the seniors. Middle-aged and seniors do not differ in this respect.
Hypothesis 6	There is a positive relationship between social support from near colleagues and the extent of professional knowledge. The strength of the relationship is the same in each age group.
Hypothesis 7	There is a positive relationship between social support from near colleagues and the extent of meta-cognitive knowledge. The strength of the relationship is the same in each age group.
Hypothesis 8	There is a positive relationship between social support from near colleagues and the number of professional skills. The strength of the relationship is the same in each age group.
Hypothesis 9	The age of the employee and the number of organizational facilities that are available to him or her are negatively related.
Hypothesis 10	There is a positive relationship between organizational facilities and the extent of professional knowledge. The strength of the relationship is the same in each age group.
Hypothesis 11	There is a positive relationship between organizational facilities and the extent of meta-cognitive knowledge. The strength of the relationship is the same in each age group.
Hypothesis 12	There is a positive relationship between organizational facilities and the number of professional skills. The strength of the relationship is the same in each age group.
Hypothesis 13	There is a positive relationship between organizational facilities and the extent of social recognition. The strength of the relationship is the same in each age group.

Table 1: (*Continued*)

Hypothesis 14	There is a positive relationship between organizational facilities and the extent of growth and flexibility. The strength of the relationship is the same in each age group.
Hypothesis 15	There is a negative relationship between age of the employee and the degree of attention from the immediate supervisor.
Hypothesis 16	There is a positive relationship between attention from the immediate supervisor and the extent of professional knowledge. The strength of the relationship is the same in each age group.
Hypothesis 17	There is a positive relationship between attention from the immediate supervisor and the extent of meta-cognitive knowledge. The strength of the relationship is the same in each age group.
Hypothesis 18	There is a positive relationship between attention from the immediate supervisor and the number of professional skills. The strength of the relationship is the same in each age group.
Hypothesis 19	There is a positive relationship between attention from the immediate supervisor and the extent of social recognition. The strength of the relationship is the same in each age group.
Hypothesis 20	There is a positive relationship between attention from the immediate supervisor and the extent of growth and flexibility. The strength of the relationship is the same in each age group.

Research Methodology

The next two sections will provide some general information about the respondents and the data collection, and about the survey.

Sampling and Data Collection

Two entrances have been used in order to obtain a substantial sample. Both the AWVN (General Employers' Association in the Netherlands) and Atrive (A Dutch Consultancy Organization for Social Housing Associations, Governments, Housing and Healthcare Combinations) have announced the study in a newsletter for the organizational members. There was an explanation of the criteria for participation and we made samples of employees appropriate for our questionnaires. These criteria concerned the maximum size of the company, i.e. 150 employees, the minimum level of functioning of the employees (middle or higher level employees), with the target of an equal distribution over three age groups (20–34 years, 35–49 years and 50+) and a reflection of different types of existing functions in the organization.

Although the EU-definition adopts 250 employees as the size limit for SMEs we believe that companies consisting of more than 150 employees resemble larger enterprises to a considerable amount as far as their personnel management strategies are concerned. In order to really differentiate the SME sample from the sample on multinationals we wanted to be sure that the general characteristics do differ from the large-firm HRM blueprint.

The category of employees aged 35–49 roughly corresponds to a category that is indicated in the term "mid-career" (Hunt & Collins 1983; Janssen 1992). Psychogerontological research has shown that the midlife experience is not strictly bound to a particular age. Some people already experience major changes in their thirties (mid-thirty crisis) whilst others only note them when they are forty or almost fifty (Munnichs 1989: 224).

As we are particularly interested in the consequences of growing older, we have decided to make a comparison of three successive age groups of the working population, namely the *starters* (20–34 years), the *middle-aged* (35–49 years) and the *seniors* (50+). In this way the whole professional career has been covered by comparing these three age groups.

The reason for limiting selection to employees active at least at a middle level of functioning or in a middle management position and not employees from lower levels was explained earlier by Boerlijst *et al.* (1993). Looking for study data which can be generalized for application in the future, we have made allowances for the possibility that the present workers, particularly the older ones, will be difficult to compare, on one point at least, with the employees who will be populating our companies in twenty years' time.

Until twenty years ago, simple functions and simple tasks were dominant in most working organizations. As a consequence, the bulk of older employees in our existing working population has a rather low level of education. As the complexity and level of difficulty of functions will on average be higher than it is now, we have every reason to expect that the average educational level will likewise have undergone a sharp rise by the year 2010.

In contrast to the situation in multinationals, *T*-tests have demonstrated that management's attitude towards, approach to and treatment of these two categories, is not significantly different. This is why the two groups of employees have been added in order to maximize the number of respondents.

One hundred and forty seven organizations have agreed to participate, producing 435 pairs of employees and their direct supervisors. Finally, data have been gathered in 95 companies from 233 employees and 217 direct supervisor, including 194 pairs. These outcomes imply a response rate of 53.6% for the employees and 49.9% for the direct supervisors. In Table 2, the number of respondents for each branch is given. A subdivision is made for the two different data sources: the employees and the supervisors.

The division of the three age groups is 31.3% starters, 45.5% middle-aged and 23.2% seniors. Despite the effort to obtain a balanced division of the two sexes in this study, from the total sample 68.0% of the respondents is male and 32.0% is female. However, it is generally known that in the higher regions of the organization most functions are fulfilled by men. In this respect, the composition of our sample can be interpreted as an advantage because the division is in fact really representative for the actual situation in Dutch SMEs and consequently more accurate.

Table 2: Number of employees, supervisors and combined ratings, per branch.

Branch	Number of Self-Ratings	Number of Supervisor Ratings	Number of Combined Ratings
Agriculture including fishing industry	8 (3.4%)	8 (3.7%)	8 (4.1%)
Manufacturing industry	29 (12.4%)	24 (11.1%)	19 (9.8%)
Building industry and installation	4 (1.7%)	5 (2.3%)	3 (1.5%)
Retail and leisure (including restaurants)	9 (3.9%)	8 (3.7%)	6 (3.1%)
Transport and communication	5 (2.1%)	3 (1.4%)	3 (1.5%)
Business services (banks, consultancy and insurances)	26 (11.2%)	25 (11.5%)	24 (12.4%)
Public services (civil services, education and health care)	143 (61.4%)	144 (66.4%)	131 (67.5%)
Unknown	9 (3.9%)	0 (0%)	
Total	233 (100%)	217 (100%)	194 pairs (100%)

The Survey

In the survey, three types of questionnaires have been used to determine the relationship between organizational predictors and professional expertise. Questionnaire A examines the occurrence of the predictor variables and was only filled in by the employees. The employees themselves seem to be in the best position to account for the predictors. It is their evaluation of possibilities to participate in career activities that matters.

Questionnaire B pertains to the self-ratings of professional expertise and had to be filled in by the individual employee while item set C had to be filled in by the corresponding immediate supervisor. Item sets B and C are nominally identical, except for the fact that the items in the self-ratings questionnaire refer to the employee him or herself, and the ones in the questionnaires for the supervisors refer to a particular employee.

All expertise items (i.e. 78 ones) refer to attributes or behaviors typically attributed to experts or outstanding performers in various field. Each item refers to one of the five dimensions. In administering the questionnaire to the supervisors, all items were adjusted into sentences like: "In that period my employee fulfilled the role of mentor for . . ." and "I think that my employee is . . . adjusting flexibly to technological changes." A six-point rating scale was added to each item rating from 1 ("very little," "never," "not at all," "very small," "very uncertain" and so on) to 6 ("very great deal of," "very often," "extremely," "to a considerable degree," "very large," "highly suitable," "very good" and so on — wording dependent on item content) on which an individual's assumed position, relative to the item in question, could be measured.

For a full appendix of the questionnaire (the original Dutch version and its English translation) and the psychometric evaluations, from both a pilot study as well as the main study in multinationals, one is referred to an article by Van der Heijden (2000). It appeared that the construction of a seemingly valid and reliable multitrait instrument for measuring degrees of individual expert performance was successful. Both for the employees' self-ratings and for the supervisor ratings, the five scales appear to be very homogeneous. Expertise is supposedly a multi-dimensional quality. The five dimensions are not fully mutually exclusive. However, they represent correlated aspects of professional expertise. This is why, after appropriate analyses, the representation of the construct (i.e. the factor structure) was oblique instead of orthogonal.

A considerable overlap consists in the meaning of the scales, given the relatively high inter-scale correlations. The distinctive power of the different scales is satisfactory, however, given the higher intra-scale correlations and the outcomes of the Multitrait Multimethod Analysis and the quantitative validation studies regarding the convergent and discriminant validity. Professional expertise does involve different types of knowledge, skills and the ability to generalize, to grow and to be flexible, as well as social recognition.

In Table 3 the reliability coefficients, using Cronbach's α are given.

For each scale of professional expertise, the reliability of the supervisor ratings is higher compared with that of the self-ratings. It has been suggested (Van der Heijden 1998) that the greater directional diversity (lower inter-item coefficients) of the self-ratings does not prove that employees give a less valid, but rather a somewhat more differentiated self-image. Each item refers to different behavior, a different behavioral style or a different outcome. Moreover, we have to bear in mind the fact that supervisors, although responsible, tend to be only superficially acquainted with their employees, especially as far as the higher-level employees are concerned.

Knowledge and *meta-cognitive knowledge* of a given employee are perhaps the most difficult person-bound factors for other people to assess validly and supervisors may find this somewhat problematic. This is why we preferred to use the self-ratings for the scales of knowledge and meta-cognitive knowledge. As far as the quantity of *skills, social recognition* and *growth and flexibility* are concerned, the assessments made by the supervisor were used. Skills, apparent from overt behavior, an evaluation of the degree of acknowledgement and an estimate of a person's growth potential are attributes for which the immediate supervisor, in general, is expected to deliver the annual performance review.

Table 3: Reliability coefficients of each scale as expressed by Cronbach's α.

Scale	Number of Items	Alpha Self-Ratings	Alpha Supervisor Ratings
Knowledge	17	0.84	0.90
Meta-cognition	15	0.88	0.91
Skill requirement	12	0.83	0.88
Social recognition	15	0.84	0.89
Growth & flexibility	19	0.85	0.92

For the measurement of the variable "*social support from immediate supervisor*" four items have been used: "Is your immediate supervisor able to evaluate the value of your work and its results?", "Does your immediate supervisor regularly express an opinion on your work?", "Is your immediate supervisor in general ready to help you with the performance of your tasks?" and "Does your immediate supervisor regularly give you supportive advice?". For the first item a six-point rating scale has been used, ranging from: (1) "not at all" to (6) "very much." For the second and fourth item a six-point rating scale has been used, ranging from: (1) "never" to (6) "very often." For the third item a six-point rating scale has been used, ranging from: (1) "In my opinion, he/she shows little willingness to help me" to (6) "In my opinion, he/she is very willing to help me." The reliability-index, using Cronbach's α, for the total item-set is 0.78.

The variable "*social support from near colleagues*" was measured by exactly the same four items, with obviously "near colleagues" instead of "immediate supervisor" in the item formulation. For this variable the reliability-index, using Cronbach's α, is 0.74.

For the measurement of the variable "*organizational facilities*" four items have been used, with the following formulation for the first three items: "In your opinion, are the possibilities in your organization to participate in . . . sufficient?" The type of facilities on the dotted line being respectively social networks, training and development courses in your own job domain, and training and development courses in a different or new job domain. The fourth item phrases: "In your opinion, are the possibilities in your organization for further personal development sufficient?" For each item a three-point rating scale has been used: (1) "no," (2) "yes, in theory, but . . ." and (3) "yes." The reliability-index, using Cronbach's α, for the total item-set is 0.68.

For the measurement of the variable "*attention to the employee's further career development from the immediate supervisor*" 24 items have been used. The different constituent items being: "Have activities concerning your career, your function or your functioning been initiated by your immediate supervisor, during the last five years?", "Have activities concerning your participation in social networks been initiated by your immediate supervisor, during the last five years?", "Have activities concerning your participation in training and development programs in your *own* job domain been initiated by your immediate supervisor, during the last five years?", "Have activities concerning your participation in training and development programs in a *different* or *new* job domain been initiated by your immediate supervisor, during the last five years?", "Have activities concerning your further personal development been initiated by your immediate supervisor, during the last five years?", "Have activities concerning your mobility to other functions or positions been initiated by your immediate supervisor, during the last five years?", "Discussions on career development in the period prior to this 'employee-supervisor' relationship.", "Discussions on career development in the *short* term.", "Discussions on career development in the *medium* term.", "Discussions on career development in the *long* term.", "Discussions or agreements on the enlargement of the degree of social support.", "Discussions or agreements on the enlargement of the employees' social network.", "Discussions or agreements on the supply and transfer of information.", "Discussions or agreements on training and development programs.", "Discussions or agreements on the improvement of the work climate.", "Discussions or agreements on the modification of the job content.", "Discussions or agreements on the mobility to another function." For each item a two-point rating scale has been used: (1) "no" and (2) "yes." The reliability-index, using Cronbach's α, for the total item-set is 0.73.

Results

The first section describes the differences in the occurrence of the organizational predictor variables in the three age groups. The relationship between the predictors and professional expertise is examined in the second section.

Differences in the Occurrence of the Organizational Predictor Variables in Three Successive Career Stages

In Table 4 the group means for the organizational predictors are given.

A One-Way Analysis of Variance (ANOVA) was applied to test whether the predictor means in the three age groups are equal (Noruis 1993: 269). In the case where the F-test appeared to be significant, independent samples T-tests were carried out in order to determine which pair(s) of age groups differ significantly regarding their mean. The outcomes of the analyses of variance are depicted in Table 5.

Only the F-test for the variable *social support from near colleagues* is significant at the 0.05 level. As regards the degree of social support from near colleagues we have found that *starters* receive greater social support from their near colleagues, compared with both the *middle-aged* ($t(174) = 2.24, p < 0.05$) and the *seniors* ($t(123) = 2.21, p < 0.05$). The difference between the middle-aged and the seniors is not significant.

With these outcomes, Hypothesis 1 is not confirmed. The p-value of the F-test is 0.072. However, the trend is in accordance with our expectations. The degree of social support from one's immediate supervisor diminishes with the employee's age. Hypothesis 5 is fully confirmed. Compared with the starters, the middle-aged and the over-fifties receive less social support from their near colleagues. The degree of social support from near colleagues is roughly the same for the middle-aged and the seniors. Hypotheses 9 and 15 regarding the relationship between age on the one hand and organizational facilities and attention from immediate supervisor on the other, must be rejected. The F-tests are not significant.

Because of limitations concerning the size of this article, extensive age effects on the measurement of professional expertise will be described elsewhere in the near future (Van der Heijden, 2001a, 2002a).

Outcomes Concerning the Regression of Scales of Professional Expertise on Organizational Predictor Variables

This section describes the outcomes concerning the regression analyses with the organizational predictor variables. The technique of *hierarchical multiple regression analysis* has been used to evaluate the contribution of age and a different predictor variable, and

Table 4: Profile of the perceptions by employees regarding organizational career activities divided by age.

Predictor Variables	Total Sample		Starters (20–34 Years)		Middle-Aged (35–49 Years)		Seniors (50 Years and Older)	
	Mean	Std. Dev.	Mean	Std. Dev.	Mean	Std. Dev.	Mean	Std. Dev.
Organizational factors								
Social support from immediate supervisor	15.48	3.16	15.90	3.18	15.61	2.97	14.65	3.39
Social support from near colleagues	14.59	2.83	15.31	3.16	14.34	2.55	14.11	2.77
Organizational facilities	8.57	2.45	9.00	2.35	8.26	2.48	8.60	2.49
Attention from immediate supervisor	6.53	4.03	7.30	4.02	7.05	4.16	5.78	3.69

Table 5: Results of analysis of variance (ANOVA) for each organizational predictor variable.

Source of Variation	Sum of Squares (SS)	Degrees of Freedom (df)	Mean Square (MS)	F	Significance
Organizational factors					
Social support from immediate supervisor				2.66	0.072
Between	52.38	2	26.19		
Within	2265.79	230	9.85		
Total	2318.16	232			
Social support from near colleagues				3.54	0.031
Between	55.59	2	27.79		
Within	1782.17	227	7.55		
Total	1837.76	229			
Organizational facilities				1.89	0.154
Between	22.51	2	11.25		
Within	1324.53	222	5.79		
Total	1347.04	224			
Attention from immediate supervisor				2.30	0.103
Between	73.92	2	36.96		
Within	3359.20	209	16.07		
Total	3433.22	211			

to determine the contribution of the combination of age and the predictor variable, these being the interaction terms.

In each case the nominal scale variable "age" was inserted in the regression equation by using dummy variables, and the predictor variable was inserted as a covariate. The interaction effects were tested hierarchically. Firstly, a regression analysis without interaction was tested, and subsequently it was examined whether the interaction added a significant contribution. Where this was not the case, the results of the analysis without interaction terms are given and interpreted. Where the interaction seemed significant, the analysis including the interaction terms has been interpreted.

The outcomes of both the multitrait-multimethod analysis approach and the quantitative validation studies using LISREL, regarding the measurement of professional expertise, justified a multivariate approach for the analyses (see Van der Heijden 2000 for more details).

Social Support from Immediate Supervisor

With our data, Hypothesis 2, concerning professional knowledge, cannot be confirmed. However, because of the fact that the multiple R is almost significant ($p = 0.06$), the results are presented in Table 6 in Appendix 1. Where the degree of social recognition is the dependent variable, 4% of the variance is explained by the degree of social support the employee receives from his or her immediate supervisor. In case growth potential is the dependent, in Table 7 in Appendix 1 one can see that 9% of the variance is explained by the degree of social support from the immediate supervisor. Contrary to our expectations, social support from the immediate supervisor is significant for the development of both social recognition as well as growth potential (see Table 8 in Appendix 1).

For meta-cognitive knowledge and skills, neither significant main nor interaction effects have been found.

Social Support from Near Colleagues

Both Hypotheses 7 and 8 have to be rejected. Contrary to our expectations, neither the development of meta-cognitive knowledge nor the development of skills is significantly influenced by the amount of social support from one's near colleagues.

Where knowledge and social recognition are the dependents some unexpected interaction effects have been found. For the starters, it seems that in both cases the relationship between the degree of social support from near colleagues and the dimension of professional expertise is negative, especially when social recognition is the dependent (see Tables 9 and 10 in Appendix 1 and Figures 1 and 2 in Appendix 2). Further research is needed to understand this pattern of outcomes.

For the middle-aged the relationship is negligible in case of the knowledge dimension and positive in case of the dimension of social recognition. For the seniors, the relationship is also positive, but stronger, in both cases. Possibly, near colleagues are in a better position to give their older colleagues feedback regarding job performance. It might be that employees have to reach a certain level in their job field before colleagues can express a fundamental judgement and give profound advice. Little by little, the employee becomes able to really demonstrate his or her capabilities. In the senior stage, the professional is remarkably able to transmit knowledge and experience which enables one's colleagues to give constructive feedback.

As far as the development of growth potential is concerned, 8% of the variance is explained by the degree of social support from near colleagues (see Table 11 in Appendix 1). Apparently, the range of the feedback that is given goes further than the job that is actually performed in the everyday practice of working life. That is to say, also future demands and the capability to adapt flexibly to changes come up.

Organizational Facilities

Organizational facilities only seem to play a role as far as the development of knowledge and growth potential are concerned (see Tables 12 and 13 in Appendix 1). Hypotheses 10 and 14 are confirmed while Hypotheses 11, 12 and 13 must be rejected. Obviously, the effect of the availability of organizational facilities is dependent upon its content. A further investigation into the character of the organizational facilities can be clarifying (see Table 14 in Appendix 1).

Table 14 reveals that less than 50% of the employees have sufficient possibilities to participate in social networks. Besides, the percentages of employees indicating that they are enabled to participate in training and development courses in a different or new job domain or aimed at a further personal development, are lower compared with the percentages of employees indicating that they have possibilities for courses in their own job domain, especially in the older age groups. It seems to be the case that, in general, management is less inclined to stimulate the employability for other functions outside the employee's own well-known territory.

It is very well conceivable that both stimulating network participation as well as stimulating the career development in a broader sense would imply that one's amount of meta-cognitive knowledge, skills and social recognition would be enlarged too. Participation in social networks enables the employee to compare his or her capabilities with colleagues in order to analyze strengths and weaknesses in his or her functioning, i.e. meta-cognitive knowledge. As far as social recognition is the dependent it might be that both participation in social networks as well as participation in programs aimed at a further personal development enable the employee to attain general credibility and a public image of being up-to-date and to build up communication skills.

It is very interesting and hopeful to find that although the character of the organizational facilities is somewhat restricted, the influence upon the amount of growth potential is positive.

Attention From the Immediate Supervisor to the Employee's Further Career Development

Five percent of the variance in professional knowledge is explained by the degree of attention from the immediate supervisor to the employee's further career development (see Table 15 in Appendix 1). The strength of the relationship between the predictor and this dimension of professional expertise is the same in each age group. The same is true in case the social recognition is the dependent (see Table 16 in Appendix 1). Hypotheses 16 and 19 are fully confirmed by these outcomes.

Hypotheses 17 and 18 must be rejected. No significant relationship between on the one hand the predictor variable, and on the other hand the extent of meta-cognitive knowledge or the amount of professional skills has been found.

Yet, for the dimension of growth and flexibility an unexpected interaction effect has been found (see Table 17 in Appendix 1). In Figure 3 in Appendix 2 the regression lines for the three age groups are depicted. We may conclude that the relationship between the amount of attention from the immediate supervisor and the amount of growth potential is rather strong for the starters, much weaker for the middle-aged and negligible in the case of the seniors. This is a very remarkable and interesting outcome. After all, the middle-aged are the ones for whom career advancement is such a central theme.

Here again, we try to account for the lack of significant results by studying the character or focus of attention from the immediate supervisor. Table 18 in Appendix 1 shows the valid percentages for the different items that comprise the scale of the predictor variable.

As shown in Table 18, for the seniors scarcely any activities aimed at a broader career development are initiated. Only 22.2% of the seniors perceived any activity from their supervisor concerning their participation in training and development programs in a different or new job domain, in the previous five years. As far as one's further personal development is concerned, only 24.1% of the seniors indicated that activities have been undertaken by their immediate supervisor.

Also, 24.1% of the seniors reported that their supervisor initiated activities concerning mobility to other functions or positions, while only 14.8% of the employees indicated that actual discussions had taken place, or agreements had been made pertaining to the subject. Indeed, only 7.4% were really in a position to take advantage of activities relating to further mobility.

Where discussions regarding the career development have taken place, they almost always focus on the here-and-now of the professional life. Thirteen percent of the seniors indicate that they have discussed their short-term career development with their immediate supervisor. However, where medium and long-term career development are concerned, the percentages are, respectively, 1.9% and 5.6%.

The data suggest that the focus of attention on the part of the immediate supervisor on the seniors is too narrow, in other words, predominantly aimed at the employee's present contribution and familiar job domain. Activities, discussions and agreements aimed at broader development and professional growth, are hardly initiated at all. For the seniors in particular, attention from the immediate supervisor for these subjects is extremely important. They are the ones for whom the enlargement of growth potential is so crucial in order to stay employable (Van der Heijden 2001b; Van der Heijden 2002b). It seems that the limited focus of attention has negative consequences, precisely for the seniors more than for other age groups.

Conclusions and Discussion

Significant age differences in the number of organizational facilities and in the degree of attention to a further development on the part of the supervisor have not been found. It seemed interesting and hopeful to find these results. Yet, a further examination of the character of the organizational facilities and of the attention from the immediate supervisor indicated that the focus of attention in many cases is rather limited, that is, predominantly aimed at the here-and-now functioning and the current job domain, especially for the older employees.

Because of the fact that the p-value for the F-test regarding the age differences in the extent of social support from the immediate supervisor is 0.072, we believe that we have good reason to conclude that the extent of supervisory attention diminishes with age. In our earlier study on over-forties (Boerlijst 1994; Boerlijst *et al.* 1993) we also found evidence for the assumption that most supervisors neglect the function and functioning of their older workers.

The extent of social support from one's near colleagues appears to decrease once the mid-life stage has been reached. A possible explanation has been sought in the competition and rivalry among middle-aged employees because of the limited opportunity to progress to a higher position, owing to an increasing flattening of working organizations. Looking at the recommendations concerning learning principles and working climate for older adults (Thijssen 1996), we must conclude that this situation is not at all favorable.

In the context of one's further professional development, our data enable us to conclude that, contrary to our expectations, social support from one's immediate supervisor plays a significant role with regard to the development of social recognition and growth potential. The relationship with the extent of professional knowledge is almost significant ($p = 0.06$).

The items comprising the scale of social support from the immediate supervisor relate to the job that is actually performed in the "here-and-now" and not directly to future job requirements. However, our outcomes do suggest that supervisors in SMEs recognize the necessity to stress future growth aspects from the individual employee. The support and advice in this direction also have positive consequences for the amount of acknowledgement the employee receives. Adaptability and flexibility is widely recognized as a distinctive competence leading to competitive advantage in a world in which one has to react to changing environmental factors such as the emergence of worldwide markets and standards for better product quality, demands for faster delivery times and closer business partnerships (Gupta & Cawthon 1996).

Possibly, the lack of significant results for the other three dimensions, i.e. knowledge, meta-cognitive knowledge and skills, is a consequence of the necessity of social support mostly aimed at survival in this fast-changing environment. Profound feedback related to the current job next to the guidance of one's growth potential could enlarge one's professional expertise in the presently performed job.

Effective leaders should create a set of values and beliefs for employees and passionately pursue them, respect and support them, set an example, focus employees' efforts on challenging goals and keep them driving toward those goals, provide the resources they need to achieve their goals, communicate with them, value their diversity, encourage creativity among them and maintain their eyes on the horizon (Scarborough & Zimmer 2000; Siropolis 1990).

Social support from near colleagues seems to be especially important for the development of one's growth potential. Here again, the outcomes suggest that attention for future job requirements and for the capability to adjust flexibly to changes are part of the evaluations, feedback and advice colleagues give. The lack of significant results for meta-cognitive knowledge and skills might be attributable to the fact that thorough support and help aimed at the current job is missing due to competition and rivalry between near colleagues in the here-and-now functioning. It is interesting to find that, even these negative phenomena can not prevent that the amount of growth potential increases by working together and exchanging support and feedback with near colleagues.

For the amount of knowledge and social recognition, some remarkable interaction effects have been found. For the starters the influence of social support from near colleagues seems to be negative for both dependents, especially in case of social recognition. For the middle-aged it is negligible when knowledge is the dependent and positive in case of social recognition. For the seniors the relationship is much stronger and positive in both cases.

Possibly, employees have to reach the phase of middle-aged regarding their professional career in order to profit from constructive feedback from colleagues. We hypothesize that this is attributable to either the character of feedback in the first career stages, meaning that it is not optimal due to the earlier mentioned rivalry and competition, or just to the fact that it is very hard to express a fundamental judgment and profound advice on the functioning of the starters.

Further research is needed to investigate which train of thought can be confirmed. Of all people, the starters are precisely the ones for whom assistance and help in order to get familiar with the "ins and outs" in the organization is such a central theme.

As far as the influence of organizational facilities is concerned we have found that they only seem to play a role in the development of professional knowledge and the extent of growth potential. In our previous study (Van der Heijden 1998) we referred to the fact that management in organizations is not inclined to stimulate the employability for other functions outside the employee's own familiar territory. However, here again it appears that we have to conclude that even although, in general, the organizational possibilities are less compared to the situation in larger companies (Van der Heijden 1998), the effects upon a broader development are more favorable.

We suggested that the other dimensions might be further developed when employees would have more opportunities to participate in social networks and follow courses, both at one's current professional domain as well as in a broader sense. Next to the development of professional knowledge and skills, these activities may be fruitful with regard to the degree of one's self-insight because of the possibilities of comparison with colleagues, and with regard to the enlargement of one's capabilities and communication skills. The latter leading to an increase of one's acknowledgment too.

As regards the influential relationship between the variable attention from the immediate supervisor to further career development and professional expertise, we have found positive relationships, the same for all age groups, with respect to the dimensions of knowledge and social recognition.

For the amount of growth potential as the dependent, a rather strong relationship has been found for the starters, while it is much weaker for the middle-aged and negligible for the seniors. Because of the fact that a further career development is such an important theme for the middle-aged it is very important to conduct further research in order to understand this pattern of outcomes.

After a further exploration of the character or focus of attention we have found that, especially for seniors, the focus is too narrow, i.e. mostly aimed at the employee's current contribution and familiar job domain. It seems that the wrong focus of attention, especially for the seniors, has negative consequences. The implications of the lack of attention for a broader development in the light of one's employability can be read in two articles on the relationship between professional expertise and the amount of employability (Van der Heijden, 2001b, 2002b).

All in all, from this study one may conclude that supervisors in SMEs are to some extent interested in the long-term future of their employees. Although the amount of attention paid to aspects related to the future functioning is lower compared to that related to current requirements, the situation is not hopeless. The results are more or less in line with the ones from the previous study in multi-nationals (Van der Heijden 1998), implying that management in SMEs also has a preoccupation with instrumental leadership, i.e. aimed at the here-and-now and less future-oriented, instead of appropriate people management (see also Boerlijst *et al.* 1993). Yet, the fact that some organizational factors do positively influence the development of growth potential gives us starting-points to further guide the professional development of individual employees in small and medium-sized enterprises.

Although small firms often do not employ professional experts to manage human resource issues, are considerably less likely to be unionized (and thus have more freedom in determining their human resource strategy), and vary markedly with regard to the provision of training (see Marlow & Patton 1993), there are lots of starting-points to use HRM in a strategic way with the goal of gaining competitive advantage.

SMEs are the ideal site for the application of HRM practices. Firstly, communications are more direct, people have to work more flexibly, the hierarchy is flatter, the impact of each employee on organizational performance is clearer and the greater insecurity makes the organization more responsive to changes in market and customer demands. Secondly, the nature of change programs in small businesses appears to be much more informal and organic. Larger organizations, with relatively bureaucratic formal change programs, may have much to learn from SMEs in this respect. Rather than taking the absence of larger formal programs to be a weakness of HRM in smaller organizations, it may well be the competitive advantage of them (Bacon *et al.* 1996).

A weakness of our study is that as far as the independent variables are concerned, we only have self-report measurements. While individuals strive to achieve consistency in their self-reported response pattern, it could be that the variables pertaining to the amount of organizational facilities are clustered (see also Kasl 1978). This means that the fact that, for example, seniors score more negative on these variables could either be due to striving to achieve consistency in their response pattern, or it could indicate that they genuinely suffer from a lack of attention, as is apparent from a wide range of measurements.

In line with this it would be interesting to compare the employee's assessments with the one's made by supervisors concerning the occurrence of the predictor variables. Because we were primarily interested in how employees experience the possibilities for career activities, we submitted this part of our survey to them. However, in the context of our wish to gain more insight into the communication that takes place between an employee and his or her immediate supervisor, and the differences in experience, it seems advisable to take the opinion of the supervisor into account (see also Stoker & Van der Heijden 2001).

Another point is the choice of the three age categories as representations for three different stages in the professional career. Although in the section on the methodology a justification has been given, it is still dubious whether the boundaries have been chosen in such a way that the different career stages people go through are correctly covered. Increasingly, career research and career problems are becoming individualized. This is why real longitudinal research in the area of career psychology is essential and recommended.

Another point of concern is the fact that the public sector is over-represented (see Table 2 on the sample structure). If one leaves out the companies for which it is unknown whether they belong to the public or the profit sector (112,490 companies out of a total set of 693,600 for which it is unknown), the division between public and profit sector is 0.3% vs. 99.7%. This implies a bias in the sample. It is important in future research to compare HRM practices in public vs. profit sector in order to understand the differences in personnel management strategies more profoundly.

Concerning the operationalization of some predictor variables, like Thijssen (1996) we defend the use of more concrete measurements of participation in career development programmes and activities. More detailed information concerning the content, duration and subject of these programmes and activities could contribute considerably to our knowledge concerning the value of such programmes in the context of further professional development.

Further research is also needed to determine whether the supposed direction of the hypothesized and tested relationships have been correctly chosen. Our research is unidirectional: individual factors and their interaction with age are described as antecedents of professional expertise.

However, it is not inconceivable that reciprocal relationships exist, meaning that the degree of professional expertise influences the number of career initiatives that are within a person's reach. The degree of employability, in its turn, could also determine the amount of effort that is invested in further development from the side of the organization as well as that of the individual.

References

Alexander, P. A., Schallert, D. L., & Hare, V. C. (1991). Coming to terms: How researchers in learning and literacy talk about knowledge. *Review of Educational Research, 61*(3), 315–344.

Bacon, N., Ackers, P., Storey, J., & Coates, D. (1996). It's a small world: Managing human resources in small business. *The International Journal of Human Resource Management, 7*(1), 82–100.

Becker, G. (1993). *Human capital: A theoretical and empirical analysis, with special reference to education* (3rd ed.). Chicago: University of Chicago Press.

Bereiter, C., & Scardamalia, M. (1993). *Surpassing ourselves. An inquiry into the nature and implications of expertise*. Chicago: Open Court.

Boerlijst, J. G. (1994). The neglect of growth and development of employees over 40 in organizations: A managerial and training problem. In: J. Snel, & J. Cremer (Eds), *Work and aging* (pp. 251–271). London: Taylor & Francis.

Boerlijst, J. G., Van der Heijden, B. I. J. M., & Van Assen, A. (1993). *Veertig-plussers in de onderneming* [Over-forties in the Organization]. Van Gorcum/Stichting Management Studies, Assen.

Boerlijst, J. G., Van der Heijden, B. I. J. M., & Verhelst, N. D. (1996). The measurement of expertise. Paper presented at the Fourth Asia Pacific Conference on Giftedness, Jakarta, 4–8 August 1996. Theme: Optimizing Excellence and Human Resource Development. Organized by the University of Indonesia, Faculty of Psychology and the Indonesian Foundation on Education and Development of Gifted Children, 4–8 August 1996. The Conference is on Collaboration with the Asia-Pacific Federation of the World Council on Gifted and Talented Children.

Chanaron, J. (1998). *Managing innovation in European small and medium-sized enterprises.* Nijmegen Lectures on Innovation Management, Nijmegen Business School, Catholic University Nijmegen, Maklu, Antwerpen.

Chi, M. T. H., Glaser, R., & Farr, M. J. (Eds) (1988). *The nature of expertise.* Hillsdale, NJ: Lawrence Erlbaum.

Deshpande, S. P., & Golhar, D. Y. (1994). HRM practices in large and small manufacturing firms: A comparative study. *Journal of Small Business Management, 32*(2), 49–56.

Duberly, J. P., & Walley, P. (1995). Assessing the adoption of HRM by small and medium-sized manufacturing organizations. *The International Journal of Human Resource Management, 6*(4), 891–909.

During, W., & Kerkhof, M. (1998). Management styles and excellence: Different ways to business-success in European SMEs. In: W. During, & R. P. Oakey (Eds), *New technology-based firms in the 1990s* (Vol. IV, pp. 38–51). London: Paul Chapman Publishing Ltd.

Ericsson, K. A. (1996). *The road to excellence. The acquisition of expert performance in the arts and sciences, sports and games.* Mahwah, NJ: Lawrence Erlbaum.

Ericsson, K. A., & Lehmann, A. C. (1996). Expert and exceptional performance: Evidence on maximal adaptations on task constraints. *Annual Review of Psychology, 47,* 273–305.

Ericsson, K. A., & Smith, J. (1991). Prospects and limits of the empirical study of expertise: An introduction. In: K. A. Ericsson, & J. Smith (Eds), *Toward a general theory of expertise. Prospects and limits* (pp. 1–38). Cambridge: Cambridge University Press.

Gupta, M., & Cawthon, G. (1996). Managerial implications of flexible manufacturing for small/medium-sized enterprises. *Technovation, 16*(2), 77–83.

Hendry, C., Arthur, M. B., & Jones, A. M. (1995). *Strategy through people — adaptation and learning in the small-medium enterprise.* London: Routledge.

Hornsby, J. S., & Kuratko, D. F. (1990). Human resource management in small business: Critical issues for the 1990s. *Journal of Small Business Management, 28*(3), 9–18.

Hunt, J., & Collins, R. R. (1983). *Managers in mid career crisis.* Sydney: Wellington Lane Press.

Janssen, P. (1992). *Relatieve deprivatie in de middenloopbaanfase bij hoger opgeleide mannen. Een vergelijking tussen drie leeftijdsgroepen* [Relative deprivation in the mid-career phase in more highly educated males. A comparison of three age groups]. Ph.D. Thesis, Catholic University Nijmegen, Datawyse, Maastricht.

Kasl, S. V. (1978). Epidemiological contributions to the study of work stress. In: C. L. Cooper, & R. Payne (Eds), *Stress at work* (pp. 3–48). Chichester: Wiley.

Koch, C., & De Kok, J. (1999). *A Human-resource-based Theory of the small firm.* EIM, Zoetermeer.

Koch, C. L. Y., & Van Straten, E. (1997). *Strategische verkenning. Personeelsbeleid in enkele MKB-bedrijven. Een inventarisatie* [Strategic exploration. Personnel management in a few SMEs]. EIM, Zoetermeer.

Legge, K. (1995). *Human resource management, rhetorics and realities.* London: Macmillan.

Marlow, S., & Patton, D. (1993). Managing the employment relationship in the small firm: Possibilities for human resource management. *International Small Business Journal, 11*(4), 57–64.

Munnichs, J. M. A. (1989). Loopbaan, levensloop en de midleven-ervaring [Career, life course and the mid-life experience]. In: J. Munnichs, & G. Uildriks (Eds), *Psychogerontologie. Een inleidend leerboek over ouder worden, persoonlijkheid, zingeving, levensloop en tijd, sociale context, gezondheid en interventie* (pp. 223–228). Van Loghum Slaterus, Deventer.

Noruis, M. J. (1993). *SPSS for windows: Base system user's guide.* Release 6.0. Chicago: SPSS.

Onstenk, J. H. A. M. (1993). Het opleiden van laaggeschoolde oudere werknemers [Training semi-skilled and unskilled older workers]. In: J. J. Peters *et al.* (Eds), *Gids voor de opleidingspraktijk. Visies, modellen en technieke.* Bohn Stafleu Van Loghum, Houten.

Organisation for Economic Co-operation and Development (OECD) (1997). *Globalisation and small and medium sized enterprises (SMEs), 1.* Synthesis Report, OECD, Paris.

Plett, P. C., Lester, B. T., & Yocum, K. L. (1991). *Training for older people.* Geneva: International Labour Office.

Royal Association MKB, the Netherlands (2000). *Internet-site.*

Sarason, B. R., Shearin, E. N., Pearce, G. R., & Sarason, I. G. (1987). Interrelations of social support measures: Theoretical and practical implications. *Journal of Personality and Social Psychology, 52,* 813–832.

Scarborough, N. M., & Zimmer, T. W. (2000). *Effective small business management. An entrepreneurial approach* (6th ed.). NJ: Prentice-Hall.

Schein, E. H. (1978). *Career dynamics: Matching individual and organizational needs.* Reading, MA: Addison-Wesley.

Schultz, T. W. (1971). *Investment in human capital: The role of education and research.* New York: Free Press.

Siropolis, N. C. (1990). *Small business management. A guide to entrepreneurship* (4th ed.). Boston: Houghton Mifflin.

Stickland, R. (1996). Career self-management — can we live without it? *European Journal of Work and Organizational Psychology, 5*(4), 583–596.

Stoker, J. I., & Van der Heijden, B. I. J. M. (2001). Competence development and appraisal in organizations. *Journal of Career Development, 28*(2), 97–113.

Storey, D. J. (1992). *Developments in the management of human resources.* Princeton University Press.

Storey, D. J. (1994). *Understanding the small business sector.* London: Routledge.

Thijssen, J. G. L. (1996). *Leren, leeftijd en loopbaanperspectief. Opleidingsdeelname door oudere personeelsleden als component van human resource development* [Learning, aging and career prospect]. Ph.D. Thesis, Catholic University Brabant, Tilburg.

Trost, G. (1993). Prediction of excellence in school, university and work. In: K. Heller, F. J. Mönks, & A. H. Passow (Eds), *International handbook of research and development of giftedness and talent* (pp. 325–336). Oxford: Pergamon Press.

Van der Heijden, B. I. J. M. (1996). *Life-long expertise development: Goal of the nineties.* Book of Proceedings of the Fifth Conference on International Human Resource Management, 24–28 June 1996, Hyatt Islandia San Diego, California, USA.

Van der Heijden, B. I. J. M. (1998). *The measurement and development of professional expertise throughout the career. A retrospective study among higher level Dutch professionals.* Ph.D. Thesis, University of Twente, PrintPartners Ipskamp, Enschede.

Van der Heijden, B. I. J. M. (2000). The development and psychometric evaluation of a multi-dimensional measurement instrument of professional expertise. *High Ability Studies. The Journal of the European Council for High Ability, 11*(1), 9–39.

Van der Heijden, B. I. J. M. (2001a). Age and assessments of professional expertise. The relationship between higher level employees' age and self-assessments or supervisor ratings of professional expertise. *International Journal of Selection and Assessment, 9*(4), 309–324.

Van der Heijden, B. I. J. M. (2001b). Professional competence and its effect upon employability throughout the career. A study among Dutch middle and higher level employees in small businesses. *The International Journal of Entrepreneurship and Innovation, 2*(3), 171–181.

Van der Heijden, B. I. J. M. (2002a). Age and assessments of professional expertise in small and medium-sized enterprises. Differences between self-ratings and supervisor ratings. *International Journal of Human Resource Development and Management, 2*(3/4), 329–343.

Van der Heijden, B. I. J. M. (2002b). Prerequisites to guarantee life-long employability. *Personnel Review, 31*(1&2), 44–61.

Van der Heijden, B. I. J. M., & Boerlijst, J. G. (1997). Bekwamer door loopbaanontwikkeling; tussen ervaringsconcentratie en ervaringsfragmentatie [More competent through career development; between experience concentration and experience fragmentation]. In: H. Leenen, B. Rosendaal, & H. van der Zee (Eds), *Concurreren op deskundigheid; ervaringen, instrumenten, concepten.* Samson Bedrijfsinformatie, Alphen aan den Rijn/Diegem.

Ylinenpää, H. (1997). *Managing competence development and acquisition in small manufacturing firms.* Ph.D. Thesis, Luleå University of Technology, Luleå.

APPENDIX 1

Table 6: Hierarchical multiple regression analysis with the extent of knowledge as dependent variable and with age and social support from immediate supervisor as predictor variables ($N = 233$).

Variables	Knowledge					
	B-Coefficient	β-Coefficient	ΔR^2	Multiple *R*	R^2	*F*
1. Middle-aged	0.13^*	0.15^*				
2. Seniors	0.07 n.s.	0.07 n.s.				
			0.02			
3. Social support from immediate supervisor	0.02 ($p = 0.05$)	0.13 ($p = 0.05$)	0.02	0.18	0.03	2.55 ($p = 0.06$) (df 3,229)

Note: n.s.: not significant.
$^*p < 0.05$.

Table 7: Hierarchical multiple regression analysis with the degree of social recognition as dependent variable and with age and social support from immediate supervisor as predictor variables ($N = 194$).

Variables	Social Recognition					
	B-Coefficient	β-Coefficient	ΔR^2	Multiple *R*	R^2	*F*
1. Middle-aged	−0.09 n.s.	−0.08 n.s.				
2. Seniors	−0.20 n.s.	−0.15 n.s.				
			0.02			
3. Social support from immediate supervisor	0.03^*	0.15^*	0.02	0.21	0.04	2.90^* (df 3,190)

Note: n.s.: not significant.
$^*p < 0.05$.

Table 8: Hierarchical multiple regression analysis with the degree of growth and flexibility as dependent variable and with age and social support from immediate supervisor as predictor variables ($N = 194$).

Variables	Growth and Flexibility					
	B-Coefficient	**β-Coefficient**	**ΔR^2**	**Multiple R**	**R^2**	**F**
1. Middle-aged	−0.14 n.s.	−0.12 n.s.				
2. Seniors	−0.32**	−0.23**				
			0.04			
3. Social support from immediate supervisor	0.04**	0.20**	0.04	0.30	0.09	6.17** (df 3,190)

Note: n.s.: not significant.
**$p < 0.01$.

Table 9: Hierarchical multiple regression analysis with the extent of knowledge as dependent variable and with age and social support from near colleagues as predictor variables ($N = 230$).

Variables	Knowledge					
	B-Coefficient	**β-Coefficient**	**ΔR^2**	**Multiple R**	**R^2**	**F**
1. Middle-aged	−0.16 n.s.	−0.18 n.s.				
2. Seniors	−0.92*	−0.89*				
			0.03			
3. Social support from near colleagues	−0.01 n.s.	−0.06 n.s.	0.002			
4. Middle-aged × soc. Support from near colleagues	0.02 n.s.	0.33 n.s.				
5. Senior × social support from near colleagues	0.07**	0.96**	0.03	0.23	0.05	2.58* (df 5,224)

Note: n.s.: not significant.
*$p < 0.05$.
**$p < 0.01$.

Table 10: Hierarchical multiple regression analysis with the degree of social recognition as dependent variable and with age and social support from near colleagues as predictor variables ($N = 192$).

Variables	Social Recognition					
	B-Coefficient	β-Coefficient	ΔR^2	Multiple *R*	R^2	*F*
1. Middle-aged	−0.83 n.s.	−0.73 n.s.				
2. Seniors	−1.42**	−1.05**				
			0.04			
3. Social support from near colleagues	−0.02 n.s.	−0.09 n.s.	0.003			
4. Middle-aged × soc. support from near colleagues	0.05 n.s.	0.66 n.s.				
5. Senior × social support from near colleagues	0.08*	0.90*	0.03	0.26	0.07	2.59** (df 5,186)

Note: n.s.: not significant.
*$p < 0.05$.
**$p < 0.01$.

Table 11: Hierarchical multiple regression analysis with the degree of growth and flexibility as dependent variable and with age and social support from near colleagues as predictor variables ($N = 192$).

Variables	Growth and Flexibility					
	B-Coefficient	β-Coefficient	ΔR^2	Multiple *R*	R^2	*F*
1. Middle-aged	−0.12 n.s.	−0.10 n.s.				
2. Seniors	−0.33**	−0.23**				
			0.04			
3. Social support from near colleagues	0.04*	0.17*	0.03	0.28	0.08	5.23** (df 3,188)

Note: n.s.: not significant.
*$p < 0.05$.
**$p < 0.01$.

Table 12: Hierarchical multiple regression analysis with the extent of knowledge as dependent variable and with age and organizational facilities as predictor variables ($N = 225$).

Variables	Knowledge					
	B-Coefficient	β-Coefficient	ΔR^2	Multiple R	R^2	F
1. Middle-aged	0.15[*]	0.17[*]				
2. Seniors	0.06 n.s.	0.06 n.s.				
			0.02			
3. Organizational facilities	0.03[*]	0.15[*]	0.02			
				0.20	0.04	2.99[*] (df 3,221)

Note: n.s.: not significant.
[*]$p < 0.05$.

Table 13: Hierarchical multiple regression analysis with the degree of growth and flexibility as dependent variable and with age and organizational facilities as predictor variables ($N = 187$).

Variables	Growth and Flexibility					
	B-Coefficient	β-Coefficient	ΔR^2	Multiple R	R^2	F
1. Middle-aged	−0.12 n.s.	−0.10 n.s.				
2. Seniors	−0.35[**]	−0.25[**]				
			0.05			
3. Organizational facilities	0.04[*]	0.18[*]	0.03			
				0.29	0.08	5.42[**] (df 3,183)

Note: n.s.: not significant.
[*]$p < 0.05$.
[**]$p < 0.01$.

Table 14: Frequency distribution and response rate in middle and higher level employees, per age group, for the items comprising the scale of "organizational facilities."

Scale Items	Answer Category	Starters (N = 73)	Middle-Aged (N = 106)	Seniors (N = 53)
In your opinion, are the possibilities in your organization to participate in *social networks* sufficient?	(1) No	18 (25.0%)	32 (30.2%)	10 (18.9%)
	(2) Yes, in theory, but there are some impediments	24 (33.3%)	32 (30.2%)	19 (35.8%)
	(3) Yes	30 (41.7%)	42 (39.6%)	24 (45.3%)
In your opinion, are the possibilities in your organization to participate in training and development courses in your *own job domain* sufficient?	(1) No	12 (16.7%)	20 (19.0%)	10 (18.5%)
	(2) Yes, in theory, but there are some impediments	19 (26.4%)	25 (23.8%)	14 (25.9%)
	(3) Yes	41 (56.9%)	60 (57.1%)	30 (55.6%)
In your opinion, are the possibilities in your organization to participate in training and development courses in a *different or new job domain* sufficient?	(1) No	20 (27.8%)	48 (46.2%)	22 (40.7%)
	(2) Yes, in theory, but there are some impediments	18 (25.0%)	33 (31.7%)	14 (25.9%)
	(3) Yes	34 (47.2%)	23 (22.1%)	18 (33.3%)
In your opinion, are the possibilities in your organization for *further personal development* sufficient?	(1) No	18 (25.0%)	36 (34.6%)	18 (34.0%)
	(2) Yes, in theory, but there are some impediments	16 (22.2%)	31 (29.8%)	14 (26.4%)
	(3) Yes	38 (52.8%)	37 (35.6%)	21 (39.6%)

Table 15: Hierarchical multiple regression analysis with the extent of knowledge as dependent variable and with age and attention from immediate supervisor as predictor variables ($N = 212$).

Variables	Knowledge					
	B-Coefficient	β-Coefficient	ΔR^2	Multiple R	R^2	*F*
1. Middle-aged	0.13 n.s.	0.15 n.s.				
2. Seniors	0.08 n.s.	0.08 n.s.				
			0.02			
3. Attention from immediate supervisor	0.02*	0.17*	0.03	0.21	0.05	3.31* (df 3,208)

Note: n.s.: not significant.
*$p < 0.05$.

Table 16: Hierarchical multiple regression analysis with the degree of social recognition as dependent variable and with age and attention from immediate supervisor as predictor variables ($N = 177$).

Variables	Social Recognition					
	B-Coefficient	β-Coefficient	ΔR^2	Multiple R	R^2	*F*
1. Middle-aged	−0.09 n.s.	−0.08 n.s.				
2. Seniors	−0.20 n.s.	−0.14 n.s.				
			0.02			
3. Attention from immediate supervisor	0.03*	0.18*	0.03	0.23	0.05	3.23* (df 3,173)

Note: n.s.: not significant.
*$p < 0.05$.

Table 17: Hierarchical multiple regression analysis with the amount of growth and flexibility as dependent variable and with age and attention from immediate supervisor as predictor variables ($N = 177$).

Variables	Growth and Flexibility					
	B-Coefficient	β-Coefficient	ΔR^2	Multiple *R*	R^2	*F*
1. Middle-aged	0.04 n.s.	0.03 n.s.				
2. Seniors	−0.09 n.s.	−0.06 n.s.				
			0.002			
3. Attention from immediate supervisor	0.04*	0.27*	0.02			
4. Middle-aged × attention from supervisor	−0.03 n.s.	−0.19 n.s.				
5. Senior × attention from supervisor	−0.04 n.s.	−0.019 n.s.				
			0.009			
				0.28	0.08	2.87* (df 5,171)

Note: n.s.: not significant.
*$p < 0.05$.

Table 18: Percentages of items, per age group of the employee, comprising the scale "attention for the employee's further career development from the immediate supervisor."

Scale Items	Starters (*N* = 73)	Middle-Aged (*N* = 106)	Seniors (*N* = 54)
Have activities concerning your career, your function or your functioning been initiated by your immediate supervisor, during the last five years?	50 (68.5%)	59 (55.7%)	26 (48.1%)
Have activities concerning your participation in social networks been initiated by your immediate supervisor, during the last five years?	38 (52.1%)	42 (39.6%)	21 (38.9%)
Have activities concerning your participation in training and development programs in your *own* job domain been initiated by your immediate supervisor, during the last five years?	44 (60.3%)	63 (59.4%)	27 (50.0%)
Have activities concerning your participation in training and development programs in a *different* or *new* job domain been initiated by your immediate supervisor, during the last five years?	26 (35.6%)	22 (20.8%)	12 (22.2%)
Have activities concerning your further personal development been initiated by your immediate supervisor, during the last five years?	29 (39.7%)	27 (25.5%)	13 (24.1%)
Have activities concerning your mobility to other functions or positions been initiated by your immediate supervisor, during the last five years?	33 (45.2%)	32 (30.2%)	13 (24.1%)
Discussions on career development in the period prior to this "employee-supervisor relationship."	5 (6.8%)	6 (5.7%)	2 (3.7%)

Table 18: (*Continued*)

Scale Items	Starters (N = 73)	Middle-Aged (N = 106)	Seniors (N = 54)
Discussions on career development in the *short* term.	37 (50.7%)	34 (32.1%)	7 (13.0%)
Discussions on career development in the *medium* term.	10 (13.7%)	11 (10.4%)	1 (1.9%)
Discussions on career development in the *long* term.	1 (1.4%)	3 (2.8%)	3 (5.6%)
Discussions or agreements on enlargement of the degree of social support.	3 (4.1%)	6 (5.7%)	5 (9.3%)
Discussions or agreements on enlargement of the employees' social network.	9 (12.3%)	8 (7.5%)	7 (13.0%)
Discussions or agreements on supply and transfer of information.	14 (19.2%)	47 (44.3%)	23 (42.6%)
Discussions or agreements on training and education programs.	46 (63.0%)	65 (61.3%)	24 (44.4%)
Discussions or agreements on improvement of the work climate.	16 (21.9%)	42 (39.6%)	14 (25.9%)
Discussions or agreements on modification of the job content.	31 (42.5%)	54 (50.9%)	27 (50.0%)
Discussions or agreements on mobility to another function.	9 (12.3%)	14 (13.2%)	8 (14.8%)
Activities on enlargement of the amount of social support.	1 (1.4%)	5 (4.7%)	4 (7.4%)
Activities on enlargement of the employees' social network.	7 (9.6%)	10 (9.4%)	5(9.3%)
Activities on supply and transfer of information.	9 (12.3%)	40 (37.7%)	22 (40.7%)
Activities on creation of possibilities to participate in training and education programs.	40 (54.8%)	53 (50.0%)	18 (33.3%)
Activities on improvement of the work climate.	9 (12.3%)	29 (27.4%)	13 (24.1%)
Activities on modification of job content.	32 (43.8%)	47 (44.3%)	19 (35.2%)
Activities on mobility to another function.	15 (20.5%)	16 (15.1%)	4 (7.4%)

APPENDIX 2

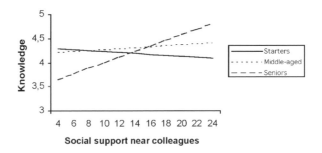

Figure 1: Interaction effect between age and the degree of social support from near colleagues, with the amount of knowledge as the dependent.

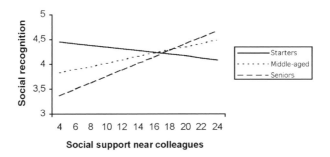

Figure 2: Interaction effect between age and the degree of social support from near colleagues, with the amount of social recognition as the dependent.

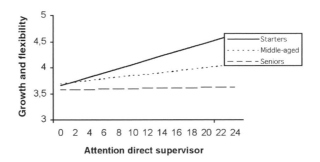

Figure 3: Interaction effects between age and the amount of attention from the immediate supervisor, with the extent of growth and flexibility as the dependent.

Chapter 14

Examining the Mental Models of Entrepreneurs from Born Global and Gradual Globalizing Firms

Paula D. Harveston, David Osborne and Ben L. Kedia

Introduction

In recent years, the field of entrepreneurship has made some strides in explaining the internationalization process of new firms (i.e. Bloodgood *et al.* 1996; McDougall *et al.* 1994; Westhead *et al.* 1998). The emergence of a large number of firms that are international *at or near inception*, and deriving a substantial proportion of sales revenue from foreign markets (i.e. born global) is driving some of this research (i.e. Cavusgil & Knight 1997; Harveston 2000; Knight 1997; Oviatt & McDougall 1994). Other new ventures that internationalize at a slower pace are considered gradually globalizing firms. The purpose of this research is to investigate the differences between the mental models of entrepreneurs leading born global and gradually globalizing firms through narrative content examination.

From an entrepreneurship theory perspective, internationalization results from the firm's search to obtain valuable resources at the lowest cost, to serve global niche markets with unique products and services, or to be founded by internationally experienced entrepreneurs (Oviatt & McDougall 1997). The firm is driven by the entrepreneur who is able to act on opportunities that others do not. The drive to be international may be felt from the very beginning in born global firms. Thus, the impact of the leader/founder of the firm should not be discounted. For instance, one entrepreneur who founded an international new venture commented that, "the advantage of starting internationally is that you establish an international spirit from the very beginning" (Mamis 1989: 38). The contribution of entrepreneurship theory is that the firm and its internationalization processes should not be viewed from a firm or industry level. Instead, the main focus of the analysis is on the entrepreneur and his or her influence on firm internationalization (McDougall *et al.* 1994). Several case studies have focused on the international activities of new ventures (e.g. Jolly *et al.* 1992; McDougall *et al.* 1994; Ray 1989).

New Technology-Based Firms in the New Millennium, Volume III
Copyright © 2004 by Elsevier Ltd.
All rights of reproduction in any form reserved.
ISBN: 0-08-044402-4

In this paper, we focus on the language of the entrepreneur in explaining the mental models of leaders of born global and gradual globalizing firms. First, we broadly discuss the previous research examining international entrepreneurship. Secondly, we focus on the role of the entrepreneurs' mental models affecting internationalization. Next, the methods section describes our approach to narrative analysis, including a description of the data and the statistical methodology. The paper closes with a review of the results, discussion, a consideration of limitations, and suggestions for future research.

Literature Review

Over the past few years the phenomena of international entrepreneurship has received increasing attention across a variety of fields including entrepreneurship, international business, marketing and strategic management. Most of the research on the international business activities of small companies has focused on exporting. Several areas of interest have been investigated including the propensity to export (Cavusgil 1994; O'Rourke 1985; Pak & Weaver 1990; Ursic & Czinkota 1984), differences between exporters and non-exporters (Bilkey 1978; Johnston & Czinkota 1982; Kedia & Chhokar 1986a, b; Langston & Teas 1976; Tookey 1964; Wiedersheim-Paul *et al.* 1978) and factors leading to export success (Christensen *et al.* 1987; Cooper & Kleinschmidt 1985; Johnston & Czinkota 1982; Tookey 1964).

In terms of propensity to export, research has mainly identified managerial characteristics as motives. For example, topics of major investigation include the manager's background (skills, travel experience, education), attitude toward international business, perceptions of risk, and costs or opportunities in international expansion. Aharoni's (1966) and Johanson & Vahlne's (1977) behavioral explanations of the management decision process serve as a theoretical foundation for this type of research (Kedia & Chhokar 1986a; Reid 1981; Ursic & Czinkota 1984; Welch & Wiedersheim-Paul 1980). Collectively, these studies suggest that in small companies a positive attitude of the owner/founder toward internationalization will be a distinguishing factor between exporters and non-exporters and will influence firm performance. Similarly, perceptions of costs, risks, and opportunities will also differentiate between exporters and non-exporters. While other factors may play important roles in internationalization, in this study we will focus solely on the entrepreneur and the mental models that affect internationalization. The following section develops hypotheses to test differences among entrepreneurs leading born global firms and gradual globalizing firms and the extensiveness of their internationalization.

Mental Models of Decision Makers

March & Simon (1958) argue that, due to a large behavioral component, strategic choices reflect the idiosyncrasies of decision-makers to a great extent. To every administrative situation, each decision-maker brings his or her own set of "givens" which reflect his or her values and principles. The importance of mental models in strategic planning and implementation have been studied in a variety of areas including cognitive strategic groups (i.e. Barr *et al.* 1992; Reger & Huff 1993; Thomas *et al.* 1993), internationalization (Murtha *et al.* 1998), marketing strategy (Menon *et al.* 1999), and new venture performance (Ensley & Pearce 2001).

Research has concluded that mental models make a difference, and for internationalization, the characteristics of the owner/founder (decision-maker) contribute to, and affect, the decision to internationalize (Bilkey 1978; Reid 1980). Indeed, more attention and greater examination of the entrepreneur and managerial characteristics have been called for in international entrepreneurship research (McDougall *et al.* 1994).

Many researchers have found that managerial mindsets affect internationalization (i.e. Bartlett & Ghoshal 1989; Bloodgood *et al.* 1996; Calof & Beamish 1994; Murtha *et al.* 1998; Wilson 1998). Bartlett & Ghoshal (1989) argue that a manager's cognitive perspectives affect the international strategic capabilities of the firm. Put simply, the mindset of the entrepreneur and the top management team affects the willingness to expand the firm's activities into international markets.

Conceptually, this line of reasoning has its foundations in the work of Perlmutter (1969). He posited that; "the more one penetrates into the living reality of an international firm, the more one finds it is necessary to give serious weight to the way executives think about doing business around the world" (p. 11). Perlmutter (1969) described three types of mindsets. Ethnocentric mindsets display a home country orientation, with a tendency to view the world from the perspective that the home country is best, while polycentric mindsets reflect a host country orientation with the view that the host country is the best. The final way of thinking reflects a geocentric mindset where the view is world-based. No single country market is best and the world is viewed as a whole.

Kothari (1978) was one of the first researchers to empirically examine Purlmutter's ideas. In a survey of Texas small manufacturers, he found that most strategies of small firms emphasize an ethnocentric focus. Other studies have used the managerial attitude toward internationalization as an indication of the managerial mindset. In general, these studies have found that managers with more positive attitudes toward international activities are more likely to engage in international activities (Ali & Swiercz 1991; Cavusgil 1980; Cooper & Kleinschmidt 1985; Lindqvist 1990; Ursic & Czinkota 1984).

There have been several studies that have directly examined the concept of managerial mindset. In linking managerial mindset to firm performance, Calof & Beamish (1994) found that firms with a geocentric orientation performed better than firms with other types of orientations (i.e. ethnocentric). Their results indicate that one of the most critical issues in the development of creating geocentric firms was international assignments and training for managers. In a similar vein, Kobrin (1994) found that the geocentric mindset of firm managers was positively associated with the extent of internationalization (percentage of sales, employees abroad, and manufacturing presence) and success in serving global markets.

Building on Kobrin's work, Murtha *et al.* (1998) developed a new way to evaluate managerial mental models and developed three new scales to measure managerial mindsets. Others have also linked geocentrism to export success (Dichtl *et al.* 1990) and higher growth (Norburn 1987). Taggart's (1998) study of the managers of Japanese, American and Continental European firms operating in the U.K. found that the polycentric mindset is vastly under-represented. He urged international firms to adopt a number of perspectives and use a variety of models when assessing international activities.

In terms of managers of born global and gradual globalizing firms and their mindsets, there is very little research. One exception is the work of Cavusgil & Knight (1997). In examining managerial mindset of born global firm managers, they further delineated the mindset concept and its subsequent link to firm performance. Their results show that managers of born global firms have a "global orientation" which is a "bundle of dispositions and competencies the sum of which appears to be positively correlated with export-marketing performance" (Cavusgil & Knight 1997: 5). In support of this finding, Kobrin (1994) had previously found that firm managers with more world-oriented mindsets (geocentric) are more likely to quickly enter international markets.

In this study, it is expected that the mental models of managers of gradual globalizing firms will be more negative toward internationalization but will be open to international opportunities if they arise. As suggested by Eriksson *et al.* (1997), managers of gradual globalizing firms are expected to have more ethnocentric mindsets and only act when they are able "to detect opportunities and reduce the uncertainties of going abroad" (p. 342). In contrast, managers of born global firms are expected to have more positive mental models toward internationalization. A recent case study found that geocentrism (more global mindset) is linked to the ability of born global firms to create their own opportunities in the international marketplace (Cavusgil & Knight 1997). Based on the preceding arguments, the following hypotheses are proposed:

Hypothesis 1. Entrepreneurs leading born global firms will have more positive mental models reflected in more positive language toward internationalization than entrepreneurs leading gradually globalizing firms.

Hypothesis 2. Drivers of internationalization (mental models) as reflected by language will be different for entrepreneurs leading born global firms and entrepreneurs leading gradually globalizing firms.

Methodology

To allow for maximum generalizability, a national sample of international entrepreneurial firms was utilized. A U.S. national sample reduces any bias due to regional economic variations. The sample was drawn from a database maintained by CorpTech. The database contained archival information on all firms in the sample and was used to compare the groups across broad categories (total sales, international sales, year the firm was founded, and number of employees) to test for non-response bias. A second consideration taken into account for sampling was the age of the business. Because the focus of this study is on born global and gradual globalizing firms and how they internationalized, it was necessary to sample firms that were founded and have generated the opportunity to internationalize in the same time frame. Thus, the sample of 1611 firms was founded within a five-year period spanning 1990–1994. All of the firms in the sample had some type of international activity that accounted for a minimum of 1% of sales to a maximum of over 70% of the firm's sales. A control group of 200 firms was also surveyed. These firms met all of the same criteria as the sample, except international sales.

The firms in the sample competed in a variety of high tech industries. The choice of a multiple vs. one industry study was made because it was anticipated that there might be too few born global firms in any single industry sector (Cavusgil & Knight 1997). A number of other studies using international entrepreneurial firms have used this approach (i.e. Burgel & Murray 2000; Jolly *et al.* 1992; Oviatt & McDougall 1995). The use of multiple industries also increased the generalizability of the study.

Survey. The major method of data gathering was a mail survey. The survey was developed inductively. Whenever possible, existing scales were used or slightly modified for the specific purpose of this study. Pre-testing was used to check the questionnaire for under-standability and content validity. The instrument was evaluated by a group of academic experts and two practitioners. This group reviewed and commented on issues such as clarity, order of questions, comprehensiveness and parsimony, and overall presentation of questionnaire.

Efforts to increase the response rate were taken including offering to send respondents an executive summary of the results (Hisrich 1975) and the questionnaire was mailed during a non-holiday period with return postage prepaid. Respondents could also fax the completed survey. The survey was addressed to members of the top management team of the firm since previous studies have found that top executives have relevant information about the strategy of the firm (Hambrick & Mason 1984) and new venture internationalization (McDougall 1989). Of the 1611 firms, twenty-six indicated that they did not want to participate in the survey and were subsequently removed from the database. Two firms had gone out of business. The final sample is composed of 231 respondents from 38 states for a response rate of 14.59%. The majority (213) of the responses was received on or before May 17 1999 and was included in all analyses. However, a number of responses (18) were received after this date (through September 21 1999) and were omitted from the analyses. In order to meet compliance with the human subjects research policy, this survey was examined and approved by the Institutional Review Board for our university.

Group Classification. Following Cavusgil & Knight (1997), respondents were asked to identify what percentage of the firm's sales came from foreign sales by the end of the third year after the firm was founded (PINT3). Firms that had greater than 25% of sales from foreign sources were classified as born global. All other firms were classified as gradually globalizing. Six firms did not respond and were dropped from further analysis. The final sample included 60 born global firms and 146 gradual globalizing firms. By the end of their third year, born global firms averaged 47.1% of sales from foreign sales with a range from 25 to 100% (s.d. = 23.85). In comparison, gradual globalizing firms had a mean of 6.91% of sales from foreign sales with a range from 0 to 22% (s.d. = 6.55).

Research Design. To exploit the content of the narrative data collected, we relied on three streams of literature. The first stream, which suggests our narrative data may contain interesting information, comes from strategic group studies that address the role of the mental models of company leaders in strategic planning and implementation (Barr *et al.* 1992; Daft & Weick 1984; Hodgkinson 1997; Hodgkinson & Johnson 1994; Osborne *et al.* 2001; Porac & Thomas 1990; Porac *et al.* 1987, 1989; Reger 1990; Reger & Huff 1993; Smircich & Stubbart 1985; Stubbart 1989; Stubbart & Ramaprasad 1988; Thomas *et al.* 1993). These studies explore the mental models of company leaders and form the basis for the "cognitive strategic groups" literature. This stream has also been shown to converge

(Nath & Gruca 1997) with earlier strategic group work based solely on non-cognitive, quantitative performance measures. Our study parallels the development of the body of strategic group literature by first having explored the quantitative characteristics of two different sets of companies (born global and gradual globalizing) and the identifying characteristics of the mental models that led to these two distinct groups.

The second stream of research that methodologically supported this study describes the growing sophistication of computer-assisted content-analysis (Holsti 1968; Kolbe & Burnett 1991; Wolf *et al.* 1993). These authors explain how computers can be effectively used to map large bodies of archival data, while avoiding some pitfalls of earlier content-analytic research designs. The third, and final, literature stream provided extensive information about using data-mining to find statistically significant quantitative patterns in qualitative data (Babbie 1992; Hair *et al.* 1992; Ramaprasad 1996; Stewart 1981). By use of word frequencies, standardization transformations, and application of metric, interdependence methods, such as factor analysis, statistical "hot-spots" could be found for further qualitative evaluation. Use of these streams of literature about strategic groups, content analysis, and statistical analysis of frequency data led to the research design we used to reveal potential differences in the mental-models applied by born global, and gradually-global firms.

Respondents were asked a series of open-ended questions about what factors had inhibited or facilitated internationalization. Each word in the narrative body was counted using a computer program. Together, more than 5,000 words were counted, and more than 1,000 of these were unique. Once the total word count was known, frequencies for each word were plotted and a subset of words was identified for further research. This reduction, which targeted high frequency non-structural words, was necessary because of practical considerations which limit PC evaluation of every word using factor analysis, and by theoretical limitations that recommend a minimum ratio of observations to variables of 2:1 for factor analysis (Hair *et al.* 1992: 227; SPSS 1995).

With as large a set of word frequencies as deemed prudent, the next stage involved dividing the total narrative body into observations. This was accomplished by entering each interview comment into a single row of a spreadsheet, labeling each row with its parent interview and group (born global and gradually global), and using a computer algorithm to provide equal numbers of rows of comments per observation. This provided fifty-seven observations that were based on equal numbers of comments, although each comment itself had a unique set of word frequencies.

With this information, a database was constructed. Its form had fifty-seven observations listed in rows, with eleven variables used as column headings. The variables used were the eleven words selected earlier because they were non-structural (not "a," "and," "the," etc.), and had relatively high raw frequency counts. Table 1 shows the basic structure of the database.

The actual database had fifty-seven rows, one for each observation. Cell entries were the standardized frequency values of the words in the respective columns. Since comparisons were to be made between born global and gradually global groups, it was important to eliminate the impact of different numbers of words in each observation because this removed differences in the amount of data provided by different interviewees. For example, one interviewee who had long, rich comments could statistically overpower another

Table 1: Illustration of database structure.

	Cost	Customer	Distribution	International	Lack	Local	Market	Marketing	Product	Regulation
1	16	20	2	10	22	5	2	26	9	15
2	3	4	4	18	18	9	6	14	5	17
3	9	16	6	11	15	5	8	8	7	19

Table 2: Frequency standardization illustration.

	Factor	Comment
1	F = word frequency count	Basic count by WordCruncher 4.6 (BYU 1989)
2	100 = a constant multiplier to eliminate numbers < 1	This transformation is a convenience that does not affect relationships
3	TWR = total words read in the observation	Total word count for each observation given by WordCruncher 4.6 (BYU 1989)
Ex.	f (market) $= 2, f$ (obs 1) $= 81$	$2 \times 100/81 = 2.47$

Table 3: Born global and gradual globalizing factor loadings.

Born Global	Gradual Globalizing	Standards
CUSTOMERS = 0.75	MARKETING = 0.73	(Hair *et al.* 1992: 239)
FOREIGN = 0.54	LACK = 0.67	+0.30 = SIGNIFICANT
INTERNATIONAL = 0.53	COSTS = 0.45	+0.40 = MORE SIGNIFICANT
		+0.50 = VERY SIGNIFICANT

interviewee whose comments were more concise. The transform used to standardize the word frequency counts was ($F \times 100$)/TWR (Table 2).

Using the completed database, several iterations of factor analysis were used to discover the underlying relationships, or factors, present in the data. While a scree plot and eigenvalue cutoffs were used, we decided to force a two-factor solution since we were looking for thematic relationships for two groups, born global and gradually global. Results showed the first runs identified words with weaker factor loadings, and permitted their culling from the original set of eleven variables. By reducing the number of variables considered to six, each with high loadings, the factor analysis met the minimum requirement of 0.5 (Stewart 1981) for sampling adequacy, as measured by the Kaiser-Myer-Olkin Test of Sampling Adequacy. Standards for loading significance at alpha = 0.01 (Hair *et al.* 1992) level were also exceeded, as shown in Table 3.

The loadings shown in Table 3 reflect the results of a VARIMAX rotation to ease interpretation of the factor outcomes. Loadings are not shown for the factor names, but the identification was verified by an earlier iteration that included born global and gradual globalizing variables in the analysis. In that run, born global and gradual globalizing variables loaded highly on separate factors, and low, or negative, on their respective companion factors.

Results

It is clear from the words of the entrepreneurs that the mental models of internationalization are different for born global and gradual globalizing firms. Three factors emerged for born global firms: Customers, international, and foreign. Leaders of born global firms emphasize the impact of the customer. Comments such as "listening to customers and customer needs" and "applications knowledge and to help customers" show that internationalization is customer-driven for born global firms. International emerged from comments such as "already have many good international contacts" and "relationships with international distributors." These comments suggest that for born global leaders internationalization is strongly affected by their informal and formal networks. The final factor is foreign. This factor is characterized by comments about "unstable foreign economies" and "foreign competition" perhaps suggesting that leaders of born global firms realize that international markets are more uncertain than domestic markets. The sum of these three factors suggests that the mental models of internationalization for entrepreneurs leading born global firms are filled with optimism, emphasis on connections, relationships, and customer relations internationally, and are balanced with the risks associated with a changing, global competitive landscape.

For leaders of gradual globalizing firms three factors emerged: cost, lack, and marketing. The first factor, cost, is emphasized by comments such as "cost of setting up office," "cost of shipping," and "cost to perform internationally." These comments were overwhelmingly negative. The second factor, lack, was also marked by negative comments. These include "lack of commitment," "lack of local reps," and "lack of distributors." This factor had the most comments of any one factor for either group. The final factor for gradual globalizing firms is marketing. Here, respondents comments included "international marketing," "local sales marketing," and "limited marketing budget." Together these three factors suggest that the mental models of internationalization for entrepreneurs leading gradual globalizing firms are filled with caution, emphasis on the time, cost, and trouble to engage internationally, and are reinforced by an overwhelming sense of the lack of knowledge, experience, and information needed to be successful.

Discussion

This research builds on the knowledge that these born global and gradual globalizing firms are distinctly different in many respects (i.e. Harveston 2000; Knight 1997). By applying ideas from strategic group research which link competitive enactment with the mental models of top management teams (Daft & Weick 1984; Hodgkinson 1997; Osborne *et al.* 2001; Smircich & Stubbart 1985; Stubbart 1989), we can gain a more comprehensive picture of the differences that identify born global and gradual globalizing firms. The key notion from this line of research is their empirical demonstration using narrative text that differences in internationalization measures reflect distinctly different mental models. We looked for distinctive differences in the mental models of the leaders of these two groups. In common with the parallel strategic groups work, we did this by analyzing narrative comments for insights about the companies studied.

The results show that leaders of born global and gradual globalizing firms emphasize significantly different topics in their discussions, and may, therefore, have different mental models with respect to competitive enactment of business strategies in the international arena (Hodgkinson & Johnson 1994; Porac & Thomas 1990; Reger & Huff 1993). The variables (words) taken together as a factor represent themes that partially describe the mental models of the speakers (Jackson 2000; Osborne *et al.* 2001; Porac *et al.* 1989;

Reger 1990). While the words in the factors/themes provide evidence of statistically significant "hot-spots" in the narrative, they lack the context needed for a better, more useful, understanding of these executives' mental models of competitive enactment (Smircich & Stubbart 1985; Stubbart & Ramaprasad 1988).

To provide the context needed to gain rich qualitative insights into the special characteristics of born global and gradual globalizing firms, this study returned to the comments, words, and phrases captured in the original study. This was accomplished by selecting all the rows of narrative containing the words in the born global theme (customers, foreign, international), and, in similar fashion, all the lines of narrative for the gradually global theme containing the words (marketing, lack, costs). To better understand how these few words in each theme can provide additional context for qualitative interpretation, the narrative extracts found for each theme appear at Appendix. These clues of important aspects of the mental models shared by born global and gradual globalizing firms led to a better understanding these two groups of firms.

As with any research, there are limitations to this study. One potential limitation is the retrospective nature of this study. Entrepreneurs were asked to report on factors leading to a decision that they may not remember very well (selective recall) depending on the time that this decision was made. Other factors such as social desirability may have introduced bias among respondents' comments. For example, entrepreneurs may want to appear to be more technologically sophisticated than they are, and therefore, talk about the attractiveness of the Internet in a socially desirable or favorable direction (i.e. higher) than actually occurs in their organizations. Mono-method bias could also be a problem. This study attempted to advance over previous studies that relied on a single respondent self-reporting for each measure by including provision for two respondents. However, two responses were not received for any of the firms participating in the survey. Therefore, the results of this study may be limited due to the self-reporting nature and single response per firm.

The ability to effectively engage internationally is an issue that affects the majority of firms including new ventures. This research shows how entrepreneurs' mental models affect the use of their scarce resources to facilitate internationalization and shows that entrepreneurial perceptions are important in affecting the internationalization process for both born global and gradual globalizing firms. Future research may wish to further examine these differences especially given the dramatic changes in worldwide economic and political environment. All firms face uncertainty in their operating environment but entrepreneurs whose firms engage in international activities face more complexity. It is hoped that this research will serve as a basis for increased discussion and future research on the international activities of new ventures. To the extent possible, this research advances previous studies by examining narrative comments to develop insights into the mental models of competitive enactment of born global and gradual globalizing firms operating in international markets.

Acknowledgments

The first author wishes to thank the Wang Center for International Business, a CIBER designated by the U.S. Dept. of Education, for research support.

References

Aharoni, Y. (1966). *The foreign investment decision process*. Boston, MA: Harvard Business School.

Ali, A., & Swiercz, P. (1991). Firm size and export behavior: Lessons from the Midwest. *Journal of Small Business Management, 29*(2), 71–78.

Babbie, E. (1992). *The practice of social research* (6th ed.). Belmont, CA: Wadsworth.

Barr, P. S., Stimpert, J. L., & Huff, A. S. (1992). Cognitive change, strategic action, and organizational renewal. *Strategic Management Journal, 13* (Summer Special Issue), 15–36.

Bartlett, C., & Ghoshal, S. (1989). *Managing across borders: The transnational solution*. Boston, MA: Harvard Business School.

Bilkey, W. J. (1978). An attempted integration of the literature on the export behavior of firms. *Journal of International Business Studies, 9*(1), 33–46.

Bloodgood, J. M., Sapienza, H. J., & Almeida, J. G. (1996). The internationalization of new high-potential U.S. ventures: Antecedents and outcomes. *Entrepreneurship, Theory & Practice, 20*, 61–76.

Brigham Young University (1989). *WordCruncher 4.5/6*. Provo, UT: Electronic Text Corp.

Burgel, O., & Murray, G. C. (2000). The international market entry choices of start-up companies in high-technology industries. *Journal of International Marketing, 8*(2), 33–62.

Calof, J. L., & Beamish, P. W. (1994). The right attitude for international success. *Business Quarterly, 59*(1), 105–110.

Cavusgil, S. T. (1980). On the internationalization process of firms. *European Research, 8*, 273–281.

Cavusgil, S. T. (1994). Born globals: A quiet revolution in Australian exporters (unpublished presentation).

Cavusgil, S. T., & Knight, G. A. (1997). Explaining an emerging phenomenon for international marketing: Global orientation and the born global firm. Working Paper, Michigan State University CIBER.

Christensen, C., daRocha, A., & Gertner, R. (1987). An empirical investigation of the factors influencing exporting success of Brazilian firms. *Journal of International Business Studies, 18*(Fall), 61–77.

Cooper, R. G., & Kleinschmidt, E. J. (1985). The impact of export strategy on export sales performance. *Journal of International Business Studies, 16*(1), 37–55.

Daft, R. L., & Weick, K. W. (1984). Toward a model of organizations as interpretive systems. *Academy of Management Review, 9*, 284–295.

Dichtl, E., Koeglmayr, H.-G., & Mueller, S. (1990). International orientation as a precondition for export success. *Journal of International Business Studies, 21*(1), 23–49.

Ensley, M. D., & Pearce, C. L. (2001). Shared cognition in top management teams: Implications for new venture performance. *Journal of Organizational Behavior, 22*, 145–160.

Eriksson, K., Johanson, J., Majkgard, A., & Sharma, D. (1997). Experiential knowledge and cost in the internationalization process. *Journal of International Business Studies*, 337–360.

Hair, J. F., Jr., Anderson, R. E., Tatham, R. L., & Black, W. C. (1992). *Multivariate data analysis, with readings*. New York: MacMillian.

Hambrick, D., & Mason, P. A. (1984). Upper echelons: The organization as a refection of its top managers. *Academy of Management Review, 9*(2), 193–206.

Harveston, P. (2000). *Synoptic vs. incremental internationalization: An examination of born global and gradual globalizing firms*. Unpublished doctoral dissertation. The University of Memphis.

Hisrich, J. (1975). Measurement of reasons for resignation of professionals: Questionnaire vs. company and consultant exit Interviews. *Journal of Applied Psychology, 60*(14), 530–532.

Hodgkinson, G. P. (1997). The cognitive analysis of competitive structures: A review and critique. *Human Relations, 50*(6), 625–654.

Hodgkinson, G. P., & Johnson, G. (1994). Exploring mental models of competitive strategists: The case for a processual approach. *Journal of Management Studies, 31*, 525–551.

Holsti, O. R. (1968). Content analysis. In: G. Lindzey, & E. Aronson (Eds), *The handbook of social psychology* (Vol. 2, pp. 596–687). Reading, MA: Addison-Westley.

Jackson, B. G. (2000). A fantasy theme analysis of Peter Senge's learning organization. *The Journal of Applied Behavioral Science, 36*(2), 193–209.

Johanson, J., & Vahlne, J. (1977). The internationalization process of the firm — A model of knowledge development and increasing foreign market commitment. *Journal of International Business Studies, 8*(1), 23–32.

Johnston, W. J., & Czinkota, M. R. (1982). Managerial motivations as determinants of industrial export behavior. In: M. R. Czinkota, & G. Tesar (Eds), *Export management: An international context*. New York: Praeger.

Jolly, V., Alahuhta, M., & Jeannet, J. (1992). Challenging the incumbents: How high technology start-ups compete globally. *Journal of Strategic Change, 1*, 71–82.

Kedia, B., & Chhokar, J. (1986a). Factors inhibiting export performance of firms: An empirical investigation. *Management International Review, 26*(4), 33–43.

Kedia, B., & Chhokar, J. (1986b). An empirical investigation of export promotion programs. *The Columbia Journal of World Business, 21*(4), 13–20.

Knight, G. O. (1997). *Emerging paradigm for international marketing: The born global firm.* Unpublished doctoral dissertation. Michigan State University.

Kobrin, S. J. (1994). Is there a relationship between a geocentric mind-set and multinational strategy? *Journal of International Business Studies, 25*(3), 493–511.

Kolbe, R. H., & Burnett, M. S. (1991). Content-analysis research: An examination of applications with directives for improving research reliability and objectivity. *Journal of Consumer Research, 18*, 243–250.

Kothari, V. (1978). Strategic approaches of small U.S. manufacturers in international markets. In: J. Susbauer (Ed.), *Academy of management best papers proceedings* (pp. 362–366).

Langston, C., & Teas, R. (1976). Commitment and characteristics of management. Paper presented at the Annual Meeting of the Midwest Business Association, St Louis, MO.

Lindqvist, M. C. (1990). Critical success factors in the process of internationalization of small hi-tech firms. In: S. Birley (Ed.), *Building European ventures* (pp. 36–60). European Foundation for Entrepreneurship Research, Elsevier.

Mamis, R. (1989). Global start up. *Inc., 11*(8), 38–47.

March, J. G., & Simon, H. A. (1958). *Organizations.* New York: Wiley.

McDougall, P. (1989). International vs. domestic entrepreneurship: New venture strategic behavior and industry structure. *Journal of Business Venturing, 4*, 387–399.

McDougall, P., Shane, S., & Oviatt, B. M. (1994). Explaining the formation of international new ventures: The limits of theories from international business research. *Journal of Business Venturing, 9*(6), 469–487.

Menon, A., Bharadwaj, S. G., Adidam, P. T., & Edison, S. W. (1999). Antecedents and consequences of marketing strategy making: A model and a test. *Journal of Marketing, 63*(2), 18–40.

Murtha, T. P., Lenway, S. A., & Baggozi, R. P. (1998). Global mind-sets and cognitive shift in a complex multinational corporation. *Strategic Management Journal, 19*(2), 97–114.

Nath, D., & Gruca, T. S. (1997). Convergence across alternative methods for forming strategic groups. *Strategic Management Journal, 18*(9), 745–760.

Norburn, D. (1987). Corporate leaders in Britain and America: A cross-national analysis. *Journal of International Business Studies, 18*, 15–32.

O'Rourke, A. (1985). Differences in exporting practices, attitudes and problems by size of firm. *American Journal of Small Business, 9*(3), 25–29.

Osborne, J. D., Stubbart, C. I., & Ramaprasad, A. (2001). Strategic groups and competitive enactment: A study of dynamic relationships between mental models and performance. *Strategic Management Journal, 22*(5), 435–454.

Oviatt, B., & McDougall, P. (1995). Global start-ups: Entrepreneurs on a worldwide stage. *Academy of Management Executive*, *9*(2), 30–43.

Oviatt, B., & McDougall, P. (1997). Challenges for internationalization process theory: The case of international new ventures. *Management International Review*, *37*(2), 85–99.

Pak, J., & Weaver, K. (1990). Differences among small manufacturing firms based on their level of export involvement: Potential, opportunistic and strategic exporters. In: T. W. Garsombke, & D. J. Garsombke (Eds), *Proceedings of USABE conference* (pp. 119–125).

Perlmutter, H. (1969). The tortuous evolution of the multinational corporation. *The Columbia Journal of World Business*, 9–18.

Porac, J. F., & Thomas, H. (1990). Taxonomic mental models in competitor definition. *Academy of Management Review*, *15*, 224–240.

Porac, J. F., Thomas, H., & Baden-Fuller, C. (1989). Competitive groups as cognitive communities: The case of Scottish knitwear manufacturers. *Journal of Management Studies*, *26*, 397–416.

Porac, J., Thomas, H., & Emme, B. (1987). Knowing the competition: The mental models of retailing strategists. In: G. Johnson (Ed.), *Strategy in retailing* (pp. 59–79). New York: Wiley.

Ramaprasad, A. (1996). A methodology for data mining. *Journal of Database Marketing*, *4*(1), 65–75.

Reger, R. K. (1990). Managerial thought structures and competitive positioning. In: A. S. Huff (Ed.), *Mapping strategic thought* (pp. 71–88). Chichester: Wiley.

Reger, R. K., & Huff, A. S. (1993). Strategic groups: A cognitive perspective. *Strategic Management Journal*, *14*, 103–124.

Reid, S. D. (1981). The decision-maker and export entry and expansion. *Journal of International Business Studies*, *12*(2), 101–112.

Smircich, L., & Stubbart, C. (1985). Strategic management in an enacted world. *Academy of Management Review*, *10*, 724–736.

SPSS, Inc. (1995). *SPSS 7.0 for Windows*. Chicago.

Stewart, D. W. (1981). The application and misapplication of factor analysis and common factor analysis. *Journal of Marketing Research*, *18*, 51–62.

Stubbart, C. I. (1989). Managerial cognition: A missing link in strategic management research. *Journal of Management Studies*, *26*, 325–347.

Stubbart, C. I., & Ramaprasad, A. (1988). Probing two chief executives' beliefs about the steel industry using cognitive maps. In: R. Lamb (Ed.), *Advances in strategic management* (Vol. 7). Greenwich, CT: JAI Press.

Taggart, J. (1998). Strategy and control in the multinational corporation: Too many recipes? *Long Range Planning*, *31*(4), 571–585.

Thomas, H., Clark, S. M., & Gioia, D. A. (1993). Strategic sense-making and organizational performance: Linkages among scanning, interpretation, action, and outcomes. *Academy of Management Journal*, *36*, 239–270.

Tookey, D. A. (1964). Factors associated with success in exporting. *The Journal of Management Studies*, *1*(1), 48–66.

Ursic, M. L., & Czinkota, M. R. (1984). An experience curve explanation for export expansion. *Journal of Business Research*, *112*(2), 159–168.

Wiedersheim-Paul, F., Olsen, H. C., & Welch, L. S. (1978). Pre-export activity: The first step in internationalization. *Journal of International Business Studies*, *9*, 47–58.

Wilson, H. (1998). International new ventures: Internationalisation in the context of prior international experience and multilateral decision making. Paper presented at the fifth International Western Academy of Management Conference, Istanbul, Turkey.

Wolf, R. A., Gephart, R. P., & Johnson, T. E. (1993). Computer-facilitated qualitative data analysis: Potential contributions to management research. *Journal of Management*, *19*, 637–660.

Appendix

Born Global and Gradual Globalizing Narrative Extracts

BORN GLOBAL narrative extracts

Customer(s) (Factor loading = 0.75)
   listening to customers and individual needs
   global customers
   applications knowledge and help to customers
   allowing international customers to know their business
   customers service

Foreign (Factor loading = 0.54)
   unstable foreign economies
   foreign competition
   foreign regulations
   foreign government subsidies and closed markets
   adoption of high tech by foreign businesses

International (Factor loading = 0.53)
   poor literature/catalog domestic and international
   already have many good international contacts
   international/large U.S. trade shows
   relationships with international distributors
   strong international
   participation in international technology standards setting group
   inability to identify correct international distributors and lack of understanding
   of national and cultural difference of our American staff
   on international business
   competitive performance by international companies
   allocating my time to look for international business
   cost of travel to international customers
   allowing international customers to know their business
   international regulatory restrictions

GRADUAL−GLOBALIZING narrative extracts

Marketing (Factor loading = 0.73)
   identifying local aggressive   lack of marketing
      sales/marketing channels   local marketing
   international marketing   local sales and marketing
   marketing   limited marketing budget
   local sales — marketing   marketing personnel

Lack (Factor loading = 0.67)
   lack of commitment   lack of strong competitive offering

lack of local representatives distance and
  lack of international salespeople
lack of distributors
lack of marketing
lack of resources
lack of capital
lack of in-house expertise
lack of capital for investment in product
  localization
lack of foreign distribution
lack of industry standards awareness &
  recognition of need was lacking
lack of marketing investment
lack of interest/education internationally
  in the Internet

lack of knowledge for product/services for
carbon free fuels
lack of interest in international markets
lack of exposure internationally; no media
advertising
lack of distribution means, as a small
company, one finds it more important to
concentrate on domestic channels
lack of internal resources is a key limiter
lack of action
lack of marketing and travel money
lack of market research
lack of international experience —
management and staff

Cost(s) (Factor loading = 0.45)
  recovery of bank costs
  cost (word listed 7 times)
  cost of setting up office
  cost of technical support
  Internet is excellent (low cost) choice
  major freight costs
  shipping costs and duties taxes
  cost of shipping
  cost of print advertising
  cost to market
  cost to perform internationally
  shipping costs & exchange rates for a
    small ticket item

low cost of fossil fuels
UN packaging requirements add to cost of
exporting
transportation costs
ability/cost of on-going maintenance &
support
ability/cost of on-site installation & training
shipping costs
cost of our services relative to international
alternatives
cost of finding distribution internationally
marketing cost
cost of entry in mark

Chapter 15

Knowledge Transfer and Policy Imperatives: Science in a Devolved Scotland

D. Jane Bower and Davena Rankin

Introduction

The newly devolved Scottish Parliament and its Executive are developing policies to address the long-term problem of a relatively low rate of business creation and growth. Sustainable prosperity in 21st Century Scotland will have to be underpinned by vigorous wealth creation. Accepting that new technologies and new organisational forms are transforming the global marketplace, and that specialisation and outsourcing possibilities are already driving the relocation of business processes to environments which most favour them, policymakers have attempted to identify and build on any competitive advantages possessed by Scotland. Conceptualisation of the processes, which will mediate business success in the emerging business environment is far from complete, particularly for new and growing firms. Nonetheless the importance of the accumulation and application of knowledge, know-how, and expertise, is a pervasive theme.

For historical reasons, which will not be detailed here, there are areas of significant weakness in the Scottish industrial base. This is particularly evident in product and service innovations in electronics and healthcare sectors. These activities are major Scottish employers, but for the most part, they do not have R&D functions located in Scotland. Partly because of this weakness, and partly because of the global perception of the importance of academic technology, a significant plank of the policies, which are emerging is the commercialisation of the academic science base (SE 2000). Scotland has considerable strength in academic science, especially in the life sciences, IT and optoelectronics. Commercialisation activities in most of its universities and research institutes have been developing for some years. Hence the Scottish Executive perceives that intensifying these activities will provide the spark which will ignite an effective dynamic in the economy.

The SME sector, its diversity, dynamics and interactions are poorly understood anywhere. However there is a growing body of research on "National Innovation Systems" (Lundvall 1992; Nelson 1993; Patel & Pavitt 1994) and technology/business networks (Gemuenden

New Technology-Based Firms in the New Millennium, Volume III
Copyright © 2004 by Elsevier Ltd.
All rights of reproduction in any form reserved.
ISBN: 0-08-044402-4

& Heydebreck 1995; Swan *et al.* 1999) which indicates that flows of knowledge, both codified and tacit, are critical to the success of technology transfer and the "cluster" concept (Porter 1990), and important for facilitating the absorption of new skills and technology by traditional firms. The reduction of barriers associated with distance for some clusters serves to highlight the persistently local character of other key features and processes of the business system — from fiscal regimes through specialist engineering services to the transfer of tacit knowledge. Understanding the dynamics of both local and distant flows of ideas, perceptions and knowledge and optimising them through appropriate policy measures will be critically important for global business competitiveness.

Management Skills in New High Technology Industries

It is not fashionable to consider that innovative scientists and engineers might be good candidates to manage high growth companies. Roberts (1991) concludes from his extensive studies of MIT spinouts, which did not achieve their growth potential until the founder engineer was replaced by a market-oriented professional manager, followed the widely accepted view that technology based firms need to "take on" staff with professional sales and marketing expertise in order to grow.

Gordon Moore offers a thought-provoking and divergent view. Moore was one of the founders of Fairchild Semiconductor and of Intel, and is regarded as one of the key founding fathers of Silicon Valley. He has discussed in detail the difficulties and the amount of time required by a group of scientists and engineers to learn how to organise themselves and their businesses (Moore & Davis 2001). However, as he points out, they not only did so, but they also evolved a novel way of doing business which gave them a uniquely effective model for transforming science into business. His observations illustrate the experiential learning curve which faces the scientist or engineer who wishes to found a business. However they also provide a powerful argument against those who insist that the only way to create a successful high technology business is to leave technically trained innovators in the laboratory and bring in "professional managers" to run the business. Moore acknowledges that there were some early stage contributions from experienced management, but his account emphasises the importance of the distinctive business and management model developed by engineers. He contrasts it with the traditional, hierarchical East Coast model and credits the success of Silicon Valley to this approach. He ascribes much of its adaptive and innovative power to its flat, flexible, highly networked characteristics. He also notes that the rich support system of venture capitalists and specialist professionals did not accumulate around the young semiconductor industry until it had begun to attract them through its success. He points out that researchers seeking to understand the success factors for cluster growth should be considering the characteristics of this early situation, not the extensive and sophisticated agglomeration of stakeholders that we see today in the mature Silicon Valley.

Moore's analysis, so apparently different from that of Roberts, leads to the conclusion that when business models are changing rapidly, traditional management and organisational skills may be as much of a hindrance as a help. The absence of ingrained habits of management and expectations of organisation may facilitate the development of new ways of doing.

In addition, as Moore contends, Silicon Valley gave birth to a new industry, whose members did not necessarily face the same sort of problems as those experienced by entrepreneurs trying to penetrate an established industry with established, shared norms and values. Although the emergence in Silicon Valley of a successful engineer-driven management culture appears to conflict with Roberts' findings, Roberts does not suggest that the MIT spinouts created a new industry with a distinctive business model and management style. They appear to have embedded themselves in the hierarchically managed East Coast business network. This may be a critical difference. On the other hand, Roberts' analysis of the MIT founders was that they did not grow their firms because they were technology-focused rather than customer- and market-focused, and this would eventually become a problem to any firm, even in a new industry where there were initially no competitors with a stronger market focus. Over time, as the market became more crowded, competition would be likely to put pressure on firms to attend closely to the effectiveness of every aspect of the business.

It is not clear from these studies whether precise comparisons can be made and a view formed about the ideal contribution of experienced managers in the context of young firms exploiting very novel technology with a new business model. Suffice it to say that the observations of Moore and of others who have documented the emergence of the microelectronics industry give encouragement to the idea that a high technology cluster can develop and thrive in a region which initially lacks entrepreneurial infrastructure and experienced management. This in turn gives credence to policies underpinning a strong academic role in Scottish cluster formation.

Spinouts from Universities

Technology-based companies which are spun out of universities by academic entrepreneurs share some of the requirements of high technology companies founded by industrial managers, but they have some distinctive problems of their own (Bower 2002). Their founders usually lack industry knowledge and contacts. The technologies are often undeveloped and generic (Howells *et al.* 1998). They may have potential applications in more than one industry, but need considerable further R&D to establish how adequately they could meet the specifications of any of these. The would-be entrepreneurs lack the expertise to identify the most promising applications and markets. Their physical location in a university at the early stages of business planning is not favourable. "Corporations and universities are not natural partners since their cultures and their missions differ" (Business-Higher Education Forum 2001: 26) moreover, the differing time horizon of the two sectors is one of several problems identified.

U.K. experience of spinning out academic ventures is not extensive. In a recent study which covered all U.K. HEIs CURDS (2001) found a total of 119 firms established in 1999/2000. Ninety two percent of these had some form of HEI ownership. In the previous five years less than 70 per year had been formed. CURDS (2001) defines academic spinoffs as enterprises in which an HEI or HEI employee(s) possess equity stakes and which have been created by the HEI or its employees to enable commercial exploitation of knowledge arising from academic research. Only 11 institutions reported income from sale of stock in spinoffs in 1999/2000, totalling £38.4 M. Three of the HEIs accounted for >80% of this income. It is impossible to predict at this point whether the recent increase in the rate of spinout formation will be accompanied by increasing returns, or whether a larger number of universities will realise returns from equity sales. Investor interest in early stage technology ventures has cooled considerably since 2000 and does not appear likely to revive in the near future.

Statistics on technology transfer from universities in the USA have been systematically collected by the Association of University Technology Managers (AUTM) for some years. They show that an increasing number of universities is devoting resources to this activity (AUTM 2002). The number of spinouts is greatest from universities with large research expenditures. In 1996 the sale of equity in spin-off companies by U.S. universities totalled $25.3 M (Bray & Lee 2000). In fact, $19m was raised from the sale of only three companies with the 13 others averaging less than $0.5 m each. Although not all universities are willing to take equity in companies, this figure does not suggest that many high-growth, high value ventures are coming out of universities. U.S. University gross income from technology licensing was about $1bn in 2000, representing 4% of the level of research expenditure (AUTM 2002), although the costs incurred in commercialisation cancelled most of this out. Both the U.K. and U.S. surveys note that few universities appear to be making profits from their technology transfer activities, which include licensing. In the few cases where significant returns have been registered, this has been the result of a lucky and unpredictable discovery of, what has become, a blockbuster drug.

These statistics give no clue as to whether there may indeed be many potential high growth ventures failing to emerge from academic institutions. They do, however indicate that, neither in the USA nor in the U.K., has a formula been established for spinning large numbers of high growth companies out of universities.

Policy Support and Cluster Emergence

Government programmes to support the development of high technology firms have made a contribution to the emergence of successful clusters. Rosenberg (2002) notes the role of public policies in the growth of Silicon Valley, for example Defense Department procurement policy requiring a minimum percentage SME involvement in contracts. However, analysis of cluster development has indicated that a highly directive, over-centralised policy approach is ineffective (Bresnahan *et al.* 2001).

In the following sections we examine how some of the policy support measures for innovation, business creation and development are being used in the current Scottish context to underpin university spinout creation. We consider the extent to which they facilitate the local interactions identified as characteristic of cluster development and analyse the extent of their articulation with functioning business networks.

Key Measures Supporting Academic Spinouts in Scotland

Among the many support measures available to Scottish academic spinouts, four are discussed here which appear to be acting synergistically to support the success of individual spinouts. They are also contributing to the growth of a strong entrepreneurial support system within Scotland. The four considered here are Connect, the Enterprise Fellowship scheme (the principle author of this paper leads the entrepreneurship training component of this programme), Proof of Concept (PoC) awards and SMART awards. The first is designed to strengthen the entrepreneurial skills of a key group of academic entrepreneurs, the second has created a growing network of early stage technology firms, investors, and professional advisers. PoC and SMART awards provide project finance towards the cost of early stages of high technology product and process developments.

Enterprise Fellowship Scheme

The Enterprise Fellowship scheme was set up in 1997 by Scottish Enterprise the Scottish Development Agency (SE) to increase the number of academic spinouts and to improve understanding of commercialisation within the academic sector. Since that date each year a small number of academic entrepreneurs (3–6 per year), who were in the process of spinning out a company based on technology they had invented, have joined a one-year programme. The programme, managed by the Royal Society of Edinburgh (RSE), has provided one year's salary while they worked on the start-up process, £5000 expenses, entrepreneurship training delivered by Glasgow Caledonian University (GCU), and mentoring by industrial, academic and Scottish Enterprise mentors. The programme was independently evaluated for Scottish Enterprise by Segal Quince Wicksteed in 2001. Their report (SE 2001) was, in general, highly favourable, finding a high level of additionality compared with other economic development programmes. A high proportion of the entrepreneurs have set up companies which have attracted substantial investment and created more than the targeted number of highly skilled jobs. Most of the entrepreneurs were postdoctoral scientists or engineers with no business experience. The main features contributing to additionality which the study found were the financial freedom to concentrate for a year on the start-up process, the opportunity provided by the training and mentors to access a wide network of support, and the ability to share experience with the group of Fellows going through the programme at the same time.

The network of support referred to is the local network of professional advisers, business angels, professional investors, economic development professionals and others whose business focus is on young, technology-based firms. This is itself a young, and immature network, but it is growing in size and sophistication due in part to the growing number of young technology firms in Scotland. Another important factor in the growth of the Scottish network has been Connect, whose activities have catalysed the formation of linkages between the stakeholder groups (see next section).

The Enterprise Fellowship programme focuses very closely on the unique needs of academic technology entrepreneurs in the Scottish context. The fellows are all highly intelligent individuals, well able to learn and apply knowledge about "hard" business issues. The areas in which they most need support in are:

(a) Learning the language and culture of technology business through interaction with relevant business and industry personnel. This process also generates the contacts they need.

(b) Acquiring a market-led mind set rather than technology-led attitude they start out with. This drives an essential change in their priorities.
(c) Keeping up their motivation and enthusiasm as they enter an unfamiliar and often threatening world.

Apart from the obviously critical element of providing fellows with some time and direct sources of information and advice to tackle the start up process, the programme as a whole provides several mechanisms for achieving (a–c). RSE, SE and GCU contributors all take a very proactive approach to introducing fellows to situations and contacts which meet their specific needs. The RSE hosts a number of events to which present and past Fellows, mentors and others involved in technology entrepreneurship make contributions. This has created a specific learning community encompassing everyone associated with the programme. Another important effect of these occasions is to strengthen the motivation of everyone associated with the programme by creating a sense of belonging to a respected community whose activities are legitimated through its affiliation with the prestige of the RSE and all that it stands for. This, and the peer support group which develops in each cohort, are powerful re-inforcers of the initial self-confidence and enthusiasm of Fellows.

The SQW report also comments on negative aspects of the Fellows' experience. These are the considerable difficulties and delays, which most Fellows have experienced in coming to a mutually acceptable agreement with their host university about intellectual property issues. A key issue which the report highlights is the unequal negotiating position of the Fellows, as employees of the university. It recommends on the one hand that an experienced third party should be involved, and on the other, that a clearer approach should be developed, with the lead in this taken by the Scottish Executive. These recommendations are under consideration and it is to be hoped that acceptable and helpful new measures will be introduced.

Connect Scotland

Connect Scotland was set up in 1996 by Ian McDonald, then a lecturer at Edinburgh University. He had been impressed by the organisation of the same name in San Diego, which effectively brought together would-be academic entrepreneurs with a network of investors and professional advisers who were interested in doing business with technology spinouts. Although Connect has received public support and is cited by policymakers as part of the support for technology entrepreneurship, it is very much the result of McDonald's personal initiative. He persuaded all the stakeholder groups, public and private, to contribute in cash and in kind to its activities from the start, and has skilfully directed it since its initiation.

Its mandate was *"to support the creation, development and growth of technology-based enterprise throughout Scotland."* It organised a programme of activities — "Enterprise workshops," "Meet the Entrepreneur," "Springboard Workshops," and an annual investment conference. These activities were designed to provide occasions for professional advisers, would-be technology entrepreneurs, venture capitalists and all the other stakeholders to formally present themselves in contexts in which they could also meet informally and learn about each other. Connect has proved very popular with a steadily growing number of stakeholders from all the groups — professional advisers, formal and informal investors, universities, and academic and industrial technology entrepreneurs (TechMaPP 2001).

In common with the SQW report on the Enterprise Fellowship programme, the TechMaPP report which evaluated Connect Scotland noted that many of the Scottish spinout companies interviewed had complained of the major difficulties they had experienced in their negotiations with the Universities over intellectual property agreements. Most had given this as a primary reason for joining Connect since they hoped it would give them access to legal advice which otherwise was proving to be prohibitively expensive. After joining, most companies found Connect an excellent support group, particularly in relation to finding finance. It was a good direct source of contacts for raising funding, and its Director, Ian McDonald, was a very good source of advice and introductions to a wider range of contacts.

Summarising the comments from companies about Connect, the TechMaPP report found that it had been "a crucial supporting environment for organisations at the start of their business lives," whose events and informal atmosphere had helped people to acquire crucial business experience and understanding in a non-threatening environment.

What the report did not comment on was the extent to which Connect had also provided a forum to educate the professional advisers and investors, and to interact effectively with technology entrepreneurs who have not yet acquired any degree of business sophistication. For the necessary conversations to take place, the professionals have just as great a need to learn how to communicate effectively with technologists in the early stages. Professionals also have to develop their skills in each specialist technology industry through the process of doing deals in that industry. In the Scottish situation there are very few of these deals coming through yet, and it is only by developing companies from the earliest stages that this skill set will be developed and refined within Scotland.

Grants in Support of Development Projects

Companies in the TechMaPP study were all high technology firms, but only a few were academic spinouts. This sample had not had much difficulty in raising initial funding from private sources. They complained of the long lead times in obtaining government grant finance. In contrast, most of the Enterprise Fellows setting up companies in 2000–2002 have expressed strong approval for PoC and SMART awards. These awards filled a difficult gap for academic entrepreneurs who lacked industry knowledge and had a way to go to convince private investors of the viability of their very early stage technologies. Although the entrepreneurial support network in Scotland becomes more sophisticated each year, it has been accompanied in the last two years by a general reduction in interest in technology ventures and withdrawal of some VC offices. Consequently the relative importance of grant finance to spinouts is probably increasing now that private seed finance for technology companies is harder to obtain. Academic entrepreneurs have been forced to start their businesses in a period of much less private willingness to invest in the riskiest stages of technology spinouts on the part of both venture capitalists and business angels.

The PoC and SMART awards both fund technological innovation projects. The Proof of Concept (PoC) fund (www.scottish-enterprise.com/proofofconceptfund, maximum award £200,000, median award ∼ £150,000), is not open to incorporated firms. It supports projects within the academic laboratory, which can lead to the development of a business or, alternatively, to patentable technology which can be licensed into a business. This technology is owned by the university in which the PoC is held.

SMART awards (www.linscot.co.uk) are available to small businesses and to individuals planning to start a business. A Stage 1 award, 75% of eligible costs, maximum £45,000, covers a technical and commercial feasibility study. If this is successfully completed a Stage 2 award can be applied for. This covers 35% of eligible costs up to a maximum award of £150,000 for both stages of SMART. Applications from the Science Base are actively encouraged. Technology developed with a SMART award belongs to the company, which holds the award. A Science Base award must be made to a company, not an institution, although the institution may hold up to 25% equity in the company.

For academic entrepreneurs these two sources of early stage finance fill awkward financial gaps. They also provide opportunities to showcase the technologies and the inventors, and provide them with a favourable profile. Award ceremonies with poster displays are attended by interested financiers and professional advisers as well as economic development professionals, and press coverage of these events adds to their effectiveness in disseminating information to the entrepreneurial support network. They may also have negative features, however, for academic entrepreneurs since they may contribute barriers to the spin out process. So long as the venture is still accumulating value within the university there may be insufficient incentive to speed the spinout process. This is considered in the following section.

Discussion

The programmes described above are providing a flexible core of support for academic spinout formation. They are also acting synergistically to reinforce the organic growth of a generic entrepreneurial support network for technology firms in Scotland. There is now an emerging community of professional advisers, investors, economic development professionals and growing, high technology firms. It is moving up a learning curve of high technology creation and growth.

There is still a long way to go, however, before it can be stated that an infrastructure for wealth creation has been built. Although significant investment has been attracted into a number of academic spinouts, including some which have benefited from some or all of these support measures, few if any have reached profitability. There is no great confidence that the current pullback in technology investment, which has affected companies at all stages of development, is about to turn around. This leaves a considerable threat hanging over all companies which have not yet attained profitability. The confidence which has underpinned the growth of the entrepreneurial network is a recent and probably fragile phenomenon. It might well be extinguished by a series of failures, or even by the prolonged absence of significant success.

Another problem area is the universities' approach towards their IPR, which was so frequently cited as a major barrier to spinning out. Universities in Scotland and in the

U.K. generally differ considerably in their approach to IPR issues and until recently there has been a dearth of information about both policies and outcomes. This is beginning to change, with the CURDS (2001) report published and others in the pipeline. The CURDS report found only 11 institutions in 1999/2000 which had income from selling equity of spinouts, and of these, eight had made on average less than £1m, and this was during a period of high interest in technology ventures. This, then, looks rather similar to the situation in the USA, where a very few universities made any income from the sale of equity in spinouts (Bray & Lee 2000, discussed above).

There is thus no evidence as yet that the proceeds from the sale of equity in spinouts are to be viewed as a reliable and significant source of income for universities. It is worth noting here that U.S. universities' policy on ownership and control is relatively hands-off (see, e.g. Bray & Lee 2000). Some are not willing to take any equity. The U.S. universities which do have a policy of taking equity typically take a 5% position in place of a licence issue fee or royalties. U.S. universities which take equity typically have large well-established Technology Transfer offices and large royalty streams. Informal discussions with spinout founders indicate that Scottish universities generally do take equity at the moment, and the amounts they expect vary widely. It appears, however, that at present most (though not all) expect to hold large equity stakes and a high measure of control over companies. The institutional view in these cases appears to be that the university should set up the company and exert control for a prolonged period.

Where the company has been set up by the university rather than by the inventors, the latter have limited influence on this. This situation may also exacerbate the difficulties of interesting private sector investors. A substantial equity position held by the institution can deter external investors and leave too little to incentivise management who face an extremely risky task. Even with less than 25% of the equity, and even if they are in principle committed to spinning it out, universities can cause delays in the spinout process through the complex maze of procedures which must be gone through before a final decision can be formalised. The institution, as a significant shareholder, has a duty of governance which carries very onerous responsibilities (HEFCE 2000). The position is greatly simplified if the academic entrepreneur sets up the company (provided that they wish to make a full commitment to this, which may not be the case — Franklin *et al.* 2001), then comes to an agreement with the institution to assign rights to it which do not include a large equity holding, rather than the institution taking the lead in forming the company. The assignation agreement can be tailored to meet the university's need to make an acceptable return, if the venture is successful, without incurring delays and costs due to unnecessary expenditure of senior management time within the institution. It is also easier by this route to distance the institution from any risks, predicted or unexpected, which might accrue to the new technology.

The SQW report recommended that IP issues should be given more careful consideration, in the light of their importance in spinout formation. There is no evidence that universities are good at managing the process of setting up entrepreneurial companies. The evidence cited here from the TechMaPP and SQW studies, as well as informal reports, points unsurprisingly to the opposite conclusion. It is to be hoped that the Scottish Executive and the universities will soon clarify the question of priorities and put in place procedures which will allow these priorities to be addressed effectively.

An interesting postscript to this was a recent article in the Financial Times (Campbell 2001), on the first round of the University Challenge Fund, set up with £70 M from

Government, The Wellcome Trust and the universities. The fund was launched in 1999, and 105 spinouts created. Only £24 m in non-Challenge funding has been attracted thus far. Only four investments have been made by venture capitalists. It is not clear whether there will be any payback — the funds are running out and there has not been much success yet in getting follow-on funding.

This raises some important issues relating to how far public policy measures are useful in supporting the academic spinout process, and at what point they risk blocking the commercialisation process. Is the technology downturn all to blame for the lack of interest in Challenge Fund companies? Do these funds allow the universities to evade commercial reality? Do they allow academics to put off acquiring a truly commercial orientation? The Enterprise Fellowship scheme and Connect have been very proactive in pushing academics into the commercial, entrepreneurial world. This has been cited as a major factor in their additionality, by evaluators and by the entrepreneurs which have used them. The question this raises is whether Scottish universities' can best support spinouts by letting them escape early and without too many onerous ties to the parent institution. Connect and the Fellowship scheme have had considerable success in fostering an entrepreneurial network which has itself developed a good articulation with wider business and industry networks. This is providing support for the commercialisation transition, which is growing steadily as the local network matures. The links to global finance, industrial partners and skilled management are beginning to be forged. Moore & Davis (2001) argue that scientists and engineers can create effective management cultures. However there is no need to rely on this alone when founding teams can catch the attention of managers looking for attractive career possibilities. The growth of the Scottish network and its linkage into global industry and business, are the preconditions for this to occur. The evidence suggests that this process is underway.

References

AUTM (2002). *AUTM licensing survey: FY 2000*. Association of University Technology Managers, UTM, Chicago.

Bower, D. J. (2002 in press). Matching technology push to market pull: Strategic choices and the academic spinout firm. *International Journal of Entrepreneurship and Innovation Management*.

Bray, M. J., & Lee, J. N. (2000). University revenues from technology transfer: Licensing fees vs. equity positions. *Journal of Business Venturing, 15*, 385–392.

Bresnahan, T., Gambardella, A., Saxenian, A., & Wallsten, S. (2001). "Old economy inputs" for "new economy" outcomes: Cluster formation in the new Silicon Valley. *Industrial and Corporate Change, 10*(4), 835–860.

Business-Higher Education Forum (2001). *Working together, creating knowledge*. Downloadable from www.acenet.edu/programs/bhef/about.cfm.

Campbell, K. (2001). From ivory tower to market. *Financial Times* (20 December).

CURDS (2001). *Higher education-business interaction survey*. A Report by the Centre for Urban and Regional Development Studies, University of Newcastle upon Tyne, Newcastle-upon-Tyne, December, downloadable from www.hefce.ac.uk.

Franklin, S. J., Wright, M., & Lockett, A. (2001). Academic and surrogate entrepreneurs in university spin-out companies. *Journal of Technology Transfer, 26*, 127–141.

Gemuenden, H. G., & Heydebreck, P. (1995). The influence of business strategy on technological network activities. *Research Policy, 24*(6), 831–850.

HEFCE (2000). *Related companies: Recommended practice guidelines.* Report 00/58 produced for HEFCE by RSM Robson Rhodes, revised December 2000, www.hefce.ac.uk.

Howells, J., Nedeva, M., & Georghiou, L. (1998). *Industry-academic links in the U.K.* PREST, University of Manchester, Manchester, December 98/70 Report.

Lundvall, B. A. (Ed.) (1992). *National innovation systems and interactive learning.* London: Pinter Publishers.

Moore, G., & Davis, K. (2001). *Learning the Silicon Valley way.* Stanford Institute for Economic Policy Research Discussion Paper No. 00–45.

Nelson, R. (Ed.) (1993). *National innovation systems: A comparative analysis.* Oxford and New York: Oxford University Press.

Patel, P., & Pavitt, K. (1994). *The nature and importance of national innovation systems.* STI Review No. 14, OECD, Paris.

Porter, M. (1990). *Competitive advantage of nations.* Free Press.

Roberts, E. B. (1991). *Entrepreneurs in high technology: Lessons from MIT and beyond.* Oxford and New York: Oxford University Press.

Rosenberg, D. (2002). *Cloning Silicon Valley: The next generation high-tech hotspots.* London: Pearson Education.

Scottish Enterprise (SE) current policy at www.scotent.co.uk.

SE (2000). *The way forward: Framework for economic development in Scotland.* Scottish Executive (June).

SE (2001). *Scottish enterprise/Royal Society of Edinburgh enterprise fellowships, interim review and evaluation.* Summary Report (June). Conducted by Segal Quince Wicksteed. Report available from Alison Graham or Iain Ross, Scottish Enterprise, Atlantic Quay, Glasgow.

Swan, J., Newell, S., & Robertson, M. (1999). Central agencies in the diffusion and design of technology: A comparison of the U.K. and Sweden. *Organization Studies, 20*(6), 905–931.

TechMaPP (2001). *Report 11 Synthesis Report Case Studies of Emerging Technologies in Scotland,* TechMaPP, Department of Business Studies, University of Edinburgh. Report commissioned by SHEFC.

Part VI

Global Issues

Chapter 16

Born Global Approach to the Internationalization of High Technology Small Firms — Antecedents and Management Challenges

Sami Saarenketo

Introduction

The competitive environment of high technology small firms (HTSF) has changed dramatically during the last decade. Nowadays, even the smallest firms have to face the challenge of globalization. An increasing number of small firms are becoming involved in international activities more rapidly and intensively than historically has been the case. The ability to engage the firm in international activities has become very important for the survival and growth of the firm. This is particularly true within the high-tech sectors where firms often have high upfront R&D costs but operate in narrow niches scattered thinly from one country to another. It is suggested that the firms have to achieve international coverage very quickly for their products. This is due to the industry dynamics, such as shortening product life cycles and discontinuous technological changes. Usually, there is a limited "window of opportunity" and if the firm wants to survive and grow it has to grasp this opportunity in time, before competitors or changes in customer preferences block it. All this sets a lot of challenges for the management of the firms.

This paper examines the internationalization process of HTSFs within the Finnish Information and Communication Technology (ICT) industry. The emphasis in the research is on the antecedents of rapid and intensive internationalization. In addition, we attempt to highlight some of the management challenges related to the process that deserve more attention. Previous research states, for example, that earlier experience, partners and networks, global vision, and risk orientation of the management team play a significant role in the firms' rapid and intensive internationalization process. However, these associations are not straightforward and there is no single best way to manage the firm's expansion into international markets. On the basis of earlier research, we have formulated a number of hypotheses for empirical testing. Our data from 124 small and medium-sized enterprises (SMEs)

New Technology-Based Firms in the New Millennium, Volume III
© 2004 Published by Elsevier Ltd.
ISBN: 0-08-044402-4

demonstrates that there are several driving forces behind the firms' internationalization process. In support of previous studies, the drivers of rapid and intensive HTSF internationalization include among others the global vision, risk behavior of the management team, partners and financing.

The paper proceeds as follows. First, the relevant literature on the internationalization process of the firm and born global phenomenon is reviewed, the emphasis being on HTSF context. Second, the hypotheses based on earlier research and theories are formulated. At the same time, we attempt to underline some of the related management challenges, which deserve more attention. Third, the empirical findings from the survey of Finnish ICT SMEs are presented. Finally, the concluding remarks are provided.

Review of the Literature

The Internationalization Process

The Internationalization process of firms has been a topic of increasing research interest for almost thirty years now. However, despite the huge amount of research, a universally accepted definition of internationalization is still lacking. Internationalization of the firm has been defined, e.g. as "the process of increasing involvement in international operations" (Welch & Luostarinen 1988: 36) and "the process of adapting firms' operations (strategy, structure, resources, etc.) to international environments" (Calof & Beamish 1995: 116). The basic logic in the process or stages-models of internationalization is that the process evolves in a slow, incremental manner (as the firm gains more experiential knowledge) towards a greater commitment in foreign markets and involves a varying number of stages. Some of the most influential developers of this perspective include Johanson & Wiedersheim-Paul (1975), Johanson & Vahlne (1977, 1990) and Luostarinen (1979). The similarity in the various models is that they are behavioral in nature building on the theory of the growth of the firm (Penrose 1959) and the behavioral theory of the firm (Cyert & March 1963). The major factors affecting the cautious behavior include lack of foreign market knowledge, high risk aversion of the managers and high perceived uncertainty among other things. Extensive reviews of international process models have been conducted, e.g. by Andersen (1993) and Melin (1992).

An alternative behavioral approach to explain the internationalization of the firm is the network approach. This view focuses on non-hierarchical systems where companies invest to build up and monitor their position in international networks (Johanson & Mattsson 1988). Two of the core parameters include the degree of internationalization of the network and an independent firm within the network. Network theory has been found to be useful when exploring SMEs (Coviello & Munro 1995), the focus being on the linkages that enable the firm's internationalization.

Foreign Direct Investment (FDI) theory is the third traditional stream to explain the internationalization of the firms. By FDI we refer to firm's fully integrated mode of international operation, such as acquisition, merger or the establishment of greenfield subsidiary. As Coviello & Martin (1999) point out, the general body of FDI theory has come forward from several different theoretical developments. Transaction cost analysis (TCA) is one of the central views adopted. This approach states that, in choosing the organizational form and location, firms are influenced by the attempt to minimize the overall transaction costs. Though this theory is widely used, it is supposed to be of more relevance for large companies with larger resources. For the young HTSFs, the capital requirements of the investment alternative may make the FDI a less feasible option, favoring exporting and cooperative modes of internationalization. It must be noted that the above-mentioned approaches to internationalization are not mutually exclusive. As noted by Coviello & McAuley (1999), internationalization is a dynamic and evolutionary process that has both behavioral and economic elements.

Until the late 1980s the studies focused more on the larger multinationals (MNCs) in mature manufacturing industries. Although the traditional (stages) theories have been useful in many cases, they have undeniably proved to be more valid in the case of the larger MNCs, neglecting the SMEs. Today, as service companies represent an increasing number of the world's population of firms the general applicability of the stages model has diminished over time. As Westhead *et al.* (2001) notice, previous studies have often failed to recognize that the reasons for internationalizing in small service firms may be markedly different from those in larger manufacturing firms. Also Coviello & McAuley (1999) state that internationalization patterns of manufacturing firms tend to follow the traditional models better than their service counterparts do. Furthermore, as the HTSFs' significant contribution to the world economy is increasingly being acknowledged — e.g. as valuable sources of innovation (Tushman & Anderson 1986) — the internationalization of these firms is gaining more interest as well.

Born Global Approach to Internationalization

To understand the born global phenomenon, one must be aware of the traditional theories on the internationalization of the firm. These views, however, offer a somewhat limited insight into the contemporary situation in the internationalizing of HTSFs. Indeed, during the last decade, the idea of how a HTSF may proceed in its internationalization was turned on its head. McKinsey's (Rennie 1993) study of Australia's high-value-added manufacturing exporters reported that numerous small and medium-sized firms (SMEs) competed successfully — virtually from their initiation — against large, established firms in the global arena. Although this phenomenon is not completely new, it is something that the traditional models are unable to fully explain and predict.

In this study, we define a born global firm, according to Knight & Cavusgil (1996), as "a firm, that has reached a share of foreign sales of at least 25% after having started export activities within three years of its birth." Thus, the rapidity and intensity of internationalization are the two key parameters. There has been a slight lack of consensus in the definition and operationalization of the phenomenon. For example, other researchers use between 2 and 5 years from start-up as a classification criterion. The definition used in this paper is, however gaining a foothold as other researchers are adopting it as well (see e.g. Madsen *et al.* 2000). Although many researchers have adopted the term Born Global to describe the new phenomenon, other names have been used. These include e.g. *Innate Exporters* (Ganitsky 1989), *International New Ventures* (Oviatt & McDougall 1994, 1997), *Global Start-ups* (Mamis

1989; Oviatt & McDougall 1995), *High Technology Start-ups* (Jolly *et al.* 1992) *Technology-Based New Firms* (Autio *et al.* 1997) and *Instant Internationals* (McAuley 1999). Despite the variety of names given, all of the concepts explain pretty much the same issues: rapid and intensive/dedicated internationalization of SMEs. As Bell *et al.* (2001: 176) states, this kind of internationalization behavior is commonplace among firms that offer their products/services to small, highly specialized global "niches" and is especially common among SMEs located in SMOPEC (small open economies) countries that face the double jeopardy of targeting narrow "niches" in limited domestic markets. However, Born Globals have been identified in all parts of the world, implying that global forces are advancing their development (Oviatt & McDougall 1994). In addition, the phenomenon is not restricted to HTSFs and knowledge-intensive industries. Born global firms have been found in a wide range of industries.

One of the most recent developments in studying the internationalization of small entrepreneurial firms is the concept of "Born-again global" firms (Bell *et al.* 2001) which may be associated with "Late global" firms (Saarenketo *et al.* 2001). These firms start their international activities very rapidly and intensively, but relatively late in the company's history. Critical incidents (Bell *et al.* 2001) such as the launch of a new product or changes in the management team may act as triggers in this kind of internationalization behavior.

HTSF Internationalization Drivers and Management Challenges

Despite the increased interest in research on the internationalization of HTSF, there is still a limited amount of evidence on what are the forces that lead to rapid and intensive internationalization of the firm. Much of the prior research is case-based and exploratory in nature. However, several suggestions of the drivers of rapid and intensive internationalization have been offered. Based on this earlier literature, we have formulated the following hypotheses for empirical testing. At the same time, we attempt to underline some of the related management challenges, which deserve more attention.

Earlier International Experience

Even in the world of "global consumers" or "standardized demands" the crossing of borders still means higher-level uncertainty and many kinds of discontinuities. This is demonstrated by e.g. language and culture, and differences in regulatory and competitive landscape (Rangan & Adner 2001). Earlier experience is of great importance and the top management team of the firm is probably its most critical (intangible) resource. Managers who have been exposed to the international professions before are likely to be more aware of the challenges and also international profit opportunities than persons who do not have such experience (Bloodgood *et al.* 1996). In other words, the managers have been able to learn from previous experiences and can add their new knowledge to these accumulated experiences (Eriksson *et al.* 1997). Therefore, the prior experience may influence the rapid and intensive internationalization of the firm. In their study Bloodgood *et al.* (1996) found that it was indeed the work experience that had a positive effect on the internationalization activity and not education. Roberts & Senturia (1996) found that none of the top managers of the firms in their U.S. sample had direct international experience from having lived abroad. Still, a majority in their sample had someone in the management team who had worked previously in or with a company that did substantial business in global markets. Autio *et al.* (2000) in their study of international growth in entrepreneurial firms contradicted previous findings by concluding that years of international experience had no effect on growth. We formulate the first two hypotheses as follows:

H1a. The earlier international experience of the management team is positively related to the rapidity of the HTSF's internationalization.

H1b. The earlier international experience of the management team is positively related to the intensity of the HTSF's internationalization.

Global Vision

"To be global one must first *think* globally" (Oviatt & McDougall 1995). This "geocentric thinking" may be enhanced by the international experience of the management. Former research has linked the success of early internationalization to having an international vision right from the firm's start-up (see e.g. case analyses by Oviatt & McDougall 1995). Simply put, global vision means that the firm sees the whole world as one big market place, no real distinction is made between domestic and global marketing, etc. The biggest challenge for the management is to communicate the vision to everyone else associated with the firm (Oviatt & McDougall 1995). Vision has to be understood by the firm's employees, customers, partners, financiers etc. in order to be put into practice. Furthermore, although the importance of the vision should not be underestimated, it cannot compensate for the deficiencies in the firm's product and/or service. In summary, we formulate the following hypotheses:

H2a. The global vision is positively related to the rapidity of the HTSF's internationalization.

H2b. The global vision is positively related to the intensity of the HTSF's internationalization.

Risk Behavior

As Bonaccorsi (1992) notes, many authors have found significant differences between exporters and non-exporters with regard to the risk perception and risk-taking propensity. Due to the risk-aversion and uncertainty-avoidance of the firms' managers, internationalization is traditionally seen to evolve in a slow and incremental manner. Rapid and intensive internationalization can indeed be considered as a high risk — high revenue strategy. As McKinsey's consultants Bates *et al.* (2001) point out there lies many perils of moving too fast: "... the faster you build a business, the less time you have to study the market, test assumptions, understand competitors, and optimize resources." However, It can be even more risky to concentrate only on domestic market activities in the conditions where the competitors and potential customers operate on a global scale. Risk behavior of the managers may be linked to their earlier experience. If a manager is internationally experienced, he/she may not perceive that high risks because of accumulated knowledge on how to run things abroad reduces the uncertainty. But even if the manager is able to capitalize on his/her previous experience, there is still the challenge in determining what is the reasonable risk level, how rapidly and intensively the firm can afford to move. In summary, we formulate the following hypotheses:

H3a. The higher risk tolerance of the managers is positively related to the rapidity of the HTSF's internationalization.

H3b. The higher risk tolerance of the managers is positively related to the intensity of the HTSF's internationalization.

R&D Activity

In the high technology industries even the small firms commonly face high upfront R&D costs and at the same time the product life cycles are getting shorter. The so-called "opportunity window" for the firm is therefore constrained; the firms cannot afford to focus on domestic markets solely but are forced to cross borders very early on. Prior research has found support for the effect of the R&D activity on the firm's internationalization. Kobrin (1991) provided evidence that technological intensity (measured as a ratio of R&D expenditures to sales revenue) was the most important structural determinant of globalization in the industries that he studied. In their study of British start-up companies in high-technology industries, Burgel & Murray (1998) found that start-ups that internationalized their activities had higher R&D-to-sales ratio. Lindqvist's (1991) findings from Swedish new technology based firms, on the other hand, demonstrated that higher R&D intensity was not related to speed and intensity of the firm's internationalization. It could be assumed, that if the R&D consumes most of the financial resources there may not be enough left for carrying out the internationalization process rapidly and intensively. Inspired by the prior controversial evidence, we formulate the following hypotheses:

H4a. The intensity of the R&D activity is positively related to the rapidity of the HTSF's internationalization.

H4b. The intensity of the R&D activity is positively related to the intensity of the HTSF's internationalization.

Market Niche

Knight & Cavusgil (1996) list a number of recent trends that have led to the emergence of born global firms. These include among others the increasing role of niche markets and greater demand for specialized or customized products. The HTSFs typically operate in rather specialized market niches. Therefore, particularly for firms initiating from small open economies, the domestic market is often too small to generate adequate cash flow. We formulate the following hypotheses:

H5a. The narrow market niche is positively related to the rapidity of the HTSF's internationalization.

H5b. The narrow market niche is positively related to the intensity of the HTSF's internationalization.

Partnerships

To "overcome the lack of resources at the time of start-up" (McDougall *et al.* 1994: 470) HTSFs frequently rely on "hybrid" governance structures in their internationalization, i.e. partnerships, strategic alliances and networks with other firms. Partnering with more established industry players may provide the firm with a variety of resources and legitimacy (Eisenhardt & Schoonhoven 1996) that can help the firm to enter international markets more rapidly and intensively. A low-cost alternative to vertical integration is offered (Bonaccorsi 1992). However, it should be noticed that if other firms are to be utilized to leverage the firm's own resource-base, a whole new set of tasks and transaction costs are introduced to the management. In fact, the challenges in managing these relationships mean that there is no guarantee to a HTSFs successful internationalization. There may be a danger that a partner behaves opportunistically and does not conform to the rules and/or contract agreed upon. For example, we have witnessed the case of large Japanese firms forming predatory alliances with American HTSFs (Oviatt & McDougall 1994) Nevertheless, this risk has to be taken due to resource deficiencies and consequently because of the unfeasibility of utilizing the FDI/internalization option. A firm may partner with firms from the home country, or firms with the host country. The latter might lead to more rapid and intensive internationalization as the firms from the host country may be better established, having better market knowledge, etc. However, as Preece *et al.* (1998) point out, focusing solely on foreign partnerships might neglect the broader impact of cooperation. In summary, we formulate the following hypotheses:

H6a. The number of partners is positively related to the rapidity of the HTSF's internationalization.

H6b. The number of partners is positively related to the intensity of the HTSF's internationalization.

External Financing

The knowledge is frequently considered as the most powerful resource of the HTSFs. However, rapid and intensive internationalization of the firm also demands also other resources such as financing. The entrepreneurs' own available funding falls short rather quickly and for the young start-up companies, the cash flow is usually an insufficient in-strument to finance the growth and internationalization process. At the same time, the firms commonly lack the guarantees for getting bank loans. Even though there has been a radical change in the financial climate for HTSFs since the spring of 2000, we believe that external equity-based financing (in the form of e.g. venture capital) may still act as a potential catalyst of the firm's internationalization. The triggering effect of financing may be even more enhanced, if the investor itself operates on a global scale with wide industry networks that may support the firm in question. Findings from Burgel & Murray's (1998) study of British start-up companies in high-technology industries demonstrated that there were no significant differences between start-ups that internationalized and those that did not with regard to access to venture or angel capital. We formulate the following hypotheses:

H7a. The existence of external financing (venture capital) is positively related to the rapidity of the HTSF's internationalization.

H7b. The existence of external financing (venture capital) is positively related to the intensity of the HTSF's internationalization.

Global Competition

In the past, the tyranny of distance meant bigger barriers for entering the global markets. Now with the development of rapid communications, transportation, etc., the HTSFs are forced to adjust to the global competitive environment either proactively or reactively. Born global firms are generally described as being more proactive in their internationalization behavior. If the firm faces global competition in its market segment, this is supposed to lead to more rapid and intensive internationalization. Coviello & Munro (1995) demonstrated that intensity of global competition in the industry was an important factor in explaining the rapid internationalization of HTSFs. According to Petersen & Pedersen (1999: 75) "the business environment of a particular industry and its implications for strategy may countervail the firm manager's inclination to cautious, incremental behavior." In the HTSFs, accelerated process of internationalization may often be triggered e.g. by an aspiration to get a "first mover advantage" and "lock-in" the potential customers before competitors (Bell 2001). It has to be stressed that the ICT sector is not actually an undifferentiated landscape (Rangan & Adner 2001) but consists of many sectors as a result of convergence of information technology and telecommunications industries. However, ICT is a particular industry where many of the market segments are in fact global by nature, which is supposed to fuel the process of the firms' internationalization. We formulate our last hypothesis as follows:

H8a. The global competition is positively related to the rapidity of the HTSF's internationalization.

H8b. The global competition is positively related to the intensity of the HTSF's internationalization.

Our preliminary theoretical model of the antecedents of rapid and intensive HTSF internationalization is presented in Figure 1.

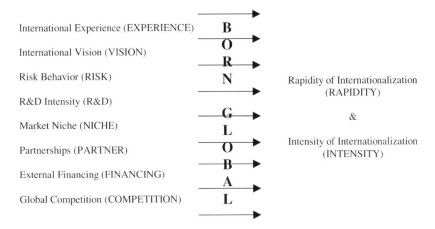

Figure 1: Antecedents of rapid and intensive HTSF internationalization.

Data and Methodology

Sample

The population of interest was defined as Finnish small and medium-sized enterprises (SMEs) providing value added services in the ICT-sector. These include content providers and software providers for service platform and management systems. Hardware manufacturers and companies providing mainly educational or consultancy services were not included in the study. Due to the rapid development of the ICT-sector and the unsuitability of standard industry classification codes, there was no single up-to-date sampling frame available for the purposes of the study. Therefore, the names and contact information of the companies were searched from multiple sources, e.g. Kompass Finland Database, The Statistical Bureau of Finland database of Finnish companies, IT magazines, and Internet sites of the companies themselves, universities, cities, science parks, incubators, venture capitalists, and industry organizations. Since the companies of interest were operating in the ICT-sector, an Internet-based questionnaire was used for data collection.

Due to the rapid development of the Infocom sector and the unsuitability of standard industry classification codes, there was no single up-to-date sampling frame available for the purposes of the study. Therefore, the names and contact information of the companies were searched from multiple sources, e.g. Kompass Finland Database, The Statistical Bureau of Finland database of Finnish companies, IT magazines, and Internet sites of the companies themselves, universities, cities, science parks, incubators, venture capitalists, and industry organizations.

In total 493 companies were identified, and contacted by telephone between November and December 2001. During this phase, 33 companies were found ineligible, and 74 companies refused to participate in the study. The 386 companies, who on the telephone agreed to participate, received the following day an e-mail message containing instructions for answering the web-based questionnaire. A reminder message was sent to those who had not returned their answer within two weeks. One hundred and twenty-four companies returned their answers, resulting in an effective response rate of 32.1% (124/386). The questionnaire was divided into two parts, the second of which handled the internation-alization of the firms. This second part was answered by 94 firms, of which 55 (58.5%) carry out foreign operations. Born globals were operationalized according to the criteria suggested by Knight & Cavusgil (1996). A Firm is a Born Global if it had a share of foreign sales of 25% or higher and started international operations within three years after its startup. According to these criteria 18 firms (32.7%) of the internationalized firms were categorized as born globals. This relatively large share of born global firms within Finnish ICT industry supports the earlier research stating that the phenomenon is context specific and more common in knowledge-intensive industries.

Measures

Independent variables

International experience (EXPERIENCE) was composed of two variables: (1) share of company's management team members (%) that have international professional experi-ence; (2) share of company's management team members (%) that have received their education/studied abroad. The experience variables formed a single principal component factor explaining the 75% of the variance. International Vision (VISION), Risk Behavior (RISK) and Global Competition (COMPETITION) all include perceptual aspects of attitudinal "mindset" of the firms' managers. Therefore, we put the 11 items (Likert-scale from 1 to 5) together in order to see whether we actually get three different dimensions. Appendix 1 shows the factor analysis (explaining the 61.62% of the variance) of variables relating to the three constructs. Two of the variables had loadings on two factors. The variable "it is important for our company to internationalize rapidly" loaded on both VISION and COMPETITION. In addition, "the market in which we operate is global by nature" had the same effect. This indicates that either the items have not been able to catch the essence of the constructs thoroughly, or that the constructs are not independent of each other. In fact, it is quite simple to assume that the two constructs are partly overlapping, i.e. if the firm has a global vision it tends to see also the competition as global more easily.

R&D Intensity (R&D) was measured as a percentage of turnover that the firms estimate they use for research and development. The R&D intensity is a widely used measure in studies of innovation and it is often used as a proxy for innovation This is due to the fact that it is found to be positively related to measures of innovative output such as patents and new product introductions (Hitt *et al.* 1997). Market Niche (NICHE) was measured by the degree to which the managers perceived the firm's clients belong to a specific precisely determined target group. Partnerships (PARTNER) were calculated in terms of the number of domestic and foreign partnerships that the firm has. External Financing (FINANCING) was measured by the degree to which managers perceived the availability of funding influenced the firm's decision to enter international markets.

Dependent variables

The rapidity of internationalization (RAPIDITY) was measured in terms of the number of years from the founding of the company to the start of its international operations. The intensity of internationalization (INTENSITY) was composed of two variables: (1) number of different country markets in which the firm has operations; (2) share of revenues (%) from international operations in the year 2001. The intensity variables were positively correlated and formed a single principal component factor accounting for 75% of the variance. The standardized factor score was saved and used as a single measure for intensity of internationalization.

The descriptive statistics in Table 1 show that Finnish SMEs within the ICT sector start their international operations rather soon, within four years on average. One third of the managers have international professional experience. The firms typically operate in eight countries and get one third of their revenues from abroad. A quarter of the turnover is put into R&D and the range of the R&D intensity in our sample varied from 1 to 150%. Most of the partnerships are formed with other SMEs, mainly domestic but also foreign.

Hypotheses Testing

Two approaches, correlation analysis and linear stepwise regression were applied to assess the proposed hypotheses. First, in order to reveal the relations between the antecedents and rapidity and intensity of internationalization, a correlation matrix was computed as shown in Table 2. The results demonstrate that R&D Intensity (R&D) was significantly associated with the rapidity of internationalization. According to our expectations, the higher the R&D

Table 1: Descriptive statistics.

Variable	Min	Max	Mean	Std. Dev.
Year of establishment	1954	2001	1994.34	6.75
Turnover	0	500	17.31	54.39
Full-time employees	0	720	29.92	76.01
Rapidity	0	30	3.62	5.79
Number of countries	0	50	8.40	10.23
Share of revenues 2001	0	100	33.53	33.94
R&D Intensity	1	150	26.72	26.42
International work experience	0	100	32.55	33.61
International education	0	100	18.31	28.04
Number of small domestic partners	0	50	5.80	8.93
Number of large domestic partners	0	10	2.56	2.53
Number of small foreing partners	0	44	3.45	6.84
Number of large foreing partners	0	10	1.55	2.15

Table 2: Correlation coefficients.

	Rapidity	**Intensity**
Rapidity	1.00	
Intensity	−0.119	1.00
International experience	−0.110	0.220
Global vision	0.087	0.376**
Risk behavior	0.132	−0.357*
Global competition	−0.119	0.192
R&D intensity	−0.313*	0.044
Standardized product	0.164	0.159
Market niche	0.127	0.166
Partnerships	0.272	0.233
External financing	−0.138	0.468**

*Correlation is significant at the 0.5 level.
**Correlation is significant at the 0.01 level.

Intensity of the firm, the faster it starts international operations. The prior research also recognizes this effect of the R&D activity on the firm's internationalization (see e.g. Kobrin 1991). None of the other measures was related to the rapidity of firm's internationalization. This may be partly due to the cross-sectional nature of the study. On the other hand, all of the internationalized firms within the sample had initiated their international activities rather quickly. Almost 70% of the internationalized firms had started their internationalization within three years from their founding. However, there were some firms that were older and had started their international operations much slower. This may have had an influence on the variable. As Autio *et al.* (2000) notice, there is more to the concept of rapidity than just the starting of internationalization (start of international operations — founding), i.e. the rapidity of a firm's subsequent international growth (once the initial commitment has been made). Regarding the intensity of internationalization, significant correlations were found in three of the variables. These were global vision, risk behavior and financing.

Linear stepwise regression was applied to test the hypotheses. Therefore, the potential problems of multicollinearity were removed. The results of the hypotheses testing with regression analysis appear in Table 3. All the hypothesized antecedents were included in the list of independent variables. The table shows the independent variables that had the significant effect on the dependent variables. Four of the hypothesized antecedents were significantly associated with the intensity of internationalization. Supporting the previous studies on born global firms and international entrepreneurship, the global vision (VISION) of the firm had a positive effect on intensity of the firm's internationalization. Thus, H2b was supported. Also perception of the risks (RISKS) was related to the intensity, i.e. the less risky the management perceives the internationalization, the more intensively the internationalization is carried out. Therefore, H3b was supported. Although there was a correlation between the R&D Intensity (R&D) and rapidity, the variable had no significant effects in the regression; therefore H4a and H4b were rejected. The partnerships (PARTNER) were

Table 3: Hypothesis testing with regression analysis (independents with significant effect).

Hypotheses	Dependent	*R*-Square Adj.	Model Sig.	Independents	Beta	Sig.
Linear stepwise regression	Intensity	0.479	0.000	Vision	0.237	0.047
				Risk	−0.352	0.002
				Partner	0.318	0.008
				Financing	0.277	0.024
Linear stepwise regression	Rapidity	0.121	0.010	Partner	0.374	0.010

positively associated to intensity of internationalization, supporting H6b. This is in line with earlier research: Born global firms with limited resources seem to employ "hybrid" structures in order to obtain leverage form external resources. Availability of external financing (FINANCING) was also associated with more intensive internationalization. Therefore, H7b was supported. This is understandable because despite the utilization of external resources the internationalization process requires capital. The hypotheses concerning the rapidity of internationalization did not perform very well. Only one hypothesized antecedent, partnerships (PARTNER) had an effect on rapidity of internationalization. However, the effect was contrary to our expectations rejecting H6a, which means the more partners the firm has, the slower it has started its international activities. Earlier international experience (EXPERIENCE) did not have a significant effect on rapidity or intensity of internationalization, thus H1a and H1b were rejected. This result is in line with the findings of Autio *et al.* (2000), who concluded that years of foreign operating experience had no effect on foreign growth. As D'Aveni (1994) points out, what its most important is not how much a firm knows going in, but how rapidly it can learn. However, it must be mentioned that one third of the managers had international professional experience. This certainly has some level of influence, although no significant effects were found in the tests. Market niche (NICHE) was not associated with either rapidity or intensity of the firms' internationalization. This may be partly due to the vagueness of our measurement item, which needs to be developed further. Global competition (COMPETITION) did not have an effect on rapidity or intensity of internationalization, therefore both H8a and H8b were rejected. We still believe the competition within the ICT industry is very global. However, the managers of the firms may not consider the competition as a decisive factor as the market is global as well; thus global competition is "taken for granted" and the variable does not make a distinct effect.

Conclusions

Our main objective in this paper was to analyze the internationalization process of HTSFs. We studied the impact of several possible antecedents on rapid and intensive internationalization of firms. Therefore, valuable insights into the triggers of rapid and intensive internationalization of the firms were offered. The management challenges related to the process were also highlighted. These include how to communicate the global vision to all the firm's associates and how to determine the reasonable risk level during the process of rapid and intensive internationalization.

Our study generated several interesting aspects on the antecedents of rapid and intensive internationalization. The intensity of internationalization was driven by many of the hypothesized antecedents. Supporting the prior research on born globals and international entrepreneurship, the global vision of the firm was associated with more intensive internationalization. Furthermore, risk orientation of the managers was also connected with intensity. If the management does not consider the risks brought about by internationalization, this will result in more dedicated internationalization. The partnerships are positively associated to intensity of internationalization. This is in line with earlier research. Born global firms with limited resources seem to employ "hybrid" structures in order to obtain leverage from external resources. The use of outside resources may also be seen as a way to manage risks and maintain greater flexibility within the firm. Availability of external financing was also associated with more intensive internationalization. This is understandable because despite the utilization of external resources internationalization requires capital. The hypotheses concerning the rapidity of internationalization did not perform very well. Only one hypothesized antecedent, partnerships had an effect on rapidity of internationalization. Furthermore, the effect was contrary to our expectations, i.e. the more partners the firm has, the slower it has started its international activities.

Our sample can be evaluated in terms of its idiosyncrasy. We focused only on single country and single industry. The majority of the firms were young and formed a rather homogeneous sample with little variations in some of the variables. Having a data set consisting of firms from a variety of industries, may have provided clearer differences. Our measurement items may have also been imperfect and therefore, may not have delineated the constructs thoroughly. Furthermore, our cross-sectional data provides only a static "snap-shot" perspective of the HTSFs' internationalization process. Therefore, our results must be looked at cautiously and their generalizability to other contexts is not straightforward.

The majority of prior research on the internationalization of the firm has focused on large multinationals (MNCs) in manufacturing industries. This bias causes a possible risk that the assistance of policy-makers and practitioners is solely geared to the needs of these firms (Westhead *et al.* 2001), neglecting the specific needs of e.g. managers in HTSFs. The born global approach to internationalization, i.e. rapid and intensive internationalization can be considered as a high risk — high revenue strategy. It is not a suitable approach for all HTSFs in all situations. Recently, we have witnessed a number of failures partly due to this strategy. For example, one Finnish firm focusing on mobile applications opened offices in London, Berlin, Paris, Rome, Singapore and Los Angeles — all within one year. As the economic climate shifted the things got out of hands and the firm went bankrupt.

This study focused on antecedents and management challenges of HTSF international-ization. There exist many avenues for further research. The natural link would be to study the performance effects of rapid and intensive internationalization more deeply. We have addressed this limitation in other papers, since several performance items were included in the survey. On the other hand, the rapid and intensive internationalization is an outcome per se. Share of foreign sales is commonly considered as a proxy for export performance. Other opportunities for further study are related to e.g. the role of knowledge and learning in the rapid and intensive internationalization process.

References

Andersen, O. (1993). On the internationalisation process of firms: A critical analysis. *Journal of International Business Studies*, *24*(2), 209–230.

Autio, E., Sapienza, H. J., & Almeida, J. G. (2000). Effects of age at tntry, knowledge intensity, and imitability on international growth. *Academy of Management Journal*, *43*, 909–924.

Autio, E., Yli-Renko, H., & Sapienza, H. (1997). Leveraging resources under threat of opportunism: Predicting networking in international growth. Paper for the First Finnish SME Research Forum, February 13–14, Turku.

Bates, M., Rizvi, S. S. H., Tewari, P., & Vardhan, D. (2001). How fast is too fast? *The McKinsey Quarterly*, *3*, 52–61.

Bell, J., McNaughton, R., & Young, S. (2001). "Born-again global" firms — an extension to the "Born Global" phenomenon. *Journal of International Management*, *7*, 173–189.

Bloodgood, J. M., Sapienza, H., & Almeida, J. G. (1996). The internationalization of new high-potential U.S. ventures: Antecedents and outcomes. *Entrepreneurship Theory and Practice* (Summer), 61–76.

Bonaccorsi, A. (1992). On the relationship between firm size and export intensity. *Journal of International Business Studies*, *23*(3), 601–635.

Burgel, O., & Murray, G. C. (1998). The international activities of British start-up companies in high-technology industries: Differences between internationalisers and non-internationalisers. Babson Frontiers for Entrepreneurship Research 1998, available on the web at the following address: http://www.babson.edu/entrep/fer/papers98/XV/XV_A/XV_A.html, accessed 10th April 2002.

Calof, J. L., & Beamish, P. W. (1995). Adapting to foreign markets: Explaining internationalization. *International Business Review*, *4*(2), 115–131.

Coviello, N. E., & Martin, K. A.-M. (1999). Internationalization of service SMEs: An integrated perspective from the engineering consulting sector. *Journal of International Marketing*, *7*(4), 42–66.

Coviello, N. E., & McAuley, A. (1999). Internationalisation and the smaller firm: Review of contemporary empirical research. *Management International Review*, *39*, 223–256.

Coviello, N. E., & Munro, H. J. (1995). Growing the entrepreneurial firm: Networking for international market development. *European Journal of Marketing*, *29*(7), 49–61.

Cyert, R. M., & March, J. G. (1963). *A behavioral theory of the firm*. Englewood Cliffs, NJ: Prentice-Hall.

D'Aveni, R. A. (1994). *Hypercompetition: Managing the dynamics of strategic maneuvering*. New York: Free Press.

Eisenhardt, K. M., & Schoonhoven, C. B. (1996). Resource-based view of strategic alliance formation: Strategic and social effects in entrepreneurial firms. *Organization Science*, *7*(2), 136–150.

Eriksson, K., Johanson, J., & Majkgård, A. (1997). Experiential knowledge and cost in the internationalization process. *Journal of International Business Studies* (2nd quarter), 337–360.

Ganitsky, J. (1989). Strategies for innate and adoptive exporters: Lessons from Israel's case. *International Marketing Review*, *6*(5), 50–65.

Johanson, J., & Mattsson, L.-G. (1988). Internationalisation in industrial systems — a network approach. In: N. Hood, & J.-E. Vahlne (Eds), *Strategies in global competition* (pp. 287–314). London: Croom Helm.

Johanson, J., & Vahlne, J.-E. (1977). The internationalization process of the firm: A model of knowledge development and increasing foreign market commitments. *Journal of International Business Studies*, *8*(1), 23–32.

Johanson, J., & Vahlne, J.-E. (1990). The mechanism of internationalization. *International Marketing Review*, *7*(4), 11–24.

Johanson, J., & Wiedersheim-Paul, F. (1975). The internationalization of the firm: Four Swedish cases. *Journal of Management Studies* (October), 305–322.

Jolly, V. K., Alahuhta, M., & Jeannet, J.-P. (1992). Challenging the incumbents: How high technology start-ups compete globally. *Journal of Strategic Change, 1*, 71–82.

Knight, G., & Cavusgil, S. T. (1996). The Born Global firm: A challenge to traditional internationalization theory. *Advances in International Marketing, 8*, 11–26.

Kobrin, S. J. (1991). An empirical analysis of the determinants of global integration. *Strategic Management Journal, 12*(Special Issue), 17–31.

Lindqvist, M. (1991). *Infant multinationals: The internationalization of young, technology-based Swedish firms*. Unpublished doctoral dissertation, Stockholm School of Economics, Stockholm.

Luostarinen, R. (1979). Internationalization of the firm. An empirical study of the internationalization of firms with small and open domestic markets with special emphasis of lateral rigidity as a behavioral characteristic in strategic decision making. Helsinki, Acta Academiae Oeconomiae Helsigiensis, Series A: 30.

Madsen, T. G., Rasmussen, E., & Servais, P. (2000). Differences and similarities between Born Globals and other types of exporters. *Globalization, the Multinational Firm, 10*, 247–265.

Mamis, R. A. (1989). Global start-up. *Inc.* (August), 38–47.

McAuley, A. (1999). Entrepreneurial instant exporters in the Scottish arts and crafts sector. *Journal of International Marketing, 7*(4), 67–82.

Melin, L. (1992). Internationalization as a strategy process. *Strategic Management Journal, 13*, 99–118.

Oviatt, B. M., & McDougall, P. P. (1994). Toward a theory of international new ventures. *Journal of International Business Studies, 25*(1), 45–64.

Oviatt, B. M., & McDougall, P. P. (1995). Global start-ups: Entrepreneurs on a worldwide stage. *Academy of Management Executive, 9*(2), 30–43.

Oviatt, B. M., & McDougall, P. P. (1997). Challenges for internationalization process theory: The case of international new ventures. *Management International Review* (Special Issue 2), 85–99.

Penrose, E. T. (1959). *The theory of the growth of the firm*. New York: Wiley.

Petersen, B., & Pedersen, T. (1999). Fast and slow resource commitment to foreign markets — what causes the difference? *Journal of International Management, 5*, 73–91.

Preece, S. B., Miles, G., & Baetz, M. C. (1998). Explaining the international intensity and global diversity of early-stage technology-based firms. *Journal of Business Venturing, 14*, 259–281.

Rangan, S., & Adner, R. (2001). Profits and the Internet: Seven misconceptions. *MIT Sloan Management Review* (Summer), 44–53.

Rennie, M. W. (1993). Global competitiveness: Born Global. *The McKinsey Quarterly, 4*, 45–52.

Roberts, E. B., & Senturia, T. A. (1996). Globalizing the emerging high-technology company. *Industrial Marketing Management, 25*, 491–506.

Saarenketo, S., Kuivalainen, O., & Puumalainen, K. (2001). Emergence of born global firms: Internationalization patterns of the Infocom SMEs as an example. Paper for the 4th McGill Conference on International Entrepreneurship, 21–23 September, University of Strathclyde, Glasgow, U.K.

Tushman, M. L., & Anderson, P. (1986). Technological discontinuities and organizational environments. *Administrative Science Quarterly, 31*, 439–465.

Welch, L. S., & Luostarinen, R. (1988). Internationalization: Evolution of a concept. *Journal of General Management, 34*(Winter), 34–57.

Westhead, P., Wright, M., & Ucbasaran, D. (2001). The internationalization of new and small firms: A resource-based view. *Journal of Business Venturing, 16*, 333–358.

Appendix

Factor Structure of Competition, Vision and Risk

Item	Competition Factor	Vision Factor	Risk Factor	Communality
It is important for us to get our product to market before our competitors do so	0.76		0.236	0.634
Our foreign competitors react rapidly to our actions	0.757	0.29	−0.268	0.649
Our competitors operate internationally	0.729		−0.134	0.633
The market in which we operate is global by nature	0.632	0.488		0.64
It is important for our company to internationalize rapidly	0.486	0.45	−0.413	0.609
The company's management sees the whole world as one big marketplace	0.475	0.331	−0.337	0.449
The role of internationalization in the company's strategy	0.262	0.83		0.761
Management's willingness to become international		0.768		0.591
High risks of international business have impeded our decision to internationalize			0.798	0.646
There is a lot of uncertainty in international operations		0.197	0.712	0.549
The risks brought about by internationalization are too great	−0.275	−0.35	0.648	0.617
Eigenvalues	3.96	1.65	1.17	
% of variance	35.97	15.02	10.63	61.62

Chapter 17

Configurations of International Knowledge-Intensive SMEs: Can the Eclectic Paradigm Provide a Sufficient Theoretical Framework?

Tamar Almor and Niron Hashai

Introduction

Born-global companies, i.e. small or medium-sized enterprises (SMEs) that are multinational from inception or from a very early phase in their life span, are receiving increasing attention (e.g. Bell 1995; Bloodgood *et al.*, 1996; Dana *et al.* 1999; Keeble *et al.* 1997; McNaughton 2000; Oviatt & McDougall 1994, 1999; Rugman & Wright 1999). Business opportunities are becoming increasingly international and national confines are disappearing, thus creating new patterns of behavior for small and medium sized businesses. Oviatt & McDougall (1999) pose that in the late 1990s about 2% of emerging businesses worldwide were global from inception and a quarter of all SMEs around the world derived a major portion of their revenues from foreign countries. Furthermore, they stated that the pace in which SMEs internationalize is increasing. Reasons for this increase can be found in the reduction in cost of international transportation and communication and shortening of the product and industry life cycle. Thus, "born-globals" are SMEs that derive most of their income from foreign markets, have started their organizational lives while exploiting international opportunities and are international in their orientation and configuration.

Although the phenomenon of "born-global" SMEs is becoming increasingly common, a comprehensive theory supporting both the internationalization process of SMEs and their international business configuration, relevant to the economic conditions of the third millennium, is still lacking (Dunning 1993a; Oviatt & McDougall 1999). It is only natural therefore that researchers increasingly are focusing their attention on theories or paradigms to explain these phenomena (Almor 2000; Coviello & McAuley 1999; Liesch & Knight 1999; Nilsen & Liesch 2000; Oviatt & McDougall 1994). Most studies that focus on the internationalization of SMEs (Coviello & McAuley 1999; Nilsen & Liesch 2000; Oviatt &

New Technology-Based Firms in the New Millennium, Volume III
Copyright © 2004 by Elsevier Ltd.
All rights of reproduction in any form reserved.
ISBN: 0-08-044402-4

McDougall 1994) claim that the traditional theories of FDI and internationalization are insufficient to explain the process. After all how can one explain the fact that SMEs, with their limited resources, are able to operate globally as though they were a large multinational corporation? Various papers suggest the development of a paradigm based on a combination of several conceptual frameworks, namely: the Eclectic Paradigm (Dunning 1977, 1988), the "stage" model (Johanson & Vahlne 1977, 1990; Welch & Luostarinen 1988) and the network perspective (Johanson & Mattsson 1988, 1992; McNaughton & Bell 1999; Sharma 1992).

This paper aims to contribute to the understanding of internationalization of SMEs by proposing a conceptual paradigm for the configuration of the internationalization of knowledge-intensive SMEs from small, open economies (SMOPEC). The paper aims to challenge the repeated need of combining schools of thought (Coviello & McAuley 1999; Nilsen & Liesch 2000; Oviatt & McDougall 1994) by offering a coherent explanation to the international configuration of SMEs based on a revised version of Dunning's Eclectic Paradigm. Surprisingly, Dunning's eclectic Paradigm has received a minor place in the extensive efforts to explain the international behavior of SMEs and even when utilized it was mainly in comparison to the stage model and the network approach (Coviello & McAuley 1999).

The reasons to focus on knowledge-intensive SMEs from SMOPECs are three-fold. First, a large percentage of newly international SMEs is characterized as knowledge-intensive and this percentage is expected to increase in the coming years (Korot & Tovstiga 1999). Second, small, developed countries exhibit a large proportion of firms with a high trans-nationality index (UNCTAD 1999), this may be explained by the fact that, by definition, these markets will be very small (or even negligible) in the home country, therefore internationalization among SMEs is prolific in SMOPEC countries (Keeble *et al.* 1997; McNaughton 2000). Third, knowledge-intensive SMEs frequently introduce a new technology that is so unique that it might be claimed that those SMEs are actually inventing their own markets and are driven to the international markets in order to exploit first mover advantages and monopolistic gains (Keeble *et al.* 1997; McNaughton 2000).

The above mentioned reasons are widely accepted throughout the relevant literature as the main triggers for knowledge-intensive SMEs from small open economies to become international from inception (or almost form inception), thus creating a context in which a large percentage of born-globals will exist.

In the next section relevant literature will be reviewed and we will show how previous research has tried to formulate a concept that is based upon the above-mentioned schools of thought. Subsequently, we will present a conceptual framework based on modifications to Dunning's Eclectic Paradigm that explains the international configuration of knowledge-intensive SMEs located in SMOPEC countries. The paradigm is then tested on data measuring internationalization processes of Israeli, knowledge-intensive SMEs. The last section of the paper will present our findings and discuss their implications.

Theoretical Framework

Classic FDI and internationalization literature can be divided into several major schools (Coviello & McAuley 1999): (1) the "economic school"; (2) the "behavioral school"; and (3) the network approach.

The "economic school" (e.g. Buckley & Casson 1976; Dunning 1977, 1988; Hirsch 1976; Williamson 1975), views internationalization as a pattern of investment in foreign markets explained by rational economic analysis of ownership, location and internalization. According to this approach firms choose their optimal structure for each stage of production by evaluating the cost of economic transactions. The firm's organizational structure and its choice of location and ownership of specific value-chain activities (Porter 1980) are determined by criteria of cost minimization.

The "behavioral school" views internationalization as an ongoing process of evolution whereby the firm increases its international involvement as a function of heightened knowledge and market commitment (Aharoni 1966; Johanson & Vahlne 1977, 1990; Johanson & Wiedersheim-Paul 1975).

A third approach (Johanson & Mattsson 1988, 1992; McNaughton & Bell 1999; Sharma 1992) views the firm's internationalization as an entrepreneurial process based on an institutional and social network that supports the firm in terms of access to information, human capital, finance and so on (Vatne 1995). The organization relies on multilateral interdependencies for its internationalization process (Axelsson & Johanson 1992; Forsgren 1989), which enable the firm to find suitable international partners.

Each of the three approaches adds different dimensions to the concept of internationalization. The "economic school" analyzes the content of internationalization by focusing on choice of location and ownership of specific value-chain activities. The "behavioral school" analyzes the internationalization process while the network approach looks at the role informal inter-firm relationships play in the internationalization process. This paper focuses on the content and configuration, rather than on the process, of SME internationalization. Therefore, we utilized the "economic school" approach and specifically Dunning's Eclectic Paradigm (Dunning 1977, 1980, 1988, 1993a, b) in order to study the internationalization of knowledge-intensive SMEs.

Because of the unique characteristics of knowledge-intensive SMEs we need to modify and adapt this paradigm in order to utilize it in explaining internationalization of these specific SMEs.

Explaining SME International Configuration by Means of the Eclectic Paradigm

Dunning's Eclectic Paradigm is based on three cumulative conditions that are necessary to become a successful MNE: ownership advantages, location advantages and internalization advantages (thus, the OLI paradigm).

Theory suggests that the configuration of international, knowledge-intensive SMEs is driven by conditions that differ from those ruling traditional industries, thus creating strategic options for the SME that may differ from the traditional options. Table 1 presents a framework for examining the configuration of international knowledge-intensive SMEs. The framework is based on relating the OLI paradigm to the major primary value activity chain (VAC) of the SME. In this paper we focus on four major generic stages of the value-added chain: R&D, production, marketing & sales and after-sales services. Table 1 explains how each stage of the VAC is related to the OLI paradigm. Below, we discuss the

Table 1: The eclectic-paradigm and its relation to the value added chain (VAC) of knowledge-intensive SMEs.

VAC	Ownership Advantage	Location Advantage	Internalization Advantage
R&D	Basis for SCA, needs to be incorporated within the VAC	Home country has a comparative advantage in R&D	Internalization is needed as the whole VAC is dependent upon proprietary R&D
Production	Not necessarily. May be irrelevant (intangibles) or of minor importance	Not necessarily due to diseconomies of scale and high labor costs in the home country	Usually not. Outsourcing may enable to exploit economies of scale and learning economies
Marketing and sales	Client base is an important SCA, thus marketing will be a part of the firm's VAC	Small and distant home country is in a location disadvantage compared to the host country	Usually firms are too small to exploit internalization advantages, but knowledge-intensive products require in-house marketing and sales force
After-sales service	Incorporation within the VAC allows to develop special supplier-customer relations, increase customers loyalty and to support propriety know-how	Small and distant home country is in a location disadvantage compared to the host country	Internalization is important as it protects against diffusion of knowledge, but may be a disadvantage if one needs to serve mass markets

four stages of the VAC of the international SME in terms of ownership advantage, location advantage, and internalization advantage.

Ownership advantage in the Eclectic Paradigm relates to the possession of intangible assets or core know-how that gives the firm a competitive advantage over its competitors and that can be exported. The main competitive advantage of knowledge-intensive SMEs lies in their core know-how to develop, market and service knowledge-intensive products. Knowledge-intensive SMEs frequently create sustainable competitive advantages (SCAs) around their unique R&D (Amin & Thrift 1994). The resource-based theory views owner-ship of technology as one of the most important bases for the development of sustainable competitive advantages. As Dunning himself acknowledges (Dunning 1993b), proprietary technology is a resource around which the profit earning potential of a firm is developed (Grant 1998) and entry barriers are created (Wernerfelt 1984). Thus, unique know-how and proprietary technology are exactly the *ownership-specific advantage* that Dunning refers to, or more specifically the advantage that stems from ownership of assets (Oa) (Dunning 1988, 1993b).

Marketing, sales and after-sales services allow the SME to build up long term rela-tions with a client base and to receive feedback regarding their technology. Moreover, as technology is the core-capability of the firm, which forms the basis for its SCAs, tight supplier-customer relations may lead to further innovations that the firm will prefer to keep as proprietary for itself. Moreover, as knowledge-intensive products require more frequent interactions between the supplier and its customers relatively to low-tech firms (Almor & Hirsch 1995), marketing and after-sales services are crucial factors in building customers loyalty and for the development of a strong clientele base. This again is consistent with the resource-base view of the firm (Wernerfelt 1984). It is expected therefore that SMEs will prefer to incorporate R&D, marketing and after-sales services in their VAC, in order to protect and upgrade proprietary knowledge and its client base which are their core asset advantages (Oa).

In contrast to the classical application of the OLI paradigm, production usually does not provide a basis for SCAs as it either involves copying the developed knowledge, such as software, to a medium through which it can be transported, or embedding it in a chip, mother-board etc., as a part of a complete system. In the first case the production skills do not provide a basis for SCAs as they are very common and can be easily provided by others. In the second case, substantial economies of scale are involved which do not allow the SME to incorporate the production process. In either case, there will not be any ownership advantages for the firm concerning the production process, therefore SMEs may not incorporate this function.

Thus we hypothesize that:

(1) Knowledge-intensive SMEs will incorporate R&D, marketing, sales and after-sales ser-
     vices in their value-added chain more frequently than production activities and perceive
     their R&D, marketing, sales and after-sales services as their SCAs.

Location advantages relate to: (1) advantages existing in the host country such as access to cheaper country-specific inputs (raw materials, labor etc.); as well as to (2) the required proximity to markets and to consumers. We assume that a location advantage exists in the home country regarding the R&D stage of knowledge-intensive products (at least in developed SMOPEC countries such as Sweden, Finland, Israel and Ireland). This comparative advantage stems from the local network and cluster of knowledge-intensive companies of which the SME is a part (Keeble *et al.* 1997; Porter 1990), as well as from the fact that the development of these products requires skilled, professional, labor, which the home country is endowed with.

A second aspect of the location advantage concerns the distance of SMOPEC firms from their target market. Although not discussed extensively by Dunning, location advantage not only concerns host country advantages, but also distance from the target market, thereby creating a *location disadvantage* for the SME. The major target markets of knowledge-intensive SMEs are the U.S., the Pacific Rim and South East Asia and the European Union. Most SMEs will be located at a distance from, at least part, of their target markets.

Knowledge-intensive SMEs frequently sell their products outside their home country, as the home country usually does not provide a large enough market for the technology developed and manufactured by SME. Therefore, The SMEs are located at a distance from their target markets, thus needing to "pay" a certain premium for distance. This premium takes the form of higher transportation costs compared to potential local competitors, slower response times, problematic after sales support, as well as higher interaction costs with potential clients. In addition such SMEs will encounter cultural differences (Hofstede 1980; Hymer 1976; Kogut & Singh 1988) as well as higher information flow costs (Buckley & Casson 1976; Casson 1994, 2000) in comparison to local competitors.

Clearly, all firms operating internationally have to cope with the above-mentioned disadvantages. While the larger multinational enterprise (MNE) is able to cope with such problems through expansion of its marketing and after sales services infrastructure by, for instance, acquiring local firms (in the major host countries) which own an existing marketing and services infrastructure, this option is beyond the means of most SMEs not only because of scarce financial resources, but mostly because of lack of managerial resources and adequate international experience (Buckley 1989; Gomes-Casseres 1997)

Location disadvantage can be solved partly by creating strategic alliances with local companies, by choosing to sell products OEM (thereby foregoing contact with the final user of the technology), by focusing on one or two large markets where the SME will establish a local presence or by serving a global niche that has relatively few customers (Gomes-Casseres 1997; Keeble *et al.* 1997; McNaughton 2000). We expect that the SME will employ at least one of the above strategies to enhance its proximity to the target market. As some of the mechanisms to compensate for the location disadvantages are related to the SME's internalization decision, we will return to this dilemma later in the section.

Location of production will be determined according to a wide range of rather contradicting considerations, such as: economies of scale, labor costs, local networks, knowledge and managerial capabilities. Not all of these considerations can be generalized, but some observations can be made. SMEs from SMOPEC countries may find it difficult to utilize economies of scale, as they are comparatively limited in production resources (both capital and labor), in comparison to firms from large countries. Moreover, as labor costs are comparatively high in countries such as Sweden, Finland, Ireland and Israel, there will not be any location advantages in those countries for placement of production facilities. On the other hand, many SMEs may be too small to move their production to "cheap labor" developing countries and will therefore manufacture in the home country even though this does not provide a location advantage. In addition, many SMEs manufacture products that are in the early stages of the Product Life Cycle (Vernon 1966) need to be manufactured in close proximity to other functions such as R&D.

Other SMEs that manufacture intangible products (software, internet applications, telecommunication etc.) will be indifferent to the location of production as manufacturing costs are relatively negligible for their operations. Therefore, we hypothesize that:

(2) R&D will be located in the home country in order to exploit local advantages; marketing and after-sales services will be located outside the home country in order to overcome location disadvantages. No preferences will be found for the location of production.

Internalization advantages relate to the ability to create superior profits by using the advantages that the firm owns, rather than by selling them or licensing them to third, international parties. The distinction between ownership and internalization is not always clear cut (Ietto-Gillies 1997) and interrelations between the two exist. Nevertheless, the internalization advantage of the R&D function seems to need little explanation. As stated above, R&D knowledge forms the basis for the firm's SCAs and will therefore need to be internalized and kept proprietary in order to strengthen the sustainability of these advantages.

Once knowledge is developed, production mainly involves embedding the knowledge in a tangible device such as a CD-Rom disc, motherboard or other electronic means. Sometimes these costs are negligible, for instance when creating a copy of software. Other times they involve large economies of scale and learning economies, as in the case of producing an electronic chip. In either case, there is no compelling reason to internalize production. Thus, while production may be incorporated in the value-added chain, it will not necessarily be internalized, but may very well be performed by independent firms.

Internalization of marketing and after-sales services on the other hand is more complicated. Internalization of those functions is required due to the knowledge content of the SMEs products that requires both high skill and specific marketing expertise and due to the need to protect proprietary know-how. On the other hand internalization of marketing and after-sales services means that the market is served by the firm itself. This is quite a demanding task for SMEs, having usually scarce managerial and financial resources. One way to internalize marketing and after-sales services is to target a market niche that consists of few customers. Indeed, many SMEs will try to create profits by focusing on relatively few transactions that provide maximum value per transaction, by selling to industrial customers or to specific market niches. However, if the firm's client base consists of a large group of customers, spread around the globe, internalization becomes almost impossible.

When serving industrial customers or specific market niches, the absolute number of clients is much smaller than when the firm serves mass-market consumers. If the firm serves a small number of industrial customers or is targeting market niches, the need for a substantial marketing, distribution and after-sales services infrastructure is reduced and a modest marketing entity may enable the firm to operate successfully. When the SME focuses on a small number of clients with similar characteristics and similar product demand (a global niche), the SME may even serve those clients directly from its home base. The SME will provide sales and after-sales services by sending engineers and other professionals to the client on an individual basis, thereby creating a "virtual MNE" and reducing the need for a sizeable marketing and services infrastructure.

By definition, most knowledge-intensive SMEs will find it quite difficult to serve a large customer base, as they need to have intensive interactions with their clients, because of the nature of knowledge-intensive products (Almor & Hirsch 1995). Moreover, SMEs are often limited in resources such as managerial skills, international experience, human resources and finance. Therefore, the resource constraining such firms is marketing, sales and after sales services infrastructure.[1] Thus, we hypothesize that:

(3a) Knowledge-intensive SMEs will internalize R&D more frequently than production.
(3b) Firms that have internalized marketing and after-sales services will mainly serve in-dustrial customers or market niches and not large and dispersed group of clients.

Empirical Analysis: Methodology and Data

This study focuses on knowledge-intensive SMEs that exhibit a high degree of international behavior. The SMOPEC country chosen for the study was Israel. Israel is regarded an international leader in terms of its contribution of hi-tech to overall exports as well as in terms of the high number of Israeli SMEs traded on the American NASDAQ stock exchange (Almor 2000).

According to Israel's Central Bureau of Statistics, 69% of Israel's industrial exports in 1999 were accounted for by knowledge-intensive industries (when excluding diamonds).

A second indicator of Israel's proclivity regarding knowledge-intensive industries exists in the sheer number of Israeli firms traded on NASDAQ. In 1999, approximately 120 Israeli firms were traded on NASDAQ, of which about 80% were defined as knowledge-intensive. Although Israel has a population of only six million people and is smaller in size compared to the Netherlands, it is home to the second largest group of non-U.S. companies traded on NASDAQ. In 1995, the number of NASDAQ-listed Israeli firms nearly equaled the number of all other foreign firms combined, excluding Canadian companies (Blass & Yafeh 1998).

The research sample was defined according to the following criteria:

(1) Knowledge-intensity. Various definitions exist to classify knowledge-intensive indus-tries (e.g. Almor & Hirsch 1995). In this study we employed share of investment in R&D as a percentage of the firm's total sales as a means of classification. Firms investing 5% or more in R&D were classified as knowledge-intensive.
(2) Small and medium sized enterprises (SMEs). Various definitions can be found in the literature regarding size. In this study "number of employees" was used as a measure of size. Buckley (1997) and Storey (1994) used the term SMEs in reference to enterprises employing less than 500 employees. However, the size threshold may vary (Dana *et al.* 1999) per country and by industry. SME definitions are based on firms operating in *local* markets. In this study we focused on *international* enterprises that are small to medium sized, i.e. small and medium sized as compared to regular MNEs. Therefore

---

[1] As opposed to traditional industries where the resource in constraint is often production infrastructure.

Table 2: Descriptive statistics of 56 knowledge-intensive, international SMEs located in Israel (data are for the year 1999).

Variable	Percentage of SMEs in the Sample	Median	S.D.	Range
Descriptive variables				
Year SME was established	–	1990	7.2	1950–1996
No. of employees	–	163	266.6	15–1050
Sales	–	$25M	$61M	$0–338M
R&D/sales	–	0.16	0.36	0.03–2.46
Serving industrial customers	87	–	–	–
Operating in niche markets	76	–	–	–
NASDAQ IPO	67	–	–	–
EU IPO	27	–	–	–
Israel IPO	6	–	–	–
Measures of internationalization				
Country 1st product sold: Israel	29	–	–	–
Country 1st product sold: U.S.	33	–	–	–
Country 1st product sold: EU	18	–	–	–
No. of countries served	–	30	18.7	1–86
% Israel based employees	–	70	25	1–100
% U.S. based employees	–	13	17	0–84
% sales in Israel	–	2	20	0–99
% sales in USA	–	35	27	0–100
% sales in EU	–	25	21	0–0.5

we limited ourselves to firms with less than 1000 employees. In the final sample only 9 firms employed more than 500 employees. Furthermore, the median value of number of employees was 163.[2]

(3) International behavior. In order to identify SMEs that generate a significant part of their sales outside Israel, we focused on those that traded outside Israel. One reason to trade a company outside one's home country is to signal internationality to clients, business partners and investors alike (Almor 2000). Therefore, we propose that Israeli SMEs traded outside Israel will be international in their business behavior.

Initially, 120 Israeli companies that were traded on NASDAQ (U.S.) and/or on various EU stock exchanges during the year 2000, were identified. All firms that were not defined as knowledge-intensive or were larger than 1000 employees were excluded from this list.

---

[2] Another parameter that is often used for classify SMEs is turnover (Cavusgil, 1984; Stray *et al.* 2001). Annual sales of the firms in the current sample do not exceed $350 Million.

The remaining firms in the sample were approached and requested to take part in a face-to-face interview, which allowed the researchers to gather data by means of a structured questionnaire. The final sample included 56 knowledge-intensive, international SMEs.

The firms in the sample were classified into four industries: (1) software (38%); (2) information and telecommunications — ICT (20%); (3) electronics (27%) and (4) "other," which included pharmaceutics, bio-technology and medical technologies (15%). Identification of the industries and classification of the firms according to industry were based on self-categorization of the firms.

Descriptive data show that the firms have a unique profile. Table 2 present percentages, median values, standard deviations and range of the variables and measures.

The data in Table 2 illustrates that firms are small in terms of the number of employees, as well as sales, they are innovative, and serve customers all over the world. The firms mainly operate in niche markets serving industrial customers.

The investigated SMEs hardly sell products in their home market. The majority of the firms in the sample started out in international markets, selling their first product outside the home country. However, their employees, as well as the senior management, are based in the home country. These descriptive data clearly show that the SMEs in the sample have a divergent configuration from classical MNEs, which have usually started in their home country, have a larger and more dispersed customer base and have production operations, personnel and management scattered around the world (Dunning 1980, 1993b, 1995; Johanson & Vahlne 1977, 1990). This divergence leads us to examine the international configuration of knowledge-intensive SMEs from SMOPECs and to ask if it is similar to that of classical MNEs.

Results

The hypotheses that were detailed earlier were tested on the sample described above. The first hypothesis concerned ownership of functions providing the basis for SCAs. It was hypothesized that knowledge-intensive SMEs will incorporate R&D, marketing, sales and after-sales services in their value-added chain more frequently than production activities.

The Cochran-Mantel-Heanszel statistic, a non-parametric measure, was used to test the hypothesis. The Cochran-Mantel-Heanszel statistic serves for testing hypotheses regarding equality of two matched distributions, measured on a categorical (nominal) scale.

Results presented in Table 3 show that R&D, marketing & sales and after-sales services are incorporated significantly more frequently than production, thus supporting hypothesis 1.

The ownership advantage findings are further supported by the firm's perception of themselves. Sixty eight percent of the firms identified their "unique technology" as one of their three most important sustainable competitive advantages. Seventy seven percent of the SMEs thought that "understanding needs of client" is among their three most important strategic capabilities and 72% of the firms regarded their "innovation" to be as such.

The second hypothesis concerned location of functions. It was hypothesized that R&D will be located in the home country and that marketing and after-sales services will be located outside the home country, it was further hypothesized that production will not present a specific pattern of location.

Table 3: Cochran-Mantel-Haenszel statistics for 56 knowledge-intensive, international SMEs located in Israel.

Variable	df	Value	Probability
Full ownership of functions	1	7.14	0.0075
Comparison of location of functions	2	37.88	<0.0001
Location R&D — production	1	2.57	0.109
Location R&D — marketing, sales and after-sales services	1	28.13	<0.0001
Location production — marketing, sales and after-sales services	1	20	<0.0001
Internalization of R&D — production	1	15.70	<0.0001

The Cochran-Mantel-Heanszel (CMH) statistic showed that a significant difference exists in patterns of location between R&D, marketing & sales and after-sales services and production (CMH value = 37.88). R&D was located in the home country significantly more frequently than marketing, sales and after-sales services. Marketing and sales as well as after-sales services were located significantly more frequently outside the home country, compared to R&D and to production. No significant difference was found between location of production and R&D, but significance was just above the required norm ($p = 0.109$).

Hypothesis 3a stated that SMEs prefer to internalize R&D more frequently than production. The Cochran-Mantel-Heanszel statistic showed that production was significantly less frequently internalized than R&D, thus supporting hypothesis 3a (see Table 3).

Hypothesis 3b stated that those firms that have internalized marketing and after-sales services mainly serve industrial customers or market niches and not large and dispersed groups of clients. Cross tabulations and chi-square tests showed that most firms in the sample serve industrial customers (87% of the sample) and market niches (76% of the sample), however, no significant relationship was found between internalization of marketing, sales and after-sales services and type of customers.

Discussion

This paper examines the content of internationalization strategies of knowledge-intensive SMEs. A theoretical framework, providing an explanation for the international configuration of SMEs and derived from Dunning's Eclectic Paradigm, was presented and tested on a sample of knowledge-intensive SMEs located in a small, open economy and traded on NASDAQ.

The study focuses on knowledge-intensive SMEs, a fast growing group of firms that seems to behave differently from firms situated in traditional industries. Differences seem to exist in the degree of internationality relative to size and age, in the way the firms are configured and in the basis for their Sustainable Competitive Advantages (SCAs). The basic differences led us to hypothesize that the international configuration of knowledge-intensive SMEs, when analyzed in terms of the Eclectic Paradigm, will be based on R&D as well as marketing and after-sales services, rather than on production. As R&D, marketing and after-sales services were expected to be the core competencies of knowledge-intensive SMEs, their ownership, location and internalization modes were expected to affect the internationalization content of such firms.

The model proposed in this paper and studied empirically shows that the firms in the sample have R&D facilities that are fully internalized, providing ownership advantages of knowledge, which form the basis for the firm's SCAs. The R&D function is located in the home country, thereby exploiting location advantages, such as local knowledge networks as well as comparative advantage resulting from the home country's endowment of skilled labour.

Production on the other hand, is not necessarily part of the value-added chain (VAC), and even if it is part of the VAC, it is not necessarily internalized and is not located according to principles guiding traditional industry. The fact that production is of lesser importance in its contribution to the firm's SCAs makes the ownership, location and internalization of production a minor issue for knowledge-intensive SMEs.

Marketing and sales, as well as after-sales services are viewed as providing an additional basis SCAs. Interactions with clients, as well as installation and maintenance of their technologies allow the SMEs to receive feedback and improve their technology and thus increase their SCA. Moreover, most knowledge-intensive SMEs market their products to a few, large customers. It may take years before an industrial potential customer becomes an actual customer. Thus, protection of the client base is vital for a SME's survival and competitiveness. Moreover, the relatively high sophistication level of products requires a highly skilled marketing and services force that needs to be instructed and updated periodically. Therefore, knowledge-intensive SMEs mostly incorporate marketing and sales as well as after-sales services within their VAC and provide these value activities mostly provided in-house. In contrast to the R&D function, however, these functions are usually located outside the home country, so as to diminish location disadvantages created by distance. Here one can note the influence of the SMEs' origin from a small, open economy with a negligible home market. Creation of some kind of international presence, in terms of marketing and after-sales services, is essential in order to compete with local competitors that do not have the burden of geographical and cultural distance from their market.

The findings in this paper address two phenomena: (1) SMEs taking part in the "international game arena"; (2) knowledge-intensive SMEs, that develop and market products which are, at least partly, intangible, compete internationally in ways that are different from classical MNEs, by internalizing R&D, marketing and after sales services and approaching global niches. These findings conform to other studies related to the internationalization of knowledge-intensive SMEs in North America (Gomes-Casseres 1997; McNaughton 2000) and the U.K. (Keeble *et al.* 1997; Stray *et al.* 2001) and thus reduce the odds that this study's results are affected by the sample (all Israeli firms).

Generalization of the findings has theoretical implications as well as implications for practitioners. Theoretically, the findings imply that traditional internationalization models have to be re-examined and tested for validity when analyzing behavior of small and medium sized enterprises that compete in the international market. Basically, we found that the

Eclectic Paradigm can be employed, albeit needing correction, or rather modifications, for the different nature of such firms. This indicates that this model, once adapted to specific characteristics of knowledge-intensive SMEs can still serve as a useful analytical tool. The question is if, for instance, the Upssala model is valid when explaining internationalization processes of knowledge-intensive SMEs, taking into account that 71% of all firms in the sample sold their first product outside the home country. The relevance of the Upssala model to the internationalization process of SMEs is debated in the relevant literature (Keeble *et al.* 1997; McNaughton 2000; Stray *et al.* 2001) but not examined in this paper.

For practitioners, the findings imply that internationalization is not limited to big firms anymore. However, content of internationalization is unique for SMEs. Mostly, SMEs seem to serve few, large clients who are dispersed around the world, and who are mostly organizations themselves. These clients are served through marketing and after-sales activities that are located close to them. R&D activities on the other hand, are mostly located in the home country, thereby exploiting comparative advantages existing in that country, such as clusters of hi-tech companies and skilled labor endowment. These findings imply that organizational configuration is geographically dispersed (R&D usually located in the home country and marketing and after sale services in the host country), that most SMEs operate within global niches and may be successful doing so.

Coviello & McAuley (1999) were right to raise the question whether the content and process of SMEs internationalization is truly different from that of larger firms. The current paper offers a partial answer to the "content" part of this question and more specifically to the question: does the international behavior of knowledge-intensive SMEs differ from that of larger knowledge-intensive firms? After all, the large MNEs of today (e.g. Microsoft, HP, IBM etc.) are different from the traditional MNEs as well. Clearly the ownership advantages of knowledge-intensive, large MNEs are similar to those of SMEs and lie in their R&D, marketing and after-sales services. As many of the large knowledge-intensive MNEs come from large countries with a large home market they usually can rely on their local sales to account for a substantial part of their turnover, thus their location disadvantage is less critical.[3]

The major difference, however, is the ability of large MNEs to develop a marketing and services infrastructure in host markets through direct investments (whether greenfield or through acquisitions) by exploiting their superior financial and managerial resources, their monopolistic power as well as their brand name. The international configuration of the SMEs discussed in this paper is unique, as they have to compensate for their distance from world's largest markets and for their limitation (in terms of time and cost) in building a substantial marketing and services infrastructure in their host markets. Internalization of marketing and after-sales services is a necessity, but can be utilized only to target niche markets (or industrial customers).

Although beyond the scope of this paper, it further raises the question of firm's growth. How, if at all, can such a SME become a MNE in the traditional sense of the term? Just like traditional SMEs that focus on a specific niche, these SMEs face a similar problem when pursuing aggressive growth strategies that will eventually, force them to abandon their niche for more lucrative markets.

---

[3] Even though some leading multinationals such as Nokia and Eriksson come from SMOPEC countries as well.

A better understanding of this issue may be derived from a systematic analysis of the preferred international business modes of knowledge-intensive SMEs. Thus, future studies may focus on the role of International Strategic Alliances (ISAs) in the internationalization of knowledge-intensive SMEs and on the contradictory role ISAs may play. Whereas ISAs may compensate for the cost and difficulty of creating marketing and after-sales services infrastructures in distant markets (Gomes-Casseres 1997), ISAs also threaten the ability of knowledge-intensive SMEs to protect their proprietary know-how. Is this conflict inevitable? Will any knowledge-intensive SME need to risk its core know-how and client base if it wishes to survive in the global market place? Are there any particular strategies that enable the SME to protect its know-how while leveraging on the marketing and sales infrastructure of larger MNEs? May international business networks provide a mechanism to reduce the need for formal ISAs? All these are critical questions that should be addressed in future studies.

References

Aharoni, Y. (1966). *The foreign investment decision process.* Boston: Harvard University.

Amin, A., & Thrift, N. (1994). Living in the global. In: A. Amin, & N. Thrift (Eds), *Globalization, institutions and regional development in Europe.* Oxford: Oxford University Press.

Almor, T. (2000). Born global: The case of small and medium sized, knowledge-intensive, Israeli firms. In: T. Almor, & N. Hashai (Eds), *FDI, international trade and the economics of peacemaking.* College of Management — Academic Studies Division: Rishon LeZion, Israel.

Almor, T., & Hirsch, S. (1995). Outsiders' response to Europe 1992: Theoretical considerations and empirical evidence. *Journal of International Business Studies, 26*(2), 223–238.

Axelsson, B., & Johanson J. (1992). Foreign market entry — the textbook view vs. the network view. In: B. Axelson, & G. Easton (Eds), *Industrial networks: A new view of reality.* London: Routledge.

Bell, J. D. (1995). The internationalisation of small computer software firms. *European Journal of Marketing, 29*(8), 60–75.

Blass, A., & Yafeh, Y. (1998). Vagabond shoes longing to stray — Why Israeli firms list in New York — Causes and implications. Discussion Paper No. 98.02. Bank of Israel, Research department.

Buckley, P. J. (1989). Foreign direct investment by small and medium-sized enterprises: The theoretical background. *Small Business Economics, 1,* 89–100.

Buckley, P. J. (1997). International technology transfer by small and medium-sized enterprises. *Small Business Economics, 9,* 67–78.

Buckley P. J., & Casson M. (1976). *The future of the multinational enterprise.* London: Macmillan.

Casson, M. (1994). Why are firms hierarchical. *International Journal of the Economics of Business, 1*(1), 47–76.

Casson, M. (2000). *The economics of international business — A new research agenda.* Cheltenham, UK: Edward Elgar.

Coviello, N., & McAuley, A. (1999). Internationalisation and the smaller firm: A review of contemporary empirical research. *Management International Review, 39*(3), 223–256.

Dana, L. P., Eternad, H., & Wright, R. W. (1999). Theoretical foundations of international entrepreneurship. In: A. M. Rugman, & R. W. Wright (Eds), *Research in global strategic management. International entrepreneurship: Globalization of emerging businesses.* Stamford, CT: JAI Press.

Dunning, J. H. (1977). Trade, location of economic activity and the MNE: A search for an eclectic paradigm. In: B. Ohlin, P. O. Hesselborn, & P. M. Wijkman (Eds), *The international allocation of economic activities, proceedings of a Nobel symposium in Stockholm.* London: Macmillan.

Dunning, J. H. (1980). The location of foreign direct investment activity, country characteristics and experience effects. *Journal of International Business Studies*, 9, 9–22.

Dunning, J. H. (1988). The Eclectic Paradigm of international production a restatement and some Possible Extensions. *Journal of International Business Studies*, *19*(1), 1–31.

Dunning, J. H. (1993a). The changing dynamics of international production. In: J. H. Dunning (Ed.), *The globalization of business*. London and New York: Routledge.

Dunning, J. H. (1993b). *Multinational enterprises and the global economy*. Wokingham, UK: Addison-Wesley.

Dunning, J. H. (1995). Reappraising the eclectic paradigm in an age of alliance capitalism. *Journal of International Business Studies*, *26*(3), 461–481.

Forsgren, M. (1989). *Managing the internationalization process: The Swedish case*. London: Routledge.

Gomes-Casseres, B. (1997). Alliance strategies of small firms. *Small Business Economics*, *9*(1), 33–44.

Grant, R. M. (1998). *Contemporary strategy analysis* (Third ed.). Oxford, UK: Blackwell.

Hirsch, S. (1976). An international trade and investment theory of the firm. *Oxford Economic Papers*, 28, 258–270.

Hofstede, G. (1980). *Culture's consequences: International differences in work-related values*. Beverly Hills, CA: Sage.

Hymer, S. H. (1976). *The international operations of national firms: A study of direct foreign investment*. Unpublished Ph.D. thesis. Cambridge, MA: MIT Press.

Ietto-Gillies, G. (1997). Alternative approaches to the explanation of international production. In: R. John (Ed.), *Global business strategy*. London: International Thomson Business Press.

Johanson, J., & Mattsson L. (1988). Internationalisation in industrial systems — A network approach. In: N. Hood, & J.-E. Vahlne (Eds), *Strategies in global competition*. London: Croom Helm.

Johanson, J., & Mattsson, L. (1992). Network positions and strategic action — An analytical framework. In: B. Axelsson, & G. Easton (Eds), *Industrial networks: A new view of reality*. London: Routledge.

Johanson, J., & Vahlne, J. E. (1977). The internationalization process of the firm — a model of knowledge development and increasing foreign market commitment. *Journal of International Business Studies*, *8*(1), 23–32.

Johanson, J., & Vahlne, J. E. (1990). The mechanism of internationalization. *International Marketing Review*, *7*(4), 11–24.

Johanson, J., & Wiedersheim-Paul, F. (1975). The Internationalisation of the firm: Four Swedish cases. *Journal of Management Studies*, *12*, 305–322.

Keeble, D., Lawson, C., Lawton Smith, H., Moore, B., & Wilkinson, F. (1997). Internationalisation processes, networking and local embeddedness in technology-intensive small firms. *Small Business Economics*, *11*, 327–342.

Kogut, B., & Singh, H. (1988). The effect of country culture on the choice of entry mode. *Journal of International Business Studies*, *19*(3), 411–423.

Korot, L., & Tovstiga, G. (1999). Profiling the twenty-first-century knowledge enterprise. In: A. M. Rugman, & R. W. Wright (Eds), *Research in global strategic management. International entrepreneurship: Globalization of emerging businesses*. Stamford, CT: JAI Press.

Liesch, P. W., & Knight, G. (1999). Information internalization and hurdle rates in small and medium enterprise internationalization. *Journal of International Business Studies*, *30*(2), 383–395.

McNaughton, R. B., & Bell, J. D. (1999). Brokering networks of small firms to generate social capital for growth and internationalization. In: A. M. Rugman, & R. W. Wright (Eds), *Research in global strategic management. International*. Stamford, CT: JAI Press.

McNaughton, R. B. (2000). Determinants of time-span to foreign market entry. *Journal of Euromarketing*, *9*(2), 99–112.

Nilsen, F. I., & Liesch D. W. (2000). International market entry of small knowledge-based firms: Towards a synthesis of economic and behavioral approaches. Paper presented at the Academy of International Business Annual Conference 2000. Phoenix, Arizona.

Oviatt, B. M., & McDougall, P. P. (1994). Toward a theory of international new ventures. *Journal of International Business Studies*, *25*(1), 45–64.

Oviatt, B. M., & McDougall, P. P. (1999). A framework for understanding accelerated international entrepreneurship. In: A. M. Rugman, & R. W. Wright (Eds), *Research in global strategic management. International entrepreneurship: Globalization of emerging businesses.* Stamford, CT: JAI Press.

Porter, M. E. (1980). *Competitive strategy: Techniques for analyzing industries.* New York: Free Press.

Porter, M. E. (1990). *The competitive advantage of nations.* New York: Free Press.

Rugman, A. M., & Wright, R. W. (Eds) (1999). *Research in global strategic management. International entrepreneurship: Globalization of emerging businesses.* Stamford, CT: JAI Press.

Sharma, D. (1992). International business research: Issues and trends. *Scandinavian International Business Review*, *1*(3), 3–8.

Storey, D. I. (1994). *Understanding the small business sector.* London: Routledge.

Stray, S., Bridgewater, S., & Murray, G. (2001). The internationalisation process of small, technology-based firms: Market selection, mode choice and degree of internationalisation. *Journal of International Global Marketing*, *15*(1), 7–29.

UNCTAD (1999). *World investment report.* Geneva: United Nations.

Vatne, E. (1995). Local resource mobilization and internationalization strategies in small and medium sized enterprises. *Environment and Planning*, *27*, 63–80.

Vernon, R. (1966). International investment and international trade in the product cycle. *Quarterly Journal of Economics*, *80*, 190–207.

Welch, D. J., & Luostarinen, R. (1988). Internationalization: Evolution of a concept. *Journal of General Management*, *14*(2), 36–64.

Wernerfelt, B. (1984). A resource-based view of the firm. *Strategic Management Journal*, *5*, 171–180.

Williamson, O. E. (1975). *Markets and hierarchies: Analysis and anti-trust applications.* New York: Free Press.

Chapter 18

Internationalisation of High Technology Small Firms in Portugal

António Teixeira and Manuel Laranja

Introduction

High Technology Small Firms (HTSFs) occupy a special place in the studies of Innovation, Economic Growth and Science & Technology Policy Research. This type of companies may assume the role of the intrepid Schumpeterian entrepreneurs, sometimes pointed out as the protagonists and the heroes of a new wave of creative destruction based upon some new radical technology. In reality these firms may be responsible for the introduction of new technological innovations, but these, need not to be radically new or cause new waves in the economic system. By introducing leading edge technologies into the market HTSFs may play an important role of diffusion at national and international levels. In particular, when these firms establish links with universities and with the technological and scientific national environment, they become key-actors in the dissemination and transfer of advanced technological know-how. Also HTSFs are often key players in industrial networks, contributing to technology transfer and innovation induction in downstream sectors.

In the past few years, perhaps as a result of the economic growth that the country has experienced, more and more Portuguese HTSFs become internationalised right from the beginning or even from the inception. This apparent new behaviour reflects a significant change in the local technical entrepreneurial mentality.

In our empirical analysis we identify and classify these firms as the "third generation" of Portuguese HTSFs, displaying a more favourable attitude towards international entrepreneurship and a sharper understanding of global market competitive factors, when compared with the two previous generations. The rapid internationalisation of HTSFs in Portugal may also be linked to the processes of structural adjustment that European countries experience as a result of the EU and the Euro currency zone.

The paper begins with an overview of the theoretical perspective on HTSFs role within a national system of innovation framework. The third section links the phenomenon of internationalisation of HTSFs, to the literature on internationalisation processes. The methodology in the empirical research is summarised in the fourth section. Next, a HTSFs classification in

New Technology-Based Firms in the New Millennium, Volume III
ISBN: 0-08-044402-4

"generations" is introduced, aimed to embrace the results of previous studies on Portuguese High Technology Small Firms.

The sixth section presents the findings of our empirical analysis, which suggests the crucial importance for these new HTSFs of having an international strategic orientation, right from the genesis of the company. Finally, the paper will draw some conclusions and some proposals for future research.

Theoretical Framework

One of the first studies on the HTSFs phenomenon in the U.S., summarised in Roberts (1991), which focuses the follow-up of spin-off companies from MIT in Boston, and from other governmental research institutes, as well as from other companies, which began to emerge from the 1960s onwards. According to Autio (1997) these studies model a sequential life-cycle approach to HTSFs' creation and development, according to which companies pass through a number of different stages. The initial stage corresponds to the generation of a business idea and the final stage would result in a sustained growth corresponding to the commercial success of the initial idea. This approach is in line with to the classic economic perspective of the firm as an input-output unit that operates in a well-defined and known market.

The analysis of the innovative role of HTSFs within such a theoretical context, strongly marked by the innovation linear model may result in a reduced perspective of HTSFs as recent companies that are just established to market new technologies (Autio & Yli-Renko 1998).

On the other hand, if we take the Mark I model of a neo-Shumpeterian evolutionary approach, we find that only a minority of technology based small companies actually introduce radical innovations, and achieve high growth and profitability (Autio & Yli-Renko 1998; Laranja & Fontes 1998). The neo-Shumpeterian approach should not stop in the Mark I or Mark II regimes, but must perhaps go further to suggest HTBFs introduction of new products and new services which may also be the result of other technology, market and appropriately regimes reflecting complex multiple interdependencies between users, carriers and developers of technological innovation (Andersen & Lundvall 1988).

According to the "systemic perspective" (Autio 1997) HTSFs are seen as "components" of the innovation system (as defined by Edquist 1997). Hence, they interact with other organizations or institutions with the purpose to develop and exchange different types of knowledge, information and other resources. As defined by Edquist (1997), the organisations composing the Innovation System are enterprises, universities, research institutes as well as financial and government institutions. General norms, expected attitudes and rules enveloping the education system, the fiscal system, unions, quality and technical compliance also have a positive or negative influence in overall innovative behaviour. On the basis of this approach HTSFs are therefore seen as important components of a wider system. Their role and contribution must be analysed within the context of National Systems of Innovation. This raises questions about the precise nature and types of influence and contribution HTSFs may have on the overall development of innovation. How do HTSFs influence the knowledge flows and the processes of accumulation of the technological knowledge on their customers? It is difficult to find evidence of the type of contributions that HTSFs may cause. However, we suggest that there is a so-called catalyst effect and an inducer effect. These effects are the result of the introduction of new technologies in the national context; on of the example set by this entrepreneur mentality that these firms could inspire; of the increase in national market sophistication regarding users of new technologies; and of the influence upon knowledge flows and the processes of accumulation of the technological knowledge within their customers. We, therefore, contend that HTSFs contribute essentially to increase the ability of their customers to the accumulative technological knowledge about their processes, products or services, allowing them to integrate this knowledge with that of other actors.

Internationalization of HTSFs: A New Phenomenon

Conventional wisdom about internationalization of small and medium enterprises suggests that because of resource constraints, small firms are unlikely to internationalize. The theory that supports this kind of approach is known as the Process Theory of Internationalization (PTI) (Autio & Burgel 1999; Oviatt & McDougall 1994, 1997) and follows the traditions of behavioural theory of the firm. Process theorists adopt an approach to internationalization, based upon "psychic distance" and experiential learning (Jones 1999). Johansen and Vahle of Uppsala University in Sweden developed this model in the seventies. In the model, it is assumed that the firm gradually increases its commitments to foreign activities as knowledge about foreign markets and operations is acquired through experience. The idea is that as market uncertainty is reduced, companies will begin to invest more abroad, and therefore, the process is rather incremental and essentially reactive (Lindqvist 1997). The company acts in response to external changes, such as competitive pressures or a saturated domestic market, but it exhibits a passive attitude towards the search for export opportunities (Autio & Burgel 1999).

In comparison, the deliberate pro-active behavior of a company is to search and identify opportunities in international markets, thus facilitating early internationalization. In reality, however, knowledge-intensive sectors are likely to have to deal with narrow windows of opportunity and with markets that may be highly competitive. Therefore, the internationalization process must start early, possibly at the genesis of the company. This approach is different from that used by process theorists, and the role of the entrepreneur may be a key-factor for the engagement in the early processes of internationalization (Simões & Dominguinhos 2001; Teixeira 2000). Currently, this perspective is known as the "New Venture Internationalization Theory" (NVI) (Autio & Burgel 1999; Oviatt & McDougall 1994, 1997). This theory supports the notion that fast internationalization requires a learning process that allows an early and fast accumulation of experiential knowledge enabling some structural transformation and routine transformation to allow international activities as early as possibly.

The two theories are similar as regards to the pattern of the internationalization process. However, they differ in terms of a fast initiation vs. a slow and late initiation (Autio *et al.* 1999). A deeper analysis (Autio & Burgel 1999) reveals a complementarity between both theories in a sense that NVI theory is more appropriate for the early internationalization stages and PTI theory is more appropriate for the later stages.

When we study the internationalization of small companies, such as HTSFs, that are niche-oriented and deal with high risks of imitation, it is necessary to take account of a conceptual effort. The diversity of HTSFs sectors will bias the generic approach to the internationalization process based on export development. Essentially, this approach focuses on downstream marketing activities and does not include upstream activities such as licensing, R&D under contract, consultancy, contract manufacture, technical services and other knowledge intensive activities (Jones 1999). Furthermore, according to the PTI theory, limited resources and international experience suggest the inevitability of an incremental and surely defensive approach to the internationalization process (Fontes & Coombs 1997; Lindqvist 1997). However, the "technology factor" (Lindqvist 1997) provides a completely different view on the internationalization process, because it requires speed to take advantage of the opportunity window.

The speedy internationalization process of HTSFs is a present phenomenon, as a result of shorter life cycles; competitive advantage and innovations; and the rapid development of the information and telecommunications technologies at the global level. The recent British-German survey research, sponsored by AGF (Burgel *et al.* 2000), concludes that international activities are common in most German and British HTSFs. Companies that undertake international activities, right from their genesis, are often denominated "Born Globals" (Autio *et al.* 1999; Oviatt & McDougall 1994, 1997; Simões & Dominguinhos 2001).

It is possible to identify resources that facilitate early internationalization. For example, the international orientation of the entrepreneur/founder, that some researchers call "accumulation of relational capital" and "visionary capability" (Simões & Dominguinhos 2001). In general, the organizational value of a small company, is related to the entrepreneur's personal characteristics. However, it is crucial that HTSF possess a network of motivations (Jones 1999) and contacts that could enable technology transfer at various stages of the innovation process. These companies often need to link to external sources of expertise and advice. These external sources are key resources that enable early internationalization.

To study the internationalization process of HTSFs, it is therefore necessary to have a model that attributes a preponderant role to the founding-entrepreneurs, particularly with regards to the international orientation forged prior to the launch of the firm. Perhaps, we may then identify a role similar to that of the intrepid Schumpeterian entrepreneur, not in terms of the contribution to the introduction of radical innovations that change techno-economic paradigms, but in the sense, that there is a "creative accumulation" that may lead to change in the development of business processes at a more general level in the economy.

Methodology

Because of the lack of studies concerning the internationalisation process of Portuguese HTSFs, we adopt an exploratory approach. The sample included HTSFs in the information technology sector consisting of a database of 279 firms. This was considered possible given the absence of systematic official data registering these kind of companies, subsequently we identified a subset of internationalized HTSFs in this database, that provided evidence of knowledge intensive business activities in foreign markets. This step was only possible, through effectively collaborating with experts in various public support institutions and private enterprise associations.

Twenty-four firms were selected from the database ($n = 279$), for semi-structured interviews, which were conducted in late 1999 early 2000, all with the founding entrepreneurs. We believe the selected HTSFs are representative of the various regions where these firms usually agglomerate, namely: Lisbon, Oporto, Braga, small poles in Aveiro, Coimbra and Setúbal.

Our qualitative methodology used the "grounded theory" approach (Strauss & Corbin 1990), consisting of a systematic group of techniques and procedures for analyzing the data collected. The operationalisation of variables resulted in 19 categories each one with 5 subcategories. This data was codified in order to conduct frequency analysis and cross tabulations between variables.

Three Generations of HTSFs in Portugal

In Portugal, the HTSFs phenomenon appear to be facing adverse local specific conditions. Laranja & Fontes (1998), highlight that a relatively low local "mass" in research and development activities associated with electronic and information technologies and other technologies like the biotech, is probably one of the main factors preventing the formation and development of these firms in higher numbers. Other adverse conditions include the low level of collaborations between research establishments and private companies. On the other hand "dynamic complementarities" (Rothwell 1989), between HTSFs and local large companies are also almost marginal in Portugal (Laranja & Fontes 1998). In addition, as in many other countries, Portugal also has low levels of academic entrepreneurship. Public support programmes especially dedicated to HTSFs are absent, and there is little interest from local venture capital companies. Despite these adversities, HTSFs in Portugal do exist and manage to survive. The economic growth that the country experienced, particularly during the second half of the eighties to the present period has probably opened up many new opportunities, that were readily taken up by new firms in high technology areas.

Moreover, previous studies (Laranja & Marques 1994), have suggested that during the late eighties, growth for domestic demand of electronic and information technologies provided some market-pull effect in the entrepreneurs' decision to start-up. These HTSFs developed products and services based upon new technological knowledge first introduced elsewhere (abroad). Their concern was to explore a niche by configuring relatively new systems and adapting them to the local market conditions. This however, required high-qualified human resources and cumulative knowledge of advanced technologies with specific needs in the local market. HTSFs of the nineties continued, in general, to proceed much the same way.

However, a small number of HTSFs started to adopt the same niche-market orientation but are now aiming at opportunities in international markets.

On the basis of the evidence presented (Fontes 1995a, b; Laranja 1994; Laranja & Marques 1994; Laranja *et al.* 1997; Pereira 1997; Simões 1997) we suggest that there are three "generations" of Portuguese HTSFs, with different characteristics.

The "first generation" of Portuguese HTSFs corresponds to companies launched in the sixties and seventies including manufacturing companies such as HOVIONE in (pharmaceuticals, founded in 1959) or PROJECONTROL in industrial electronics (Pereira 1997). Companies in the "first generation" were essentially established by engineers coming out of large established firms, although in a few cases one could also identify spin-offs of local R&D institutes. Information about this first generation is relatively scarce as many companies founded at that time did not survive until the 1990s. The Portuguese telecomm equipment group CENTREL, originated in the seventies as a buy-out of STANDARD ELÉCTRICA (a subsidiary of the British Plessey), is a good example of a source of spin-offs. This group, working for the local telecomm operator under projectionist practices, is the source of first generation HTBFs such as PROJECONTROL, SETCOM, CEL-CACHAPUZ (and later second generation, PARAREDE). This "first generation" of HTSFs had an important role in creating linkages for technology acquisition, learning and transfer to the market. Looking back we can see that these first efforts resulted in the formation of local manufacturing activities in electronics, in line with the classic model (Roberts 1991; Rothwell 1989).

The "second generation" of HTSFs corresponds to a period from the second half of the eighties to the middle of the nineties. This is the period during which the integration of Portugal in the European Union takes place. During this period there was a strong in flow of structural funds to R&D, enabling the creation and/or further development of technological infrastructures, which were dedicated to technology transfer. During this period, as a result of European funds, there was a strong increase in the number of post-graduates (Masters or Ph.D.), many of whom were going abroad to foreign universities and facing increasing difficulties in finding teaching vacancies at local Universities. This may have obliged some of these new highly qualified people to adopt a defensive entrepreneurial attitude. During this period 5 incubation and business innovation centers, were also established. AITEC, an innovation center effectively operating as a Seed Capital Investor and strongly linked to INESC (Institute of Engineering and Computer Systems), which played an important role in the emergence of new HTSFs during this period in Portugal and contributed to the launch of more than 60 firms (Pereira 1997). Many of the firms were created with the support of AITEC-INESC, that had begun to emerge during the beginning of the second half of the eighties, and are now market leaders in their own fields as it is the case of NOVABASE in professional services for software and information systems.

In "second generation" companies, relative to the first, one may find more cases of university spin-offs. However, existing empirical data (Laranja 1994) suggests that, relative to the overall number of companies (or products) in their generation, university spin-off initiatives are a minority. Also, studies carried out during this period (Fontes & Coombs 1995; Laranja & Fontes 1998; Laranja & Marques 1994) lends support to the view that companies founded in this period had an important "technology transfer" role. By and large, the technologies introduced by these companies were first developed and market-introduced abroad. However, the need for small adaptations that resulted in specific products, created opportunities for what Laranja & Fontes classified as "creative adaptation" (Laranja & Fontes 1998).

This differs from a simple acquisition and transfer role. It means that the "service content" of product and technologies that HTBFs introduce in the market, contributes to induce innovation capabilities in downstream clients. This role is less obvious under the lens of a classic Shumpeterian Mark I behaviour, but nevertheless an important one in the context of technology lagging countries.

Finally the "third generation" corresponds to a period from the end of the nineties onwards. A significant proportion of this new generation results from growth and restructuring of HTBFs of previous generations. The remaining roughly corresponds to companies created during the past five years, entering the so-called digital economy. We could point out examples such as CRITICAL SOFTWARE, an academic spin-off that develops activities in critical information systems or the company CHIPIDEA, other academic spin-off, that develops activities in the design and concept of integrated circuits. This new generation of HTSFs, continues to have a defensive approach to innovation, associated with small market niche behaviour. However, some of these third generation HTSFs benefit from the accumulated experience of entrepreneurs involved in the previous generations. As for companies of previous generations, we can identify in this new generation the same "challenging" role in the national technological context (Fontes 1995b, 1996). However, these new companies may also have an important additional role of positively influencing the prevailing national entrepreneurial mentality with regards to the search for opportunities in international markets. This change of entrepreneurial mentality would be necessary to remove inferiority complexes, relative to international competitors, particularly in businesses involving advanced technologies. In fact, internationalization is often understood by this generation of HTSFs as a qualitative jump related to the products and technologies they wish to offer. Also another important difference is that these companies show an unusual concern for strategic planning over the long term.

Results of Empirical Analysis

First, it is important to emphasize that Portuguese HTSFs with international activities, relative to total HTBFs (all three generations) are in minority. Unlike what it was found in the case of the U.K. and Germany (Burgel *et al.* 2000), where the majority of firms have a presence in international markets, in Portugal only 24 firms out of our sample of HTSFs ($n = 279$) have a presence in international markets.

However, for firms that are internationalized, around half have more than 20% of volume sales in international markets, and a quarter have close to 100% of sales with customers abroad. Compared to other kinds of SMEs in Portugal, this international intensity is somewhat unusual. Our data suggests that the dimension of the company does not influence the internationalization intensity. The statistical test calculated for the crosstabulation between the number of workers and the percentage of sales abroad, suggests a weak relationship between the two variables. Also, the empirical analysis that internationalized third generation companies display a pro-active behavior in their search for a window of opportunity in wider markets. In general, third generation companies have a clear international orientation, assumed right from their genesis. Founding-entrepreneurs appear to have a definition of the business concept that already takes a view to compete in the global market. Moreover, this international orientation of the founders is reflected

Table 1: International orientation vs. established year.

International Orientation	Established Since ...		
	Before 1985 (%)	**1985–1995 (%)**	**1996–1999 (%)**
"Since the genesis of the company"	0	63	100
"We try national market first"	0	31	0
"We don't have this"	100	6	0
Total ($N = 24$)	$N = 4$	$N = 16$	$N = 4$

Note: $\chi^2 = 20.57, p < 0.001$, Cramer's $V = 0.66$.

in the formation process of the company. Generally these companies start by building the resources required for access to international markets.

We also found that in (Table 1), the most recent HTSFs appear to have a clear international orientation from the genesis, contrasting with older firms (established before 1985) and display a more conservative behavior.

It may be that the unique knowledge of very specialised technologies for those later HTSFs (for example, software for "critical" systems, call centres management software, etc.) gives them higher possibilities to compete in the international markets. On the other hand, to sustain early entry in international markets, it is not enough to possess a competitive and innovative product or service for sustained internationalisation, but it is necessary to possess technological competences that allow the development of incremental innovations in product or service. It may be also necessary to have other "complementary" assets and competencies. Our interviews suggest that the international process is directly related to the entrepreneurs' personal contacts. In many cases, the technological and management competencies of these firms are precisely those acquired by founding-entrepreneurs in their previous international experiences and/or contacts with in multinationals (Oviatt & McDougall 1994).

In line with the literature review by Kandasaami (1998), our findings lend some support to the view that entrepreneurs' personal characteristics such as age, educational level, academic qualifications, frequency of wide periods abroad and knowledge of foreign languages largely condition the motivation for the internationalisation process. To evaluate the speed of internationalisation, we also measured the time elapsed from company formation to the full accomplishment of the first international activity. In our sample, we detected that 42% of the companies immediately go international, 21% take about two years, and the remaining 37%, needed 3 years or more to venture into foreign markets. These last 58% claim that this amount of time was crucial for the development of their products or services.

Because HTSFs are usually owned by highly qualified technical people that lack business management and marketing capabilities, we would normally expect a parochial approach to internationalisation. However, our evidence suggests a wider diversity of situations. More than 75% of HTSFs chose markets that were "economically close," (EU and other European countries) and "historically and culturally close," (Brazil and African countries of Portuguese language). These findings are in line with the theory of gradual internationalisation as proposed by the Uppsala model. Knowledge of familiar foreign markets is used in order to accumulate experience of international activities and increase international involvement gradually.

However, we also found six firms that aimed at the United States as their preferential market. Thus some firms are more inline with the NVI approach regarding speedier internationalisation. These HTSFs favour the U.S. market, partially because of the entrepreneurs' international contacts, but also because it is a market with sophisticated customers, demanding higher innovative capabilities. The form of entry to the U.S. market is through the establishment of subsidiaries or joint ventures.

With regards to the European market it was not possible to distinguish a preferential entry mode. This may have been because different modes are used in accordance to differences in different European countries. In the case of Brazil, HTSFs are more likely to be fervent but we detect a growing tendency for the establishment of integrated joint-ventures in the U.S. African markets are quite interesting for Portuguese HTSFs that develop application software and the preferential entry mode tends to be through direct exports, using foreign agents or distributors.

In our research, we also tried to analyse the impact of venture capital for the development of Portuguese HTSFs. More than 75% of all HTSFs did not receive any venture capital financing. Furthermore, HTSFs' founders claim that they are not interested in having venture capital partners. This seems to be a European trend, as it is suggested by Burgel *et al.* (2000), that only 10% of HTSFs in U.K. and Germany were financed by Venture Capital investors.

The traditional "low-tech" image of Portugal is known to cause difficulties to the internationalisation process of HTSFs and other kinds of SMEs. Our data suggests that indeed, the image of Portugal as a country with concentrated technological intensity is, in some cases, a barrier to internationalisation. To resolve this problem, the most common strategy is to use a "European" technological image to hide the Portuguese origin. HTSFs that explicitly announce their Portuguese origin, normally make extra efforts to build reputation and credibility with regards to innovation and excellence.

In line with the systemic perspective, we also analysed the interactions with academic institutions. Most HTSFs have connections with universities or investigation centres and are classified as "highly satisfactory." One of the most remarkable examples is that of HTSFs in Braga (north region), which have created strong linkages to the local Computer and Engineering Faculty of the University of Minho. Nevertheless, in this region, more than a third of these HTSFs are "unsatisfied" with the university interactions.

Conclusions

Our empirical analysis suggests that some of the more recent HTSFs have an international strategic orientation, right from the genesis of the company. The findings also suggest that Portuguese HTSFs in general, and internationalised HTSFs in particular, have potential to play a dynamic role as components of the local innovation system. In our view HTSFs contribute to the local innovation system through their influence upon the flows of knowledge accumulation and technology transfer. Further research, however, would be needed to better characterise the influence of these companies on the National System of Innovation. We believe that because they are better able to enter and compete in global technological markets, they probably add some local dynamism in different components of the National System of Innovation. The perceptions of HTSFs founders also provide some evidence with regards to the need of these companies to effectively collaborate with the universities.

For the purposes of further research we would suggest the importance of having a national HTBFs database that would enable a more longitudinal approach to the study of the apparent different "waves" or generations of HTSFs in Portugal. This type of database would allow every type of HTSF to be identified, possibly facilitating the design of appropriate specific government support policies.

Another issue raised from our analysis is that the low-tech image associated to the country can cause difficulties for the HTSFs' internationalisation process. It is nevertheless very important to use the relative success of those HTSFs already internationalised and helping to change this image, thus contributing to the open opportunities for other national companies. The observations outlined above suggest that Portuguese internationalised HTSFs follow the general European trend. All HTSFs seem to have a distinctive "technological factor" that appeals to international niches. Our findings suggest that the incremental export process of internationalisation is not the only mode of internationalisation for HTSFs. In some cases an effort must be made to conciliate "process" theory with the "venture" theory. Furthermore, the international intensity of Portuguese HTSFs is not size related and like the European HTSFs does not present any relationship with the number of employees or financial actives. However, it is intimately related with the international experience attained by the founders in previous occupations or business ventures, and has a decisive influence for the speed of foreign market entry.

The pattern of foreign market selection for entry that we found is very diversified and geared towards Portugal's geographically and historically closest markets. Although our findings give some support to behaviourist theories, they also embrace the systemic perspective of the internationalisation process. Another implication of this paper for future research is the possible contribution that these companies may be making to change the prevailing conservative entrepreneurial mentality, particularly with regards to the example set for future generations of Portuguese technical entrepreneurs. This would be, without a doubt one of the greater contributions of HTSFs to the improvement of the National System of Innovation, tendering these companies part of the solution required for the future development of the Portuguese economy.

References

Andersen, E. S., & Lundvall, B. Å. (1988). Small systems of innovation facing technological revolutions: An analytical framework. In: C. Freeman, & B. Å. Lundvall (Eds), *Small countries facing the technological revolution*. London: Printer Publishers.

Autio, E. (1997). New, technology-based firms in innovation networks: Symplectic and generative impacts. *Research Policy, 26*, 263–281.

Autio, E., & Burgel, O. (1999). Internationalisation experience, knowledge resource endowment, perceived cost of internationalisation, and growth in technology-intensive new firms. Paper presented on the 1999 World Conference on International Entrepreneurship, August, Singapore.

Autio, E., Sapienza, H., & Almeida, J. (1999). Effects of time to internationalization, knowledge intensity and imitability on growth. *Academy of Management Journal, 41.*

Autio, E., & Yli-Renko, H. (1998). New, technology-based firms in small open economies — an analysis based on the Finnish experience. *Research Policy, 26*(9), 973–987.

Burgel, O., Murray, G., Fier, A., & Licht, G. (2000). Research report — *The rapid internationalization of high-tech young firms in Germany and in the United Kingdom*. London Business School & Zentrum für Europäische Wirtschaftsforshung GmbH. Study supported by the Anglo-German Foundation for the Study of Industrial Society.

Fontes, M. (1995a). Upgrading national systems of innovation. New technology based firms, a vehicle of technology transfer and absorption at country level. Paper presented on 3rd International ASEAT Conference "Managing Technology in the 21st Century", September 1995, Manchester School of Management, Manchester, U.K.

Fontes, M. (1995b). Acquiring the inputs for new technology based firms creation: Technology, funds and market demand — a case study of Portugal. Paper presented on 3rd Annual High Technology Small Firms Conference, Manchester School of Management, Manchester, U.K.

Fontes, M. (1996). 'Upgrading' national systems of innovation. New technology based firms. In: R. Coombs, K. Green, A. Richards, & V. Walsh (Eds), *Technological change and organization.* Cheltenham, UK: Edward Elgar.

Fontes, M., & Coombs, R. (1995). New technology based firms and technology acquisition in Portugal: Firm's adaptive responses to a less favourable environment. *Technovation, 15*(8), 497–509.

Fontes, M., & Coombs, R. (1997). The coincidence of technology and market objectives in the internationalisation of new technology based firms. *International Small Business Journal, 15*(4), 14–35.

Jones, M. V. (1999). The internationalisation of small high-technology firms. *Journal of International Marketing, 7*(4), 15–41.

Kandasaami, S. (1998). Internationalisation of small and medium sized born-global firms: A conceptual model. Paper presented on 1998 International Council For Small Business (ICSB) — Singapore Conference.

Laranja, M. (1994). *Innovation trajectories: The case of small technology based firms in Portugal.* Documento apresentado na Universidade das Nações Unidas, Instituto para as novas tecnologias, Maastricht.

Laranja, M., & Fontes, M. (1998). Creative adaptation: The role of new technology based firms in Portugal. *Research Policy, 26*(9), 1023–1036.

Laranja, M., & Marques, J. (1994). As Tecnologias de Informação e Electrónica em Portugal: Importância, Realidade e Perspectivas. Direcção Geral da Indústria — DGI, Estudos DGI, Análise Industrial, Ano III, no. 3, Lisboa.

Laranja, M., Simões, V. C., & Fontes, M. (1997). Inovação Tecnológica — Experiência das Empresas Portuguesas, Texto Editora, Lisboa.

Lindqvist, M. (1997). Infant multinationals: Internationalisation of small technology based firms. In: D. Jones-Evans, & M. Klofsten (Eds), *Technology, innovation and enterprise — the European experience.* London: Macmillan.

Oviatt, B. M., & McDougall, P. P. (1994). Toward a theory of international new ventures. *Journal of International Business Studies, 25*(1), 45–64.

Oviatt, B. M., & McDougall, P. P. (1997). Challenges for internationalization process theory: The case of international new ventures. *Management International Review, 37*(2), 85–99.

Pereira, J. M. (1997). Estudo Sobre o mercado de utilização de centros de incubação de pequenas empresas industriais de base tecnológica. Relatório Final. Estudo no. 1751A preparado por Tecninvest para o Gabinete do Gestor do PEDIP.

Roberts, E. B. (1991). *Entrepreneurs in high technology: Lessons from MIT and beyond.* Oxford: Oxford University Press.

Rothwell, R. (1989). Small firms, innovation and industrial change. *Small Business Economics, 1,* 51–64.

Simões, V. C. (1997). Inovação e Gestão em PME. Gabinete de Estudos e Prospectiva Económica (GEPE) do Ministério da Economia, Lisboa.

Simões, V. C., & Dominguinhos, P. M. (2001). Portuguese born globals: An exploratory study. Paper presented at the 27th EIBA Conference, December, Paris.

Strauss, A., & Corbin, J. (1990). *Basics of qualitative research — grounded theory procedures and techniques.* London: Sage.

Teixeira, A. (2000). A Internacionalização das Novas Empresas de Base Tecnológica em Portugal, Dissertação de Mestrado, Universidade de Aveiro.